# Nanomaterials for Sustainable Energy

ACS SYMPOSIUM SERIES **1213**

# Nanomaterials for Sustainable Energy

**Jingbo Louise Liu**, Editor

**Sajid Bashir**, Editor

*Texas A&M University-Kingsville*
*Kingsville, Texas*

**Sponsored by the**
**ACS Division of of Colloid and Surface Chemistry**

American Chemical Society, Washington, DC

Distributed in print by Oxford University Press

**Library of Congress Cataloging-in-Publication Data**

Names: Liu, Jingbo Louise, editor. | Bashir, Sajid, 1967- editor. | American Chemical Society. Division of Colloid and Surface Chemistry.
Title: Nanomaterials for sustainable energy / Jingbo Louise Liu, editor, Sajid Bashir, editor, Texas A&M University-Kingsville, Kingsville, Texas ; sponsored by the ACS Division of Colloid and Surface Chemistry.
Description: Washington, DC : American Chemical Society, [2015] | [Oxford] : Distributed in print by Oxford University Press | Series: ACS symposium series ; 1213 | Includes bibliographical references and index.
Identifiers: LCCN 2015044189 (print) | LCCN 2015044602 (ebook) | ISBN 9780841231160 | ISBN 9780841231153 ()
Subjects: LCSH: Renewable energy sources--Research. | Power resources--Research. | Nanostructured materials.
Classification: LCC TJ808.6 .N36 2015 (print) | LCC TJ808.6 (ebook) | DDC 621.042028/4--dc23
LC record available at http://lccn.loc.gov/2015044189

The paper used in this publication meets the minimum requirements of American National Standard for Information Sciences—Permanence of Paper for Printed Library Materials, ANSI Z39.48n1984.



Distributed in print by Oxford University Press

# Foreword

The ACS Symposium Series was first published in 1974 to provide a mechanism for publishing symposia quickly in book form. The purpose of the series is to publish timely, comprehensive books developed from the ACS sponsored symposia based on current scientific research. Occasionally, books are developed from symposia sponsored by other organizations when the topic is of keen interest to the chemistry audience.

Before agreeing to publish a book, the proposed table of contents is reviewed for appropriate and comprehensive coverage and for interest to the audience. Some papers may be excluded to better focus the book; others may be added to provide comprehensiveness. When appropriate, overview or introductory chapters are added. Drafts of chapters are peer-reviewed prior to final acceptance or rejection, and manuscripts are prepared in camera-ready format.

As a rule, only original research papers and original review papers are included in the volumes. Verbatim reproductions of previous published papers are not accepted.

**ACS Books Department**

# Contents

## Section IV



## Indexes

# Preface: Nanomaterials for Energy, A Look Backwards

The American Chemical Society (ACS) is one of the major venues where experts and leaders discuss contemporary and current ideas in Science and Technology as was the case at the 246th ACS Meeting and Exposition that was held between September 8-12, 2013, Indianapolis, Indiana. The focus of the National Conference was on Chemistry in Motion, Industry and Nanoscience.

The current ACS Symposium Series monograph was born out of the meeting based on oral arguments presented at the symposium on Nanostructured Materials for Next-Generation Energy Storage and Conversion. The gestation period of 1125 days, longer than the 660 days it takes to create a baby Elephant. Fifty technical talks from leaders around the world based at teaching or research-intensive universities, industry and federal laboratories in Colloid and Surface Chemistry areas presented their findings and here after thoughtful reflection, modification and renewal expand upon their original treatise presented at the conference. The symposium is a crucible for ideas related to applications related to energy production, modelling and storage and their potential impact in society. Yet, energy generation, storage and application is timeless in that every generation has faced similar challenges and adopted with contemporary solutions.

This is seen even at a cursory glance at the literature in that in the last decade over 4000 articles in the technical periodicals related to fuel or solar cells with polymeric and ceramic materials encompassing the major sub category. Within new polymers, electrolytes are prominent, since fuel or solar cells improvements need advancement of anode/cathode electrode catalyst but also in membranes. The latter requires a through grounding in polymer and nanoscience to control structure and dimensionality and the book is organized along these ideas. The motivation is driven by societal needs and technical advancements in synthesis, design and characterization. Increases in raw material (oil, coal, natural gas) always lead to a reevaluation towards renewable and sustainable energy. The approach of the current generation is to investigate advanced responsive materials including polymers, ceramics and bi-functionalized materials for sustainable and green energy generation, transport, storage and usage.

Raw material prices are a major influence, but another influence is portable energy usage, such as mobile devices, utilization of renewable energy in the form, of wind, water and solar and transportation in the form of electric vehicles (EVs), unattended aerial vehicles (UAVs) and pulsed power transportation require offline storage modalities from biosensors to load-leveling power stations including double-later electrochemical capacitors, dielectric capacitors and fuel cells.

Fuel cells are a class of green energy devices based on proton-exchange membrane that are in continual development as can be observed from their published power densities. These improvements have focused on specific components, such as polyelectrolytes membranes, materials to limit cathode flooding, or increased retention at high temperatures, or materials that increase proton conductivity whilst keeping costs down. The use of hybrid materials may achieve both aims of retaining water whilst keeping high conductivity and high operational lifespans, such as amphiphilic incorporation into proton exchange membranes through incorporation of solid acids. The end goal is to achieve high thermal stability, tolerance to reactant migration/cross-over and water absorption. The last goal may also be achieved through incorporation of metal/metal oxide nanomaterials into the membrane or electrode, such as Pt or carbon nanotubes or coupling of metal organic framework architecture into hydrogen storage as a feedstock for fuel cells.

Fuel cells require methanol, or gasoline, a limitation not shared by photovoltaics that utilize direct sunlight and convert that into electrical energy, analogous to plants that utilize sunlight for generation of chemical energy via photosynthesis. Although solar energy is not a major contributor to the US energy economy, the cost per kilowatt has dropped over the last five years due to use of organic polymers utilizing bi continuous hybrid cell design with (p-type) organic polymers and (n-type) inorganic materials as semiconductors taking design concepts from organic based semiconductors and light emitting diodes. Newer designs exhibit high mobility and conversion efficiency through incorporation of single cells design, matching optical properties of the harvesting surface to natural light spectra and the open-circuit voltage. These are some ideas touched-upon or elaborated upon by our select authors and from this Stephen Creager presents his views on the future of renewable energy storage devices.

This book's focus has been described in broad terms in the proceeding chapters. Specifically we have ten authors who have selected one area of developed, related to energy and storage. In the area of smart nanomaterials (Yilmaz et al.; Chapter 1) and Anodic Oxide nanomaterials and their applications in energy generation (Thornton and DeWitt; Chapter 2), Energy Generation (Mott and Maenosono; Chapter 3), cover the first third thematically. In the middle sections related to Catalysis (Mao and Ahmed; Chapter 4), Electrolytes (Marechal et al.; Chapter 5), Next Generation of Nafion-type membranes (Leddy; Chapter 6), are elaborated and in the latter third, topics related to Electrochemical based Energy Storage (Smimova and Kolla; Chapter 7), Metal-organic framework based storage (Zhou et al.; Chapter 8), Organic Solar Cells (Yamamoto et al.; Chapter 9), a standalone biosafety chapter (Liu et al.*; Chapter 10) and a post-face looking forward (Creager; Chapter 11).

[*] Three years ago, when we held this symposium, the biological implication of components utilized in these devices was not a reality, due to the limited presence. However, with the push for EV, greater efficiencies in DSSCs and PEM-FCs, I [Dr. Bashir] asked Louise Liu to add a penultimate chapter discussing

the biological implications of accidental release of metal-organic framework based catalysts, which may be used in fuel cells and gas storage devices. Her analyses indicate a complex picture that these materials may be toxic (inhibitory to cell proliferation) or non-toxic (stimulatory) depending on their mode of interactions. Materials that exhibit oxidative stress tend to be toxic to human cells as well as microorganism. The former scenario is a disadvantage, whereas the latter scenario is an advantage, since accidental release into the water table of such metal-organic framework agents would behave as disinfectants (an distinct advantage), however if ingested, they would promote oxidative stress and apoptosis in human cells (a distinct disadvantage). More policy work is required in this important area; furthermore the nitric oxide (NO) and optical density (OD) assays may be a methodical approach to assess metal-organic framework toxicity in eukaryotic and prokaryotic cells respectively.

All ten authors are leading exponents in their respective fields and have been directly or indirectly involved with the symposium or monograph through a rigorous peer review process.

The success of the symposium and the publication of the proceedings could not have been possible without the effort and support of the ACS Ramanathan Nagarajan, Program Chair, Division of Colloid and Surface Chemistry (COLL) whom I share a great respect and admiration for. In addition, we would be remiss not to thank the support of the session moderators and presiders (listed alphabetically Drs. Christopher Brooks, Lea-Der Chen, Anne Co, and Amit Verma) and other organizers of the program.

We also acknowledge Mr. Tim Marney, Bob Hauserman, Ms. Kat Squibb, and Ms. Arlene Furman, for their tireless and timely involvement in editing and producing this book. In addition, special thanks are due to the faculty at the Department of Chemistry, Dean of our College of Arts& Sciences and the Provost for their encouragement and lastly but not least the speakers, authors, manuscript reviewers, and ACS program coordinators and officials for their contributions.

Who is this volume intended is our driving force in its gestation and creation. This monograph presents an up-to-date and eminently readable collection of papers in energy conversion, storage, and modelling, which would be of great interest to practitioners, directors, fund and program managers, researchers, and engineers in the field of energy generation and storage. It would also be the invaluable asset to students, and novices coming into the field, as well as seasoned veterans because this monograph based on the ideas discussed in the symposium will highlight leading edge frontier research in polymeric and nanomaterials including electrolytes at American and International research units, to bring the technical and general community in the direction of polymers and advanced materials for energy generation, storage, and conversion in fuel cells and photovoltaics and indirectly in batteries.

An abbreviation and author biograph sections terminate this thematic edition. Specific funding agencies are listed at the end of each chapter and we wish to thank them all for their support, particularly the American Chemical Society – Petroleum Research Fund, the Air Force Office of Scientific Research, The U.S. Department of Education Promoting Postbaccalaureate Opportunities for

Hispanic Americans Program, the U. S National Science Foundation and Texas A&M University-Kingsville University Research Award.

The thematic focus, depth, breadth, technical omissions, errors are as always ours human failing and not attributed to the ACS or Authors for which we will always " to strive, to seek, to find, [and to serve better] and not to yield" (in Ulysses, By Alfred, Lord Tennyson).

We thank you the reader for your future contribution as "A bringer of new things. ... For some three suns to store and ... yearning in desire. To follow knowledge like a sinking star" (in Ulysses, By Alfred, Lord Tennyson) towards the next generation of invention and scholarly creativity.

Sincerely,

Jingbo Louise Liu (MS: Heilongjiang University, Ph.D.: USTB, and Advanced Training: The University of Calgary), JSPS Invitation Fellow, FaST Fellow, DEBI Fellow and LBNL ALS User affiliate.

Sajid Bashir, (MA: SUNY Buffalo, Ph.D.: Warwick University, and Advanced Training: Cornell University), CChem, CSci, FRSChem.

**Jingbo Louise Liu**
Department of Chemistry, Texas A&M University-Kingsville
700 University Blvd., MSC 161
Kingsville, Texas, 78363-8202, United States
jingbo.liu@tamu.edu (e-mail)

**Sajid Bashir**
Department of Chemistry, Texas A&M University-Kingsville
700 University Blvd., MSC 161
Kingsville, Texas, 78363-8202, United States
br9@tamuk.edu (e-mail)

# Section I

## Chapter 1

# Nanomaterials Fueling the World

**Shuyang Shaun Pan,[1] Lynne Tan Xin Lin,[2] Vasileios Komvokis,[3] Artrease Spann,[1] Melissa Clough,[4] and Bilge Yilmaz*,[1]**

**[1]BASF Refinery Catalysts, 25 Middlesex-Essex Tpk., Iselin, New Jersey 08830, United States**
**[2]BASF Refinery Catalysts, 7 Temasek Blvd, 038987, Singapore**
**[3]BASF Refinery Catalysts, SK8 6QG Cheadle, United Kingdom**
**[4]BASF Refinery Catalysts, 1111 Bagby St., 2600, Houston, Texas 77002, United States**
***E-mail: bilge.yilmaz@basf.com**

Materials that store energy in forms that can be practically released and used are called fuels. In its colloquial form, this definition is typically used for materials storing energy in the form of chemical energy that could be released through combustion. Fuels produced by refining crude oil deliver the majority of energy consumed for transportation (e.g., gasoline, diesel, jet fuel etc.). Globally, the main industrial conversion process in a typical fuels refinery is fluid catalytic cracking (FCC) units around the world currently process around 15 million barrels of crude oil per day. At the heart of the FCC process is the catalyst, which dictates the productivity and yield slates from this unit operation by defining the catalytic activity and selectivity. Herein, we will discuss the critical role that nanomaterials play for meeting the transportation fuels demand of today, as well as the larger role of overcoming future challenges due to changes in feedstock supply, product demand and environmental concerns.

## Introduction

Today, a considerable portion of the energy used for transportation worldwide is delivered by liquid transportation fuels (e.g., gasoline, diesel, jet fuel etc.), which are produced in fluid catalytic cracking (FCC) units. Due to its pivotal role in FCC, the catalyst brings the needed activity and selectivity and therefore dictates the productivity and yield slates from this unit operation. Hence, it can be stated that FCC catalysts fuel our industrialized society and have a direct impact on the global economy.

Fluid catalytic cracking catalyst contains multiple sub-micron sized building blocks that define its catalytic activity and selectivity, as well as provide structural integrity and hydrothermal stability. The major components are the nanoporous zeolite, matrix, filler and binder. These components vary significantly in type and quantity. The catalyst can be made in a simplistic way by incorporating these components in a single particle by physically combining them during particle shaping (e.g., spray drying) to meet the functions required. A more innovative and sophisticated method for FCC catalyst preparation is the *in situ* process, where the nanoporous zeolite and the active matrix are synthesized directly within a kaolin microsphere. As the *in situ* method does not require additional fillers or binder, it is considered to be a more efficient manufacturing route. Nanostructures that provide the superior activity and selectivity to the catalyst are grown by *in situ* transformation of the starting materials. The *in situ* synthesis route also allows for the targeted arrangement of zeolite and matrix on the nanoscale, which results in tailored positioning of different types of active sites and interconnected pore hierarchy to maximize the selectivity of the FCC catalyst towards the desired fuels.

Catalyst design to meet the FCC operation targets/requirements defines the commercial outcome. The changing global energy/fuels landscape in terms of available input and desired ouput has a direct influence on the catalyst design, since the catalyst has to be modified on the nanoscale to combat these changes. Here, we will discuss how the effective use of nanomaterials is transforming the way we fuel our industrialized society to meet the challenges due to changes in feedstock supply (resid feeds, tights oils, biomass derived feeds, etc.), product demand (diesel vs. gasoline) and environmental concerns. This chapter is a concise account of how a maturc field continously evolves by the utilization of nanomaterials to meet the challenges it faces.

## Nanomaterials To Address Changes in Feedstock Supply

As the global energy demand increases, various unconventional feedstocks that could not be exploited in the past due to operational challenges associated with their utilization, are now being considered as attractive new possibilities for refineries. Resid feeds, tight oils and biomass-derived feeds constitute three main types of these opportunity feedstocks. Below is a summary of how nanomaterials facilitate the necessary changes in catalyst properties and performance for effective processing of these new feedstocks.

## Processing Resid Feeds

Due to the increasing demand for fuels and the relatively low availability of light crudes (in parts of the world where access to tight/shale oil is limited), some refineries are forced to process more residue feedstock in FCC units, resulting in a significant increase of the contaminant metals that negatively impact the FCC catalyst (*1*, *2*). The processing of residue feedstock is also due to economic considerations in an effort to stay competitive by processing lower cost opprtunitiy crudes, such as those which are heavier and more contaminated in nature. Globally, around 35 % of FCC units process predominantly vacuum gasoil (VGO) feedstock and 65 % process feedstock containing residue. In some regions, the tendency to process residue is even greater due to economics and feedstock availability. To improve the profitability of the FCC unit, increased flexibility to process residue feedstock is desired. This flexibility can be achieved through the use of advanced FCC catalysts with improved metals tolerance and diffusion characteristics.

To maximize refinery profitability, FCC catalysts need to be tailored to meet the specific requirements of each unit. The customized catalyst should have a carefully designed pore architecture, matrix type, zeolite surface area, rare earth on zeolite (RE/Z), as well as contaminant metals passivating components.

Zeolite Y is the type of zeolite used in FCC catalysts. It has a regular repeating pattern giving the characteristic measured as the unit cell size (UCS). The ratio of framework alumina-to-silica in a zeolite defines its cracking characteristics. Zeolites are deactivated by the loss of alumina from the structure at high temperatures in the presence of steam, such as the conditions found in the FCC regenerator. A controlled dealumination procedure applied by the catalyst manufacturer can be used to give the catalyst good performance characteristics. These are known as ultra-stable Y (USY) or reduced unit cell size catalysts. Zeolites are responsible for most of the FCC cracking activity. Product selectivity is significantly impacted by both the feedstock's diffusion through the zeolite Y pore channel system, whose average diameter is approximately 7.4 Å, and the feedstock's cracking on the zeolite acid sites.

Matrix is defined as the non-zeolite part of the cracking catalyst. It serves both physical and catalytic functions. Physical functions include providing support for the zeolite, providing particle integrity and attrition resistance, acting as a heat transfer medium and allowing for the free flow of feed and products in and out of the catalyst, respectively. Catalytic functions include heavy oil upgrading and passivation of feedstock contaminant metals.

There are two known industrial manufacturing processes of FCC catalysts: the "incorporated" and the "*in situ*" manufacturing routes. In the incorporated manufacturing process, zeolite and matrix are made separately and then combined together with inert fillers using a binder to form catalyst particles. A certain amount of binder is required to sustain the physical integrity of the catalyst particle. This requirement constrains the flexibility to produce catalyst with high matrix and at the same time moderate-to-high zeolite levels. In the *in situ* manufacturing process, a catalyst microsphere is manufactured and then zeolite is grown directly onto the microsphere from nutrients provided both by portions of the microsphere and supplemented by separate nutrient addition. The chemical bond between the microsphere and zeolite eliminates the need for a separate binder (used in the incorporated manufacturing process); therefore, compared to the incorporated manufacturing process, higher zeolite levels are possible, without compromising the catalyst's physical integrity.

Fluid catalytic cracking catalysts with the following features are required to increase the flexibility to process residue feeds:

- Coke selective matrix for bottoms upgrading and metals tolerance;
- Optimized zeolite and matrix content to maximize either conversion or distillate yield;
- Moderate to high zeolite content to provide selective cracking;
- Low fresh catalyst sodium (Na) content to minimize sodium- vanadium (Na-V) zeolite deactivation (*3*).

The additional production of hydrogen ($H_2$) and coke resulting from processing high metals content and high Conradson carbon residue (CCR) feeds has a significant impact on FCC unit operation, as most units operate against gas and coke handling limits. The maximum tolerable coke yield/production may be constrained by regenerator operating limits, such as air blower limits. Coke is defined as any carbonaceous, high molecular weight, non-volatile residue formed from cracking. There are four types of contribution to FCC coke: contaminant metals, feed additive, cat-to-oil (strippable) and catalytic coke formation (*1*). Contaminant metals coke results from reactions due to feed metals that accumulated on the catalyst particle acting as dehydrogenation catalysts (e.g. nickel (Ni) and vanadium (V)) that remove hydrogen from hydrocarbon molecules, thereby increasing the tendency to form coke. Catalytic coke is formed by thermal and catalytic reactions.

The formation of coke on the catalyst results in catalyst deactivation due to the blocking of active acid sites. Thus, the coke must be burned off the catalyst in the regenerator to restore its activity. Heat generated from burning coke increases the regenerator temperature. This can contribute to more severe catalyst deactivation, higher fresh catalyst consumption, reduced equipment operating life, and lower conversion due to reduced cat to oil (C/O). Thus, it is always very important to control the production of coke, which is particularly challenging when processing resid feeds.

The risk of excessive $H_2$ and coke production associated with processing resid feeds can be mitigated by the use of advanced FCC catalyst technologies that are tailored on the nanoscale based on operational needs. One effective strategy to improve coke selectivity has been the use of catalysts with open pore architecture with exposed zeolite, such as the distributed matrix structures (DMS ) technology platform (*3*). In DMS, the matrix is designed to provide enhanced diffusion of the feed molecules to pre-cracking sites located on the external, exposed surface of highly dispersed zeolite crystals (Figures 1 & 2).

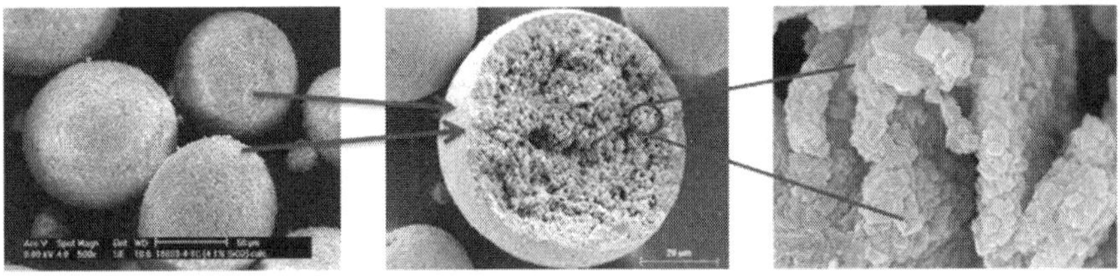

*Figure 1. Scanning Electron Microscopy (SEM) images of FCC catalyst particles from in situ synthesis where macropores are covered with nanoporous zeolite growth, leading to enhanced zeolite accessibility and activity. (see color insert)*

The large hydrocarbon molecules in the feed can also crack on the exposed zeolite surface itself, rather than just on the active amorphous matrix material. This difference provides the potential for improved selectivities with the reduced coke formation characteristic of zeolite cracking. The secondary diffusion pathway of the cracked products to the internal crystalline zeolite surface is also minimized, resulting in less overcracking (i.e. undesirable conversion of gasoline to gasoline to liquefied petroleum gas (LPG), dry gas and coke). The net result is higher bottoms upgrading with lower delta-coke (weight percent (wt %) coke yield/cat to oil), leading to increased yields of valuable liquid products.

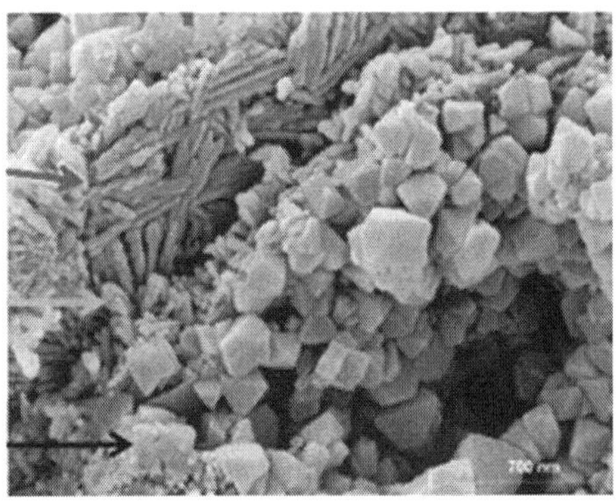


*Figure 2. Scanning Electron Microscopy (SEM) images depicting different matrix and zeolite layers in different colors. The exposed zeolite layer covering the macropore walls lead to selective cracking of large hydrocarbon molecules. (see color insert)*

The contaminant metals in residue feed that need to be controlled in the FCC include vanadium (V) and nickel (Ni), with iron (Fe) and calcium (Ca) also present in high levels in some crudes (*1*). Sodium (Na) is typically reduced to low levels by crude desalting. The potential detrimental effects of these metals on FCC performance are summarized below:

- Nickel (Ni) – Dehydrogenation activity leading to increased $H_2$ and coke
- Vanadium (V) – Catalyst deactivation, with some dehydrogenation activity

- Iron (Fe) – Surface blockage of some catalyst types resulting in lower conversion
- Calcium (Ca) – Involved in catalyst deactivation and iron poisoning
- Sodium (Na) – Involved in catalyst deactivation

When processing residue feeds, usually the feed contains a significant amount of V, which in the presence of steam and high temperatures, deactivates the catalyst by damaging the zeolite structure. The mechanism of FCC catalyst deactivation by V follows the steps below (*2*):

1) Vanadium is deposited onto the catalyst and is oxidized in the FCC regenerator
2) The oxidized form undergoes further reactions to form several highly mobile types of vanadic acids
3) These vanadic acids remove $Na^+$ from the zeolitic exchange sites to form sodium vanadate
4) The sodium vanadate hydrolyses to sodium hydroxide ($Na^+OH^-$)
5) The hydroxyl group ($OH^-$) then attacks the silica-oxygen zeolite framework leading to zeolite collapse/destruction/catalyst deactivation

In order to mitigate the V and Na effects, FCC catalyst suppliers focus on both the trapping or passivating of V coming in from the feed and the reduction of residual $Na^+$ on the fresh catalyst from the manufacturing process. FCC catalyst vendors minimize the amount of residual $Na^+$ on zeolite by reducing it to ultra-low levels via a unique combination of calcination and ion exchange steps, which improves resistance to Na-V zeolite deactivation.

To passivate V, catalyst manufacturers focus on adding vanadium trapping components in the catalyst formulation either inside the catalyst particle or as a separate particle. Another way to mitigate vanadium poisoning is by increasing zeolite surface area and decreasing the zeolite crystal size. Higher amounts of zeolite and its increased accessibility due to smaller crystallite size provide a higher amount of available active centers and consequently it takes more vanadium to deactivate those acid sites. Moving to catalyst with smaller zeolite crystallite size has also been reported to increase the yield selectivity through improvement in the diffusivities of the reactants and products. However, this requires the right silicon (Si) to aluminium (Al) ratio so that the catalyst is stable in the presence of residue feeds and under the hydrothermal conditions of the FCC (*4*, *5*).

It is generally accepted that the dehydrogenation activity of metals can be expressed in terms of Equivalent nickel (Ni) as Ni + V/4 + iron (Fe)/10 + 5 copper (Cu) - 4/3 Antimony (Sb). Typically, resid feeds contain higher amounts of Ni, V and Fe. Thus, it is especially important to mitigate the dehydrogenation effects of feedstock contaminant Ni. This is done within FCC catalysts by incorporating a specialty crystalline alumina into the matrix to trap the Ni. By examining spent FCC catalyst from refiners processing resid feeds with electron microscopy, it is generally observed that while V is homogeneously distributed throughout the particle, Ni mainly deposits and accumulates on the outer stages of the catalyst as shown in Figure 3 (*6*, *7*).

Conventional manufacturing techniques, practiced by most FCC catalyst suppliers, result in the specialty Ni-trapping alumina being more or less evenly dispersed throughout the particle. This results in a large portion of the specialty alumina being located in the interior of the particle, which renders it unavailable to react with Ni and in essence wasted. By using novel catalyst technologies, such as the multi-stage reaction catalyst (MSRC) concept, the spatial distribution of the specialty sub-micron-sized crystalline alumina within the particle is optimized to maximize its efficiency in Ni trapping and this enhanced efficiency leads to improved catalyst performance (*8*). The result is a two-stage catalyst. The inner-stage of the catalyst has a tailored nanoporous structure to allow enhanced diffusion of heavy molecules and selective pre-cracking on the exposed zeolite surface. The outer-stage is also based on the same technology platform, but is enriched with specialty crystalline alumina to trap Ni where it enters and deposits on the catalyst (*8*).

When Ni on equilibrium catalyst ($E_{cat}$) is above 1,000 parts-per-million (ppm), antimony (Sb), which acts as a Ni passivator, can also be injected into the feed. Efficacy of this Ni passivation strategy can be monitored by the absorber off-gas hydrogen ($H_2$)-to-methane ($CH_4$) ratio. It is proposed that Sb forms an alloy with Ni, mitigating its dehydrogenation activity on catalyst surface (*10*).

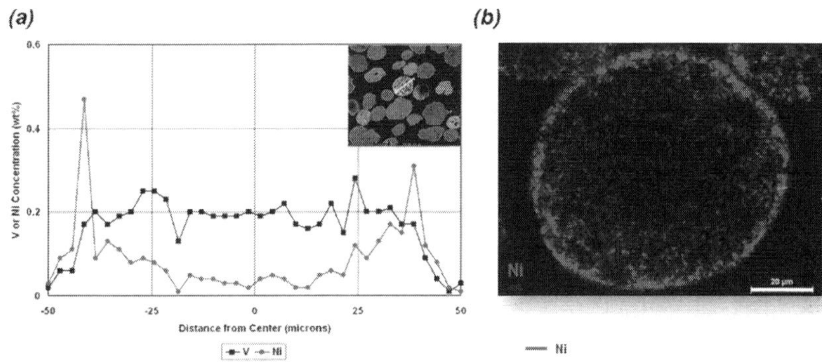


*Figure 3. (a) Deposition profiles for contaminant metals Ni and V on FCC catalysts measured by electron-probe microanalysis. (b) Nickel distribution on an FCC catalyst particle by energy dispersive X-ray spectroscopic analysis. (see color insert)*

## Tight Oils

The most dramatic change in the global energy landscape in recent years has been due to the impact of tight oil. Tight oil can be defined as the light crude oil contained in petroleum-bearing formations of low permeability, which are typically shale or tight sandstone. Due to their increased availability, tight oils are representing more and more of the feedstock diet for fluid catalytic cracking (FCCUs) units in recent years. Tight oils are produced from shale plays with many reserves in the United States and represent unique challenges for the refining industry. Significant reserves are also found in China, Russia, and Argentina. Feed properties of tight oils differ from conventional crudes; tight oils are generally lighter, have higher American petroleum institute (API) gravities, produce less coke, and have lower contaminant metals such as nickel and vanadium, but typically come with higher amounts of contaminants such as iron and calcium. Refiners around the United States are looking at processing tight oils, and many already have. The figure below graphically demonstrates the number of refiners in the United States, separated by petroleum administration for defense districts (PADDs), who have processed tight oils.

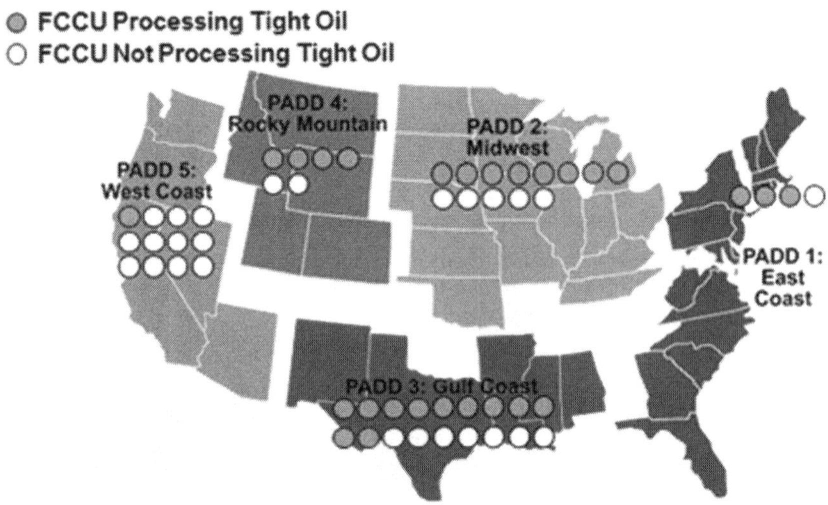


*Figure 4. A representation of the number of refineries that have moved to processing tight oils, either exclusively or partially, separated by PADD region. (see color insert)*

Not only are nanomaterials necessary for the extraction and pre-FCC upgrading of tight oils, nanomaterials are also a necessity for their processing in an FCC unit. The unique architecture of the FCC catalyst is vital to this process, with pores, channels, zeolite structures, and reactive acid sites that perform the highly selective chemistry on nanoscale. The pore opening of Y-zeolite, which provides much of the activity of the FCC catalyst, is approximately 7.4 Å wide, allowing for size-selective cracking of feed molecules into fuels such as gasoline. To deal with swings in feed qualities between conventional and opportunity crudes, a tailored catalyst approach must be followed and entails the selection of appropriate nano-scale components in the FCC catalyst.

For instance, the processing of tight oil in an FCC, because of its high paraffinicity and high API gravity, could lead to heat balance issues. To combat this, an optimum catalyst would be less coke selective, providing heat to the FCC unit. The pore architecture of the catalyst, mainly in the mesopore regime (20 to 200 angstroms), is the variable by which this limitation could be overcome. Another constraint of processing tight oils is gasoline octane. In order for the gasoline off the FCC to be valuable, it must meet the gasoline octane specifications set forth by the refinery. Gasoline from tight oil processing may be octane short.

To overcome this, the amount of rare earth cations on the catalyst can be tuned and reduced. Varying rare earth content on the catalyst changes the coordination at the active sites of the catalyst. For instance, lanthanum can stabilize three active sites as a 3+ cation, resulting in a direct influence on Unit Cell Size (UCS) of the catalyst on the angstrom scale. Rare earth elements stabilize the active sites on the zeolite, resulting in enhanced activity. This is depicted in the graphic below (Figure 5), showing the extremes of high and low UCS. Lower rare earth oxide (REO) levels will allow for higher gasoline octane, minimizing that constraint of tight oil processing.

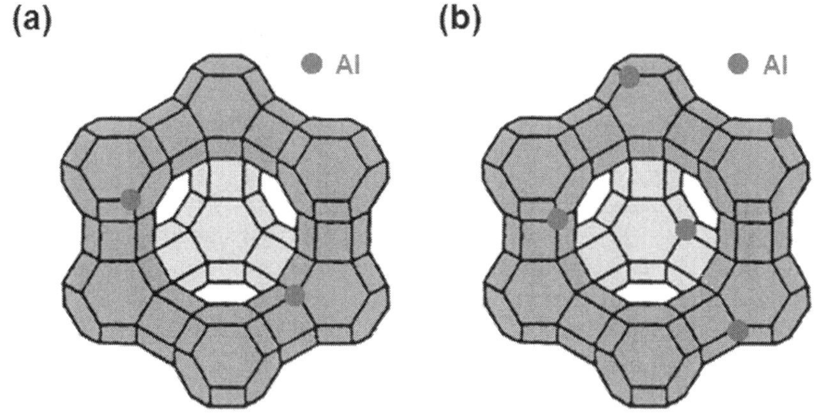


*Figure 5. Depiction of lower (a) and higher (b) unit cell size on zeolite Y framework within an FCC catalyst. Blue spots represent aluminum sites. (see color insert)*

Tight oils also bring other variables that are not commonly seen in conventional crudes: higher iron, calcium and sodium. Iron and calcium can act synergistically with other species, such as silica, to form low-melting eutectics, which can cause a loss in surface area, and subsequently, a loss in catalyst activity for some catalyst technologies. The catalyst structure must be optimized to withstand iron and calcium poisoning. With unsuitable catalyst technologies (i.e., closed pore architectures that are susceptible to blockage; and surface chemistries that favor such eutectic formation) iron containing formations can also block the surface pores of the catalyst, effectively closing off the catalyst interior to further reactions. Therefore, the surface porosity and catalyst architecture to maintain accessibility are important when choosing/desigining catalysts for processing tight oils. Case studies have shown a loss in conversion and bottoms upgrading relating to iron upsets due to unsuitable catalyst; this means less gasoline and diesel output for the refinery, and represents a problem that can effectively be mitigated via catalyst design at the nanoscale. Another issue that can arise with the increase in iron levels in feeds is the iron nodule formation on the surface of the FCC catalyst. This is a physical phenomenon seen on circulating equilibrium catalyst ($E_{cat}$) and can affect the circulation of catalyst due to a change in apparent bulk density. This physical phenomenon is depicted below in Figure 6, in which both fresh (left) and normal (middle) $E_{cat}$ are compared against iron-poisoned $E_{cat}$ (right). Nodulated surface formation on $E_{cat}$ can be as thick as 1 micrometer, but do not inherently translate to surface blockage and performance debits.

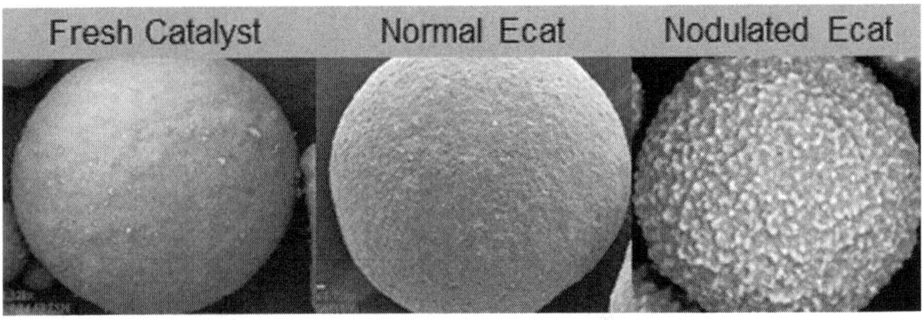


*Figure 6. Depiction of iron nodulation formation on the surface of FCC catalyst. (see color insert)*

Sodium is an alkali metal which neutralizes the acid cracking sites on the catalyst. One way to mitigate the impact of sodium is to reduce the sodium level in the fresh FCC catalyst, as previously discussed. The low sodium on fresh catalysts also reduces the detrimental effect of vanadium, as vanadium interacts with sodium synergistically (*2*). Innovative catalyst technology platforms with an open pore architecture and highly accessible zeolite content have been shown to result in catalysts with low sodium levels (*10*).

The FCC catalyst remains a versatile variable to handle crude oil refining, and as described here, changes on the nanoscale can help combat large scale tight oil processing in a refinery.

**Processing Biomass-Derived Feeds**

In recent years, with increasing energy demand and growing environmental concerns, there has been an increasing interest in alternatives to fossil-based energy sources. Out of these alternatives one sustainable source that can be converted to mainstream liquid trainsportation fuels is biomass, which also includes edible and inedible vegetable oils and animal fats.

Processing plant biomass via fast pyrolysis technology to produce bio-oil is an important form of biomass fuel development. This method can yield up to 75 % bio-oil. While bio-oil provides a sustainable alternative to crude oil, bio-oil still has challenges that need to be addressed. Currently, bio-oil is not soluble in petroleum-based fuels and contains over 400 components. Additionally, bio-oils can polymerize with time, under high temperature, and with oxygen exposure. Burning bio-oil can also be an issue due to the high oxygen and water content. Bio-oils also contain high concentrations of acids that can cause corrosion and coking can occur via polymerization of thermally unstable compounds. With these known challenges, the use of bio-oils has been limited. The thermal stability, acid content and heating value of bio-oil need to be improved before it can reach fuel standards. Catalytic cracking has been used as a technique to upgrade the current bio-oil production (*11*).

A wide range of products can be formed via FCC upgrading of bio-oil, mainly liquid oil products, water, coke, tar, and gas – primarily carbon dioxide ($CO_2$). These products depend on the type of catalyst used and FCC operating conditions. High operating temperatures primarily yield two phases - organic and aqueous, while lower temperatures yield one homogenous phase product. Research has shown optimized liquid oil yields at temperatures above 500 °C along with a reduction in aqueous phase contents; however, these higher temperatures also lead to increases in gaseous products (*12*) and coke formation due to secondary cracking and the favoring of aromatization/polymerization, respectively (*13*). The coke formation associated with bio-oils is more severe than heavy oil and oil-sand feedstocks. In addition to the control of reaction temperature, optimization of reaction time and FCC process conditions is necessary for bio-oil upgrading with improved efficiency. Decreased residence time is advantageous for reduced coke production, but the reduced time does not allow for increased bio-oil deoxygenation. The gasoline produced from bio-oil contains a higher aromatic content than conventionally produced gasoline because of the lack of hydrogen in the FCC environment (*14*).

The type of catalyst chosen for bio-oil cracking is extremely important as the conversion and product distribution are highly dependent on the catalyst. Nanoporous catalysts are considered the most effective in bio-oil upgrading, yielding maximum organic distillates, total valuable hydrocarbons, and aromatics, while producing minimum coke (*15*). Silica-alumina catalysts with an open pore architecture effectively minimize char, while catalyst containing high acidic zeolites yield the least amount of tar. At operating conditions above 450 °C, dealumination occurs more readily caused by high oxygen concentration and high water content. The coke formation that occurs on an FCC catalyst from bio-oil cracking can result in irreversible deactivation. Despite processing at low temperatures, coking can remain much higher and more significant than cracking petroleum-based feedstocks (*15*) Reactor plugging issues caused by char and coke formation occur even in low temperature reactions (*11*).

## Nanomaterials To Address Changes in Fuels Demand

Recent years have brought a major shift in the world's fuel demand for gasoline and diesel range products. While global gasoline consumption has shown minor increases, the demand for distillates has been very strong and is mainly driven by the transportation sector (*16*). The outlook for the years 2015 to 2035 shows an incremental growth in diesel demand above the 2015 level at more than three times higher than that of gasoline (Figure 7) (*16*). As a result, more and more FCC units target increasing their diesel output in order to optimize refining margins.

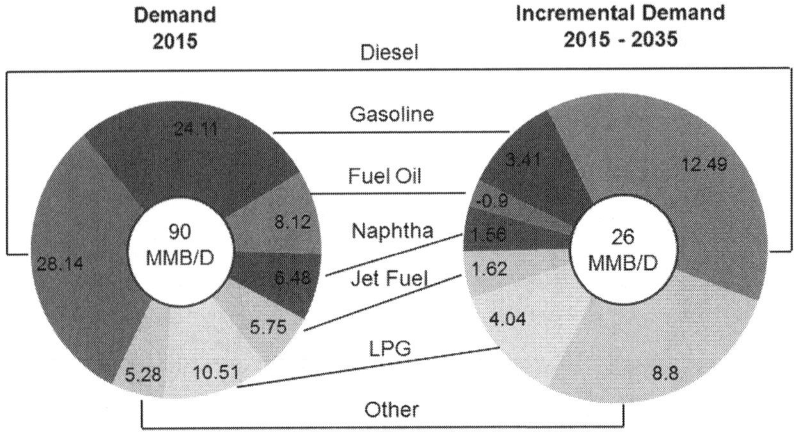


*Figure 7. Current demand and incremental demand growth for fuels. (see color insert)*

Aside from operational optimizations, such as undercutting gasoline, lowering riser temperature, slurry recycling and improvements to hardware, the catalyst has a profound impact on product distribution. If, for example, the activity in the FCC unit is adjusted by lowering the riser outlet temperature or by lowering catalyst additions, a catalyst not optimized for distillate mode operation could show high residual slurry yields. In the case of resid operations, lowering catalyst additions would also result in high coke and hydrogen yields if the catalyst does not feature an optimized metals passivation system.

Conventionally, low zeolite to matrix surface area ratios (Z/M ratios) are applied to maximize the distillate yield by both increasing the matrix surface area (MSA) and lowering the zeolite surface area (ZSA). However, without further optimization, a lower Z/M ratio results in accentuated matrix cracking with poor coke and gas selectivities. In addition, a matrix with low hydrothermal stability will result in low MSA and high Z/M in the equilibrium catalyst despite the fact that the fresh catalysts show low Z/M ratios.

Three key features are critical for a catalyst technology targeting distillate optimization (*17*). Firstly, the matrix needs to have good hydrothermal stability and low coke and dry gas make. Secondly, the matrix and zeolite content and distribution need to be optimized. Lastly, the pore size distribution of the catalyst needs to be optimized to improve the diffusion of reactants and products and the upgrading of bottoms to distillate range molecules.

In recent years there have been innovative technology platforms developed using the *in-situ* manufacturing route targeting distillate maximization operation (*18*, *19*). Most conventional matrix materials, such as alumina, predominantly have Lewis acid sites as measured by Fourier transform infrared (FT-IR) spectroscopy studies using pyridine as the probe molecule. These manufacturing processes not only generate Lewis acid sites, but also a small yet significant fraction of Brønsted acid sites (*18*, *19*). The role of Brønsted acid sites in zeolitic cracking is very well known. The Brønsted sites on the matrix material, although weaker than zeolitic sites, act synergistically with the Lewis sites to enhance catalytic properties.

This tailored version of the *in situ* manufacturing process for maximizing diesel yields from the refinery not only forms both the matrix material and zeolite in one single step, but also arranges them on the nanoscale to be in intimate contact with one another. The resulting structure is illustrated in Figure 8. The submicron Y-zeolite crystallites are found to be in close proximity with the *in situ* generated matrix. It is this unique synergy between the zeolite and matrix that leads to rapid transfer of reactant and feed molecules from zeolitic acid sites to matrix acid sites. This enhanced transfer helps to stabilize coke precursors produced by the matrix cracking, leading to higher distillate production with lower coke. Significant improvements in distillate yields and reduction in bottoms yields have been demonstrated by these novel catalysts (*19*).

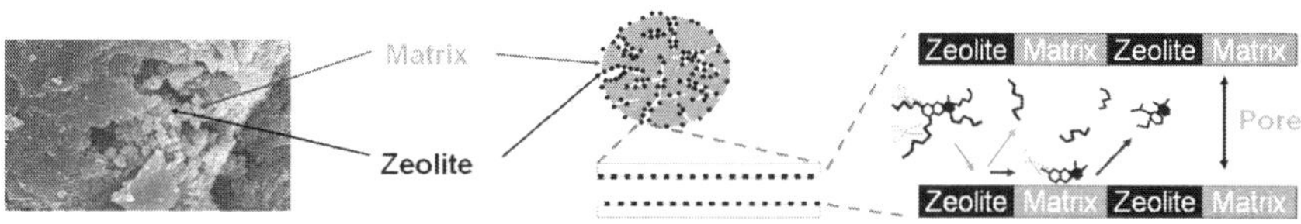


*Figure 8. SEM image and schematic depicting an FCC catalyst manufactured via in situ synthesis technology to maximize diesel yields from the refinery. (see color insert)*

## Nanomaterials To Address Environmental Concerns in FCC

The FCC has always been a critical process for delivering profits to refineries and supplying transporation fuels to our industrialized society. However, if not operated in the correct way, it can also be a major source of undesirable pollutants such as carbon monoxide (CO), nitrogen oxides ($NO_x$), sulfur oxides ($SO_x$) and particulate matter (PM). With the increasing need to monitor and reduce environmental pollutants, the use of FCC additives have become the preferred solution with minimal capital investment.

FCC units equipped with CO boilers typically have no problem meeting CO emission limits due to the complete oxidation of CO to carbon dioxide ($CO_2$), which can be safely released into the environment. In the absence of a CO boiler, which is common in full burn units, the CO needs to be oxidized in the regenerator. The extent to which this occurs depends on the regenerator design, air and spent catalyst distribution, dense bed temperature and excess oxygen present. To facilitate this oxidation, many refiners will utilize FCC additives known as CO promoters in the fresh catalyst inventory. Platinum is the most commonly used active ingredient in CO promoters although this oxidation can also be catalyzed by a wide variety of noble and base metals (*20*).

Nitrogen oxide ($NO_x$) can be formed by two mechanisms: thermal $NO_x$ produced from reaction of molecular nitrogen with oxygen and fuel $NO_x$ produced from the oxidation of nitrogen-containing coke species (*21*). Experiments have shown that the latter is the more significant contributor to total $NO_x$ in an FCC regenerator. Reducing $NO_x$ generation requires both curbing the oxidation of nitrogen compounds into $NO_x$ and reducing the $NO_x$ formed into nitrogen. This reduction can be achieved by replacing or modifying the platinum nanoparticles commonly used in CO promotors to enable selective catalysis. The use of oxygen storage compounds, which had been explored in great detail during development efforts for the catalytic converter, enabled the invention of new CO promoters that also lowers $NO_x$ emissions (*22*).

Sulfur oxide ($SO_x$) emission [($SO_x$) = sulfur dioxide ($SO_2$) + sulfur trioxide ($SO_3$)] is a function of the feed sulfur, coke yield and conversion. Hardware solutions such as wet gas scrubber and hydrotreaters are effective in reducing SOx emissions and formation, respectively, but the high cost of capital investments drove the development of $SO_x$ reducing additives. The use of $SO_x$ reduction additives was successfully commercialized in the mid-1980s with the introduction of an additive based on the magnesium aluminate [$MgAl_2O_4$] spinel structure (*23*). $SO_x$ reduction additives typically consist of a metal-based oxidation catalyst to oxidize $SO_2$ to $SO_3$ and then a magnesium-based "pick up agent" to absorb the $SO_3$ and release it as hydrogen sulfide ($H_2S$) in the reactor to be further treated. Typically, $SO_x$ reduction additives can capture sulfur associated with paraffins, naphthenes and simple aromatics. Sulfur in large, highly aromatic feed molecules is laid down as coke and ends up in the regenerator, which will then burn off to produce $SO_x$. Therefore, catalyst with high matrix activities, such as bottoms cracking catalysts, can also help to reduce $SO_x$ emissions due to their capacity to break apart larger aromatic molecules.

The amount of particulate matter (PM) released from an FCCU is dependent on the efficiency of the cyclones and the attrition resistance characteristics of the catalyst and additive inside the FCCU. To minimize the PM emissions to the environment, considerations such as reducing catalyst addition, improving the attrition resistance of the catalyst and the use of low microfines catalysts have been introduced. Catalyst addition can be reduced by using higher activity catalyst or activity enhancement additives, which can also further improve product slate. PM can also be controlled by capital intensive equipment such as electrostatic precipitators (ESP), but similar to $SO_x$ additives, attrition resistant catalyst research has historically been driven by the desire to avoid capital projects by the refineries. Low attrition products have been developed to reduce particle fracture and in turn reduce PM formation (*24*).

## Conclusions and Outlook

Energy used for transportation is typically delivered by liquid transportation fuels (e.g., gasoline, diesel, jet fuel etc.) that are produced by the fluid catalytic cracking (FCC) process. Catalyst is the core of this process as it dictates the overall profitability by defining the catalytic activity and selectivity. Successful catalyst design involves fine-tuning its sub-micron building blocks and engineering of the pore architecture on the nanoscale. Effective use of nanomaterials is crucial to meet the current and future challenges due to changes in feedstock supply (resid feeds, tights oils, biomass derived feeds), product demand (diesel vs. gasoline) and environmental concerns. Recent developments in this field and the progression of fluid catalytic cracking (FCC) catalysts and additives over the course of recent history demonstrate how a mature field can constinously evolve through succesful utilization of nanomaterials to overcome the challenges it faces.

## References


1. Pinto, F. V.; Escobar, A. S.; de Oliveira, G. G.; Lam, Y. L.; Cerqueira, H. S.; Louis, B.; Tessonnier, J. P.; Su, D. S.; Pereira, M. M. The Effect of Alumina on FCC Catalyst in the Presence of Nickel and Vanadium. *Appl. Catal., A* **2010**, *388*, 15–21.
2. Xu, M.; Liu, X.; Madon, R. Pathways for Y Zeolite Destruction: The Role of Sodium and Vanadium. *J. Catal.* **2002**, *207*, 237–246.
3. McLean, J. et al. *Distributed Matrix Structures – A Technology Platform for Advanced FCC Catalyst Solutions*, NPRA AM-03-38.
4. Camblor, M. A.; Corma, A.; Martinez, A.; Mocholi, F. A.; Perez Pariente, J. Catalytic Cracking of Gasoil. Benefits in Activity and Selectivity of Small Y Zeoiltecrystallites Stabilized by a Higher Silicon-to-Aluminium Ratio by Synthesis. *Appl. Catal.* **1989**, *55*, 65–74.
5. Tonetto, G.; Atias, J.; de Lasa, H. FCC Catalysts with Different Zeolite Crystallite Sizes: Acidity, Structural Properties and Reactivity. *Appl. Catal., A* **2004**, *270*, 9–25.
6. Kugler, E. L.; Leta, D. P. Nickel and Vanadium on Equilibrium Cracking Catalysts by Imaging Secondary Ion Mass Spectrometry. *J. Catal.* **1988**, *109*, 387–395.
7. Lappas, A. A.; Nalbadian, L.; Iatridis, D. K.; Voutetakis, S. S.; Vasalos, I. A. Effect of Metals Poisoning on FCC Products Yields: Studies in an FCC Short Contact Time Pilot Plant Unit. *Catal. Today* **2001**, *65*, 233–240.
8. McLean, J. B., Hoffer, B. W., Smith, G. M., Stockwell, D. M., Shackleford, A. S. *Multi-Stage Reaction Catalysts: A Breakthrough Innovation in FCC Catalyst Technology*, NPRA-AM-11-01.
9. Corma, A.; Grande, M. S.; Iglesias, M.; del Pino, C.; Rojas, R. M. Nickel Passivation on Fluidised Cracking Catalysts with Different Antimony Complexes. *Appl. Catal., A* **1992**, *85*, 61–71.
10. Pan, S; *Capturing Maximum Values for Processing Tight Oil through Optimization of FCC Catalyst Technology*; AFPM Annual Meeting, 2014.
11. Al-Sabawi, M.; Chen, J.; Ng, S. Fluid Catalytic Cracking of Biomass-Derived Oils and Their Blends with Petroleum Feedstocks: A Review. *Energy Fuels* **2012**, *26*, 5355–5372.
12. Adjaye, J. D.; Bakhshi, N. N. Catalytic Conversion of Biomass-Derived Oil to Fuels and Chemical 1: Model Compound Studies and Reaction Pathways. *Biomass Bioenergy* **1995**, *8*, 131–149.
13. Sharma, R. K.; Bakhshi, N. N. Catalytic Upgrading of Pyrolysis Oil. *Energy Fuels* **1993**, *7* (2), 306–314.
14. Guo, X. Y.; Yan, Y. J.; Li, T. C.; Ren, Z. W.; Yuan, C. M. C. Catalytic Cracking of Bio-Oil from Biomass Pyrolysis. *Chin. J. Process Eng.* **2003**, *3*, 91–95.

15. Adjaye, J. D.; Bakhshi, N. N. Production of Hydrocarbons by Catalytic Upgrading of a Fast Pyrolysis Bio-Oil. Part 2: Comparivative Catalyst Performance and Reaction Pathways. *Fuel Process. Technol.* **1995**, *45* (2), 185–202.
16. *Global Crude Refining and Fuels Outlook*; World Refining and Fuels Service, Hart Energy, 2014.
17. McLean, J. *Maximising FCC Distillate Production*, PTQ Catalysis Review, 2010.
18. Keeley, C.; Mayol, J.; Riva, S.; Komvokis, V.; Challis, S. *EPTQ Catal.* **2013**, 31–35.
19. Jollien, Y.-A.; Keeley, C.; Mayol, J.; Riva, S.; Komvokis, V. *Hydrocarbon Process.* **2013**, 63–67.
20. Vaarkamp, M., *Control Strategies for NOx and SOx Emissions frm FCCUs*, NPRA paper AM-04-21, March 2004.
21. Kelkar, C. P., Stockwell, D. M., Winkler, W. S, Tauster, S., Sexton, J. A., Cantley, G. A., Wick, J. P. *New Additive Technologies for Reduced FCC Emissions*, NPRA paper AM-02-56, March 2002.
22. *OxyClean CO Promoter Reduces NOx Emissions by 45%*, Performance Profile BF-7891, April 2010.
23. Evans, M., *Reducing FCC SOx in Partial Burn Regenerator*, Digital Refining, April 2008.
24. Farley, C., Andrews, R., McClung, R. *Controlling FCC Particulate Emissions*, NPRA paper AM-07-19, March 2007.

## Chapter 2

# Anodic Oxide Nanostructures and Their Applications in Energy Generation and Storage

**Stephen DeWitt[1] and Katsuyo Thornton*,[1,2]**

**[1]Applied Physics Program, University of Michigan, 2300 Hayward Street, Ann Arbor, Michigan 48109, United States**
**[2]Department of Materials Science and Engineering, University of Michigan, 2300 Hayward Street, Ann Arbor, Michigan 48109, United States**
***E-mail: kthorn@umich.edu.**

This chapter reviews the morphologies, growth mechanisms, and applications of anodic oxide films. Experimentally, it is possible to control the growth conditions of these films to yield several distinct morphologies, including self-ordered nanoporous and nanotubular films and more complex morphologies such as nanolace. Fundamental processes that lead to self-ordering of nanoscale features, including interfacial reactions, ionic transport, stress generation, and space charge accumulation, are discussed. The high-aspect-ratio geometries and nanoscale feature sizes of anodic oxide films are attractive for a range of applications either as an active component or as a template. Specifically, applications to solar cells, batteries, and supercapacitors are reviewed.

## Introduction

Anodic oxide films, formed by the electrochemical oxidation of a metal substrate, display a wide range of morphologies including compact barrier films, nanoporous films, nanotubular films, and more complex morphologies such as nanolace. The film grows as oxygen ions from the electrolyte react with the atoms from the metal substrate to form new oxide. The current is transported through the growing film by the combined migration of both metal and oxygen ions.

Anodization has been an important industrial process since the early 20$^{th}$ century (*1*, *2*). Traditionally, anodic films have been utilized as the dielectric layer in electrolytic capacitors and to improve the corrosion resistance, abrasion resistance, and dye absorption of metals (*1*). Since the development of highly ordered anodic nanostructures in 1995 (*3*), anodic films have found an increasing range of possible applications. Especially promising are the applications of anodic films for energy generation and storage, such as in solar cells, batteries, and supercapacitors. Other applications include sensors, membranes, and medical devices (*4*). In this chapter we present a brief summary of the anodic nanostructure morphologies, the mechanisms involved in anodic film growth, and the energy applications of anodic nanostructures.

## Experimental Morphologies

### Barrier Films

The anodic oxide films with the simplest morphology are barrier films. As shown in Figure 1a, barrier films are compact films that grow conformally on the substrate. These films can grow up to hundreds of nanometers thick, usually limited by the dielectric breakdown voltage of the oxide (*1*). Anodic barrier films can form on a number of metallic substrates including aluminum (Al), iron (Fe), hafnium (Hf), niobium (Nb), tantalum (Ta), titanium (It), tungsten (W), and zirconium (Zr) (*2*). These films grow in electrolytes that do not form soluble complexes with the metal cations in the substrate, thus preventing the dissolution of the oxide at the oxide/electrolyte interface (*1*). While the thin, compact nature of these films does not lend itself to applications in energy storage and generation, the simple geometry of barrier films provides a model system to examine the mechanisms governing anodization.

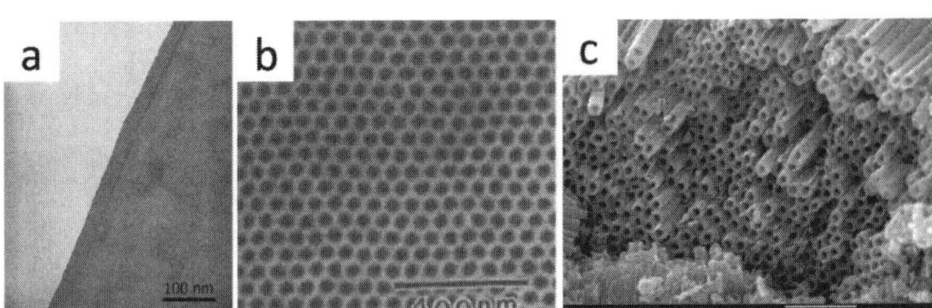


*Figure 1. Typical anodic nanostructures: (a) cross section of a barrier film (reproduced from reference (5)), (b) top-down view of a nanoporous film with the barrier layer etched away (reproduced with permission from reference 6, copyright 1997 The Electrochemical Society), and (c) nanotubular film broken in half. (Reproduced with permission from reference (7). Copyright 2006 American Chemical Society.)*

## Nanoporous Films

Nanoporous anodic films exhibit cylindrical channels that extend from the film surface down toward the film/substrate interface, as seen in Figure 1b. These films grow in electrolytes that form soluble complexes with the metal substrate cations. For example, nanoporous alumina films grow in a variety of acidic electrolytes, including chromic, oxalic, phosphoric, and sulfuric acids (*1*). Other substrates that can yield nanoporous anodic films include Nb (*8*), Ta (*9*), W (*10*), and tin (Sn) (*11*).

Although porous aluminum oxide (PAO) films have been studied since the 1850s (*12*), the first direct observation of the porous structure was by Keller, Hunter, and Robinson in 1953 (*13*). The nanopores nucleate randomly across the substrate surface and eventually self-organize into a hexagonal array perpendicular to the plane of the substrate, yielding a porous structure that is disordered near the film surface (where the pores initially formed) but becomes ordered at some depth (*6*, *14*).

In 1996, Masuda and Satoh pioneered a two-step anodization process to grow nanoporous films that are well ordered throughout the entire film (*15*). In this process, the substrate is anodized until the pores are well ordered at the base of the film. The oxide is then dissolved, leaving behind a dimpled substrate surface. The substrate is then anodized a second time, and the growth of the pores is directed by the depressions in the substrate. The pores nucleate as a well-ordered array with the optimal inter-pore spacing and grow into straight pores oriented perpendicular to the substrate surface. In addition, nano-indentation (*16*) and a focused ion beam (FIB) (*17*) have been applied to direct the growth of the pores.

The geometry of the nanopores depends on the substrate material, the electrolyte, and the applied potential/applied current, as has been systematically examined by Friedman *et al.* (*18*) For ordered anodic alumina nanopores, the inter-pore spacing can range from 20 to 500 nm (*18*, *19*), the pore diameters can range from 7 to 250 nm (*13*, *18–20*), and the barrier thickness can range from 5 to 225 nm (*20*, *21*), all of which increases approximately linearly with the applied potential (*18*, *21*). The interpore spacing and pore diameter are essentially independent of the electrolyte pH (*18*). The pore growth rate ranges between 0.1 and 16 nm/s and increases as the applied potential increases and as the pH decreases (*16–18*). Nanoporous films can be much thicker than barrier films, up to 100 μm thick (*22*). The chemical (open-circuit) dissolution rate of the oxide determines the maximum film thickness. The electrolyte slowly dissolves the walls of the growing pores, tapering them to a point. The maximum thickness is reached once the chemical dissolution at the top of the film matches the electrochemical growth at the base of the film (*23*, *24*).

The morphologies of nanoporous films on other substrates have not been examined as comprehensively. Well-ordered nanoporous films on Ta and Nb and disordered nanoporous films on W and Sn have been reported, with pore feature sizes on the order of tens of nanometers (*8–11*).

## Nanotubular Films

Nanotubular films are another commonly observed anodic nanostructure. As seen in Figure 1c, nanotubular films have a cylindrical central pore, like nanoporous films, but also have voids between the cells, separating the nanotubes from each other. As in nanoporous films, two-step anodization improves the ordering of the nanostructure (*25*).

The first anodic nanotubular films were developed by Zwilling *et al.* in 1999 using Ti and a Ti-Al alloy as the substrates (*26*). Since then nanotubular films have been grown on Zr (*27*), Hf (*28*), and Fe (*29*), as well as a variety of alloy systems (*12*). For these materials, $F^-$ or $Cl^-$ ions in the electrolyte are required to form soluble complexes with the substrate metal ions (*12*, *30*).

For nanotubular films grown on Ti, Fe, Zr, or Hf substrates, typical pore diameters range from 15 to 150 nm with wall thicknesses between 10 and 30 nm and barrier thicknesses from 20 to 80 nm (*27–29*, *31–33*). As in nanoporous alumina films, the pore diameter and barrier thickness has been shown to increase linearly with the applied potential (*31*, *33*). The nanotube wall thickness is essentially independent of the applied potential (*31*, *32*). Nanotubular titania films have been grown up to 1 mm in thickness (*34*), enabled by the use of organic electrolytes with decreased chemical dissolution of the oxide (*24*).

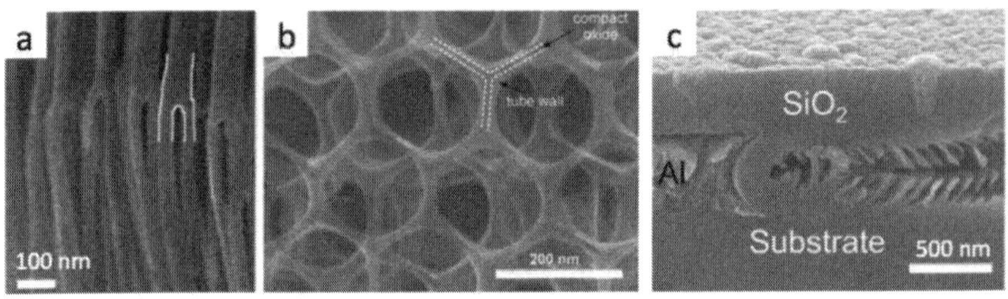


*Figure 2. Examples of more complex anodic oxide nanostructures: (a) branched nanopores (reproduced from reference (35), copyright United States Academy of Sciences), (b) nanolace (reproduced with permission from reference (36), copyright 2008 John Wiley & Sons), (c) dendritic nanopores. (reproduced with permission from reference (37), copyright 2011 The Electrochemical Society.)*

## Films with More Complex Morphologies

Beyond the standard morphologies, films with more complex morphologies have been developed, including modulated-diameter structures, branched structures, nanobamboo, perforated nanopores, nanolace, dendritic nanopores, and pores with nonhexagonal lattices. Examples of these morphologies can be seen in Figure 2. These morphologies can be generated by changing the potential during growth, by partially etching the film, by constraining the geometry during anodization, or by patterning the substrate surface before anodization. Nanoporous alumina films with periodically modulated pore diameters have been developed by switching between a low applied potential and a high applied potential (*38*). A similar approach can be applied to form nanobamboo structures (*36*). A potential decrease by a factor of $1/\sqrt{n}$ during anodization causes each pore to split into *n* branches (see Figure 2a) (*35*). By etching nanoporous films with modulated pore diameters, it is possible to generate morphologies such as periodically perforated nanopores (*39*) and nanolace (see Figure 2b) (*36*). Nanopores with horizontal dendritic structures can be formed by anodizing Al inside an $SiO_2$ mask (see Figure 2c) (*37*). Finally, FIB or nanoindentation prepatterning of the substrate to guide pore nucleation can yield pores with square or "snowflake" lattices (*17*, *38*).

## Mechanisms Governing Growth and Self-Organization

### Fundamental Processes Governing Anodic Film Growth

The range of nanostructures described in the previous section is the result of the combination of several fundamental processes: reactions at both the oxide/electrolyte and metal/oxide interfaces, ionic transport through the oxide, stress generation and its effects, and space charge accumulation. Understanding these underlying processes and their interactions would enable further optimization of the morphologies and properties of anodic oxide films.

#### *Interfacial Reactions*

At the oxide/electrolyte interface, a series of simultaneous reactions occur, involving both oxygen (O) and metal species. These reactions result in $O^{2-}$ ions entering the film from the electrolyte and in metal ion complexes such as $TiF_6^{2-}$ exiting the film (*24*, *40*). For anodic alumina, the pathways for these reactions have been studied in detail by Våland and Heusler by measuring the pH-dependent reaction fluxes of O and Al species (*40*). Anions from the electrolyte (e.g. phosphate anion, $PO_4^{3-}$, fluoride anion, $F^-$) can also react at the oxide/electrolyte interface and become incorporated into the film, comprising up to 14 wt. % of the film (*24*, *41*). In contrast to the multiple reactions at the oxide/electrolyte interface, the reaction at the metal/oxide interface is much simpler. Metal atoms from the substrate react to become ions and then enter the oxide, which may combine with incoming oxygen ions to form a new oxide (*22*).

#### *Ionic Transport*

Tracer experiments using implanted noble gas atoms indicate that anodic films grown on Al, Ti, Ta, Nb, and W substrates grow at both the oxide/electrolyte and metal/oxide interfaces, while films grown on Zr and Hf substrates only grow at the metal/oxide interface (*42–45*). Thus, for most anodization systems, both anions and cations carry the current through the film. Experiments using isotopically labeled oxygen as well as an experiment with substrates consisting of layers of Nb and either Ta or Al indicate that all of the ions in the film participate in conduction, rather than isolated defects moving through an otherwise stationary lattice and that the ionic transport occurs through a series of short jumps (*46–48*). Empirically, the ionic current (mA) follows an exponential relationship, known as the high-field transport equation (*1*),

$$i = A\exp(BE) \tag{1}$$

which relates the current density (A/cm$^2$) to the applied electric field, $\boldsymbol{E}$ (V) by means of the temperature-dependent constants $\boldsymbol{A}$ (cm$^{-1}$) and $\boldsymbol{B}$ (V cm$^{-1}$). This empirical relationship has been interpreted as describing the motion of ions over a potential energy barrier (*49–51*). In this case, $B \approx \frac{q_n a_n}{kT}$, where the subscript $\boldsymbol{n}$ denotes the ionic species (dimensionless), $\boldsymbol{q_n}$ is the charge of the ion (e/cm$^2$), $\boldsymbol{2a_n}$ is the ionic jump distance (nm), $\boldsymbol{k}$ is Boltzmann's constant ($1.38064852 \times 10^{-23}$ m$^2$ kg s$^{-2}$ K$^{-1}$), and $\boldsymbol{T}$ is the temperature (K). The total current density is then given by the sum of the partial current densities of each mobile species.

Experiments suggest that the anions and cations move through the film via a cooperative process, rather than through independent processes. The experimental values of $\boldsymbol{B}$ for anodized Al (*52, 53*) yield $\boldsymbol{q_n a_n}$ values that are much too high for a single ion hopping to an adjacent site on its sublattice (*54*). Moreover, both oxygen and the metal carry a substantial fraction of the current over a wide range of applied electric fields (*42, 44, 45*), which requires $\boldsymbol{q_n a_n}$ to be similar for $Al^{3+}$ and $O^{2-}$ ions. However, this is unlikely given different ionic charge and the similar atomic spacings of $Al^{3+}$ and $O^{2-}$ ions in anodic alumina (*47, 54*). In response to this evidence against independent transport, several models of cooperative transport have been proposed (*54–57*). In particular, the defect cluster model (*55*) and the counter-site defect model (*54*) resolve these issues by assuming that transport is facilitated by mobile defects that consist of both anions and cations, with an effective charge greater than the individual ions in the system. Further insight into the ionic transport mechanism can be achieved by a combination of atomistic simulations of ion transport and new, more accurate measurements of $\boldsymbol{i(E)}$ and the fraction of the current carried by each ion for a range of current densities.

### *Stress Generation and Its Effects*

Stresses are generated inside the film during anodization, as has been documented for multiple material systems (*58–61*). The stresses in the film can be separated into electrostriction, which is compressive and is only present while the current is applied, and the residual stresses, which remain in the film at the open circuit condition. The dependence of the residual stresses on the applied current has been studied extensively for anodized Al films (*59–61*). The residual stresses are tensile at low current densities and compressive at high current densities, with a zero-stress transition point at approximately 4 mA/cm$^2$. This compressive shift has been attributed to the increased generation of new oxide within the oxide bulk (rather than at the interfaces) as the current density increases (*61*). The stresses within the film have been linked to plastic flow from

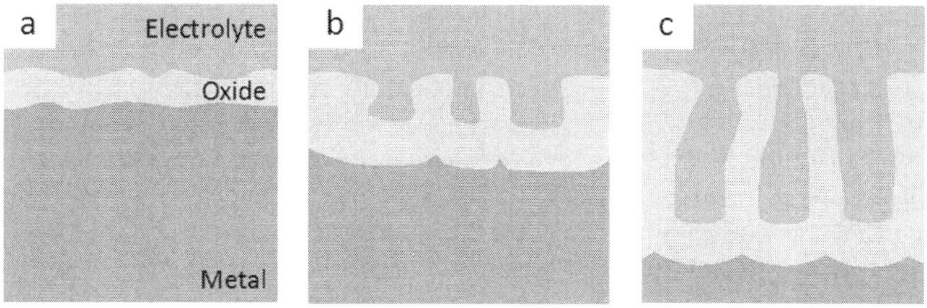


*Figure 3. Schematic diagram of the three stages of anodic nanopore growth: (a) barrier formation, (b) pore initiation, and (c) steady state growth.*

The nanopore/nanotube base to the nanopore/nanotube walls, which has been observed through the motion of W tracer atoms (*62*).

*Space Charge Accumulation*

The consideration of space charge within anodic oxides is important both during and after anodization. Simulations have shown that space charge within the oxide can significantly affect ionic transport (*54*, *63*, *64*). Space charge that is trapped in the film after anodization leads to a built-in electric field that may influence the behavior of anodic oxide nanostructures, especially in electronic applications (*65*). The charge trapped in the oxide film after anodization can be calculated from measurements of the potential along the surface of the film (*5*, *66*). The trapped charge, negative at the oxide/electrolyte interface and positive at the metal/oxide interface, is on the order of $1 \times 10^{12}$ e/cm$^2$ near each interface. The excess charge has been attributed to locally nonstoichiometric ratios between the metal and oxygen and stress-induced electron traps (*5*, *54*, *66*).

## Mechanisms and Models of Anodic Nanostructure Self-Organization

Anodic nanopores/nanotubes grow in three stages: barrier formation, pore nucleation, and steady state growth, as shown in Figure 3. During barrier formation, a nearly flat film grows on the substrate as a result of an oxide growth reaction at the metal/oxide interface and competing oxide growth and oxide dissolution reactions at the oxide/electrolyte interface. The rates of these reactions depend on the electric field. As the barrier forms, the film reaches a preferred thickness at which the growth rate at the metal/oxide interface matches the net dissolution rate at the oxide/electrolyte interface. As the barrier film approaches its preferred thickness, slight variations along the interfaces cause the current to focus on thin regions of the film. These regions grow into the substrate more quickly than neighboring areas, leading to the formation of pore nuclei (*67–69*). As the pore nuclei grow into the substrate, a larger fraction of the current passes through the pore regions, rather than the neighboring flat regions, until all of the current passes through the base of the pores. Eventually, interactions between the pores cause them to self-order and they reach the steady state growth stage, where an array of equally sized pores grows into the substrate at a constant velocity (*14*, *68–70*).

The source of the interactions that cause the ordering process is an active area of research. One proposed source of the ordering is competitive growth due to the field-dependence of the interfacial reactions. Under this theory, the preferred pore geometry is the one that leads to an electric field that maximizes the pore growth rate. Through a competitive growth process, eventually only the pores with this preferred geometry remain (*68*, *69*, *71*). However, this competition between growing pores may not be sufficient to explain the high degree of ordering and the limited window of applied potentials at which highly ordered structures can be attained (*72–75*). Pore ordering may also be guided by repulsive forces due to volume expansion during oxide formation (*74*, *75*), which is supported by experiments showing that stress applied to the substrate during anodization can suppress ordering (*76*). The pore geometry has also been proposed to result from a balance between volume expansion at the metal/oxide interface and plastic flow away from the interface (*62*, *75*). Stress-dependent reaction rates have also been proposed to guide pore ordering (*72*, *73*, *77*).

For nanotubular films, the tubes separate from each other between pore nucleation and steady state growth due to preferential chemical dissolution of $F^-$-rich oxide at the cell edges (*24*, *28*). The $F^-$ ions segregate to the outer edges of the film because they travel through the oxide more quickly than the $O^{2-}$ ions. Mechanical stresses at the junction between cells (*71*) and void formation to due ionic transport (*78*) have also been suggested as contributing factors to tube separation.

Several computational models have been developed to test the proposed mechanisms governing the ordering of anodic nanostructures. Parkhutik and Shershulsky (*68*) developed an influential analytic model of the steady-state growth of nanoporous alumina based on the field-dependence of the reactions, which produced the expected dependence of the pore separation on the applied potential. Houser and Hebert built on this work (*63*, *79*), demonstrating the importance of space charge formed due to high-field ionic transport and confirming that plastic flow could account for the observed W tracer motion (*62*). Thamida and Chang (*69*) performed linear stability analysis on the Parkhutik and Shershulsky model, giving insight into the pore nucleation process; they also examined steady-state growth by relaxing the strict geometric assumptions (i.e. hemispherical caps combined with a cylinder) of the Parkhutik and Shershulsky model. Golovin and co-workers (*72*, *73*, *77*, *80*) also applied linear stability analysis to investigate the role of stress-dependent reaction rates and $O^{2-}$ transport. Their results indicated that a mechanism to damp long-wavelength perturbations to the interface is required to form a regular array of pores, which the Parkhutik and Shershulsky model lacked. Other linear stability analysis by Hebert *et al.* (*81*) examined the role of the current efficiency for stable pore growth. Finally, in contrast to these previously mentioned simulations, which were limited to either pore nucleation or steady-state growth with a fixed geometry, Cheng and Ngan (*17*, *70*) and DeWitt and Thornton (*54*, *82*) simulated the evolution of nanopores with freely moving interfaces in two dimensions, allowing simulations of the entire growth and ordering process. Cheng and Ngan established that the field-dependent reactions alone can lead to well-ordered pore geometries (*17*, *70*). DeWitt and Thornton demonstrated that a detailed treatment of the reactions at the oxide/electrolyte interface predicts a dependence of the pore geometry on electrolyte pH that agrees with expriement (*18*) unlike the Parkhutik and Shershulsky model (*18*, *54*, *82*).

## Applications for Energy Generation and Storage

### Solar Cells

Anodic nanostructured films are promising materials for solar cells because their highly ordered, high-aspect-ratio geometry facilitates transport of photogenerated electrons and holes to their respective current collectors. Anodic nanostructured films have been investigated for use in a variety of solar cells, including dye-sensitized solar cells (DSSCs), photoelectrochemical cells (PECs), and solid-state organic solar cells.

In DSSCs, incoming photons are absorbed by dye molecules attached to the surface of the electrode, generating excited electrons. The dye transfers the excited electron to the semiconductor electrode, and then receives a replacement electron from a redox couple in the electrolyte. The photoanode is typically fabricated using compacted nanoparticulate titania ($TiO_2$), with a maximum reported efficiency of 13 % (*83*). Anodic nanotubular $TiO_2$ offers the potential for improved performance because the nanotubes provide straight electron transport paths to the current collector, without the interparticle trap sites that slow electron transport for nanoparticulate films (*4*, *84*). Beginning with the 2.9 % solar conversion efficiency first reported in 2006 (*85*), the efficiency of nanotubular $TiO_2$-based DSSCs has been steadily increasing and is currently up to 9.8 % (*86*). Nanotubular electrodes have been shown to have lower electron recombination rates than nanoparticulate electrodes, as expected, but the transport times are not improved (*87*), likely as a result of defects caused by $F^-$ anions incorporated into the film from the electrolyte (*88*). The elimination of these impurities provides a pathway to substantially improved performance of anodic nanotubular $TiO_2$ DSSCs. Porous aluminum oxide (PAO) templates have also been applied to build DSSCs with nanotubular $TiO_2$ (*89*) although the resulting devices have shown lower efficiencies than those using anodized $TiO_2$.

Anodic nanostructures have also been applied for PECs, where the energy from absorbed photons is used to form a fuel such as hydrogen gas ($H_2$). The ideal photoelectrode for a PEC absorbs a large fraction of the solar spectrum, has good charge transfer and charge transport characteristics, and is stable in aqueous electrolytes. PECs based on gallium arsenide/gallium indium phosphide ($GaAs/GaInP_2$) photoelectrodes have efficiencies of up to 16.5 %, but degrade after one day of use (*90*). $TiO_2$ photoelectrodes are stable and have sufficient charge transport properties, but the large band gap limits the theoretical efficiency to 2.2 % (*91*). However, anodic $TiO_2$ nanotubes can be doped with C to reduce the band gap, leading to efficiencies up to 8.5 % (*31*). Iron(III) oxide

($Fe_2O_3$)-based photoelectrodes have a near-ideal band gap, but suffer from poor hole transport capability (*92*). $Fe_2O_3$ nanotubes grown on PAO templates have reached efficiencies of up to 5 %, where the thin walls of the nanotubes decrease the distance the holes must travel to reach the electrolyte, reducing recombination (*92*). The photocurrent for these template nanotubes is among the highest reported photocurrents for hematite-based PECs and is much higher than reported values for anodized $Fe_2O_3$ nanotubes (*29, 93, 94*). Tandem cells pairing a DSSC with the $Fe_2O_3$ photoelectrode may provide a route the higher efficiencies required for the widespread use of PECs (*90*).

Beyond DSSCs and PECs, anodic nanostructures have also been used in organic solid-state solar cells. Anodic $TiO_2$ has been utilized as an electron acceptor in standard bulk heterojunction and double bulk heterojunction solar cells, yielding efficiencies of up to 4.1 % (*95, 96*). In addition to the straight path for electron transport provided by the nanotubular structure, the nanoscale confinement of the polymer orients its bonds for improved hole transport (*96*). This confinement effect has also been explored for organic solar cells fabricated using PAO templates (*97*).

**Li-Ion Battery Anodes**

Anodic $TiO_2$ and $Fe_2O_3$ nanostructures have been examined as lithium (Li)-ion battery anodes, and PAO has been used as an anode template. The nanotubular morphologies attainable through anodization enhance the kinetics of the anode. The thin nanotube wall enables high charge/discharge rates because it reduces the transport distance during intercalation/deintercalation (*98*). Anodic nanostructures are especially promising for microbatteries, where their three-dimensional (3D) nanostructure provides a much higher areal energy density than the traditional flat-film anodes (*99*). Microbatteries have applications in fast-growing sectors such as implantable biomedical devices, microelectromechanical systems (MEMS), radio-frequency identification (RFID) chips, and miniaturized sensors (*99*).

As a Li-ion battery anode, $TiO_2$ has three primary advantages: (1) a higher operating potential than the currently used carbon anodes, which decreases the risk of dendrite-induced catastrophic failure (*98, 99*), (2) a low volumetric change (on the order of 3 %) during lithiation and delithiation, providing the opportunity for a long cycle life (*100*), and (3) low cost and earth abundance (*101*). Anodic nanotubular $TiO_2$ battery anodes were first studied by Liu *et al.* in 2008 (*101*). Since then, reversible capacities of up to 310 mAh/g (*102*) for pure $TiO_2$ and 350 mAh/g for $TiO_2$ covered with Manganese(IV) oxide ($MnO_2$) nanosheets (*103*) have been reported, which is competitive with the theoretical maximum capacity of carbon-based anodes, 372 mAh/g (*104*). The coulombic efficiency of the $TiO_2$ anode has been shown to be nearly 100 % up to very high charge/discharge rates of 32 C and for over 1000 cycles (*102, 105*). Many investigations have focused on the crystal structure of the $TiO_2$ nanotubes, with mixed results on whether annealing improves (*101*) or degrades (*98*) the cyclability of the anode. Amorphous $TiO_2$ nanotubes can undergo an irreversible phase transformation to a face-centered-cubic phase during the first lithiation cycle, improving the capacity (*102*). Nanotubes with smaller feature sizes have been found to have better performance than those with larger feature sizes (*106*). Besides $MnO_2$, other combinations of $TiO_2$ and other materials have been investigated, including anodic nanotubes grown from a titanium cobalt (It-Co) alloy (*107*), $Fe_2O_3$ deposited into $TiO_2$ nanotubes (*108*), and either Al or Ni nanorods from a PAO template covered in $TiO_2$ (*109*, *110*).

$Fe_2O_3$ and Iron (II,III) oxide ($Fe_3O_4$) nanostructures fabricated using anodization have also been utilized as anodes for Li-ion batteries. Iron oxides are attractive materials for battery anodes due to a combination of low cost, high theoretical capacity (1005 mAh/g), and high electrical conductivity (*111*, *112*). Li-ion battery anodes constructed purely from anodized $Fe_2O_3$ nanotubes (either amorphous or annealed into $\alpha$-$Fe_2O_3$) have been shown to have poor cycability (*111*, *113*, *114*). Composite anodes that incorporate carbon species have been shown to significantly increase the capacity and cyclability of $Fe_2O_3/Fe_3O_4$ nanotubes (*111*, *113*). For example, a $Fe_2O_3/Fe_3C$/graphene composite anode has been shown to have a reversible capacity of 1118 mAh/g at 0.17 C and 503 mAh/g at 6.6 C over 1000 cycles (*113*). Copper (Cu) nanorods fabricated using a porous anodic alumina template and then coated with $Fe_3O_4$ have been shown to have a stable reversible capacity above 800 mAh/g (presumably neglecting the mass of the Cu) for at least 50 cycles at C/8 (*112*).

### Supercapacitors

Anodic nanostructures, mostly PAO templates, have also been used to fabricate supercapacitors. Supercapacitors are electrochemical energy storage devices that store energy in the electrochemical double layer and in some cases through pseudocapacitance that involves redox reactions at the electrode surface (*115*). The nanotubular and nanorod structures from anodic films improve supercapacitor performance by increasing the surface area of the electrodes and providing short, straight-line diffusion paths in the electrolyte (*116*, *117*). The active material is also in direct electrical contact with the current collector, making conductive additives and binders unnecessary (*115*). Traditional materials for supercapacitors include graphite, which has a limited specific capacitance of less than 400 F/g due to a lack of pseudocapacitance, and ruthenium(IV) oxide ($RuO_2$), which can yield a large specific capacitance but is very expensive (*115*, *118*).

Supercapacitors have been made from metal oxide, metal hydroxides, and conductive polymers using anodic nanostructures. Electrodes using anodic oxide nanostructures as the active material have specific capacitances below 220 F/g or areal capacitances below 15 $mF/cm^2$ (*8*, *115*, *125*, *126*), which are insufficient for practical applications. However, supercapacitors based on anodic-nanostructure templates exhibit much improved performance. An electrode made of PAO-templated $RuO_2$ has a reported specific capacitance of 1300 F/g (*116*). $RuO_2$-coated $TiO_2$ nanotubes have also been investigated, but the mass of the $TiO_2$ leads to a lower specific capacitance of 620 F/g (*119*). Non-Ru-based electrodes made of conductive polymers deposited into anodic nanostructure templates have also been investigated, with the best device having a specific capacitance that was nearly as high as the best $RuO_2$ electrode, 1142 F/g (*117*). A cobalt (VIII) oxide Co3$O_4$ nanotube electrode using a PAO template (*120*) has a reported specific capacitance of 574 F/g. Nickel(II) oxide (NiO) and Nickel(II) hydroxide ($Ni(OH)_2$) electrodes based on PAO templates (*121*) or deposited into $TiO_2$ nanotubes exhibited areal capacitances of up to 7,010 $mF/cm^2$ (*121–123*). $MnO_2$ electrodes templated by PAO or deposited into $TiO_2$ nanotubes have shown specific capacitances of up to 320 F/g and areal capacitances of up to 150 $mF/cm^2$ (*101*, *124*, *125*).

## Role of Fundamental Investigations To Improve Performance in Energy Applications

Improved understanding of the fundamental mechanisms governing anodization provides a path for increased performance of anodic oxide films in energy generation and storage applications. To better predict and control the growth of anodic oxide films, a more complete understanding of the general mechanisms governing the anodization process is needed, particularly for the effects of plastic flow and the cooperative mechanism governing ionic transport through the oxide. Future experimental or computational investigations may also provide more specific insights into anodic oxide films that could be directly applied to improve device performance. Four promising areas for such insights are discussed below.

First, fundamental understanding of the role of impurities from the electrolyte in directing pore morphology and film properties must be developed. The effects of $F^-$ ions from the electrolyte must be elucidated because they are responsible for the separation between nanotubes (*24*, *28*) and are believed to be recombination centers in solar cell applications (*88*). Optimized control of the incorporation and transport of $F^-$ ions during anodization may allow further control of the thickness of the nanotube walls and a decreased number of recombination centers. Second, more research is needed to understand the effect of embedded charge and the associated electric field on device performance, because it may alter the interfacial reaction rates or the charge transport through the oxide. The embedded charge can be measured experimentally using electrostatic force microscopy or scanning Kelvin probe force microscopy, as in Refs. (*5*) and (*66*), or studied computationally using the approach presented in Ref. (*54*). Third, engineering the atomic-scale structure of anodic oxide films may lead to improved device performance, particularly for battery applications. Further investigations of the role of annealing and electrochemical phase transformations may provide insight into crystal structures that facilitate the intercalation and deintercalation processes. The approach taken in Ref. (*102*) to examine phase transitions in $TiO_2$ for Li-ion battery applications, combining synchrotron X-ray characterization techniques, molecular dynamics simulations, and density functional theory calculations, is particularly promising. Fourth, the continued development of high-surface-area anodic oxide morphologies, such as nanolace and dendritic nanopores, may lead to improvements in DSSCs and supercapacitors, where the electrode surface area is a primary determinant of device performance. In DSSCs, these high surface-area-morphologies may provide the high surface area found in nanoparticulate electrodes without the detrimental interparticle recombination centers.

## Conclusion

Anodic oxide nanostructures provide a compelling mixture of high aspect ratios and highly ordered geometries with controllable features on the order of tens of nanometers. Fundamental research into the mechanisms governing anodic oxide growth may lead to new processing routes to gain further control of the properties and morphology of these nanostructures. Despite much progress in understanding these governing mechanisms, many fundamental questions remain.

Anodic oxide nanostructures have been investigated for a variety of applications, including solar cells, batteries, and supercapacitors, with promising results. However, further understanding of the formation and properties of anodic nanostructures is required to improve the performance of these devices. Such advances provide a path to commercial success of devices for energy storage and generation based on anodic nanostructures.

## References


1. Diggle, J. W.; Downie, T. C.; Goulding, C. W. Anodic oxide films on aluminum. *Chem. Rev.* **1969**, *69*, 365–405.
2. Young, L. *Anodic Oxide Films*; Academic Press: London, 1961.
3. Masuda, H.; Fukuda, K. Ordered metal nanohole arrays made by a two-step replication of honeycomb structures of anodic alumina. *Science* **1995**, *268*, 1466–1468.
4. Rani, S.; Roy, S. C.; Paulose, M.; Varghese, O. K.; Mor, G. K.; Kim, S.; Yoriya, S.; Latempa, T. J.; Grimes, C. A. Synthesis and applications of electrochemically self-assembled titania nanotube arrays. *Phys. Chem. Chem. Phys.* **2010**, *12*, 2780–2800.
5. Yasakau, K. A.; Salak, A. N.; Zheludkevich, M. L.; Ferreira, M. r. G. S. Volta Potential of Oxidized Aluminum Studied by Scanning Kelvin Probe Force Microscopy. *J. Phys. Chem. C* **2010**, *114*, 8474–8484.
6. Masuda, H.; Hasegwa, F.; Ono, S. Self-ordering of cell arrangement of anodic porous alumina formed in sulfuric acid solution. *J. Electrochem. Soc.* **1997**, *144*, L127–L130.
7. Paulose, M.; Shankar, K.; Yoriya, S.; Prakasam, H. E.; Varghese, O. K.; Mor, G. K.; LaTempa, T. J.; Fitzgerald, A.; Grimes, C. A. Anodic growth of highly ordered $TiO_2$ nanotube arrays to 134 μm in length. *J. Phys. Chem. B* **2006**, *110*, 16179–16184.
8. Lee, K.; Yang, Y.; Yang, M.; Schmuki, P. Formation of highly ordered nanochannel Nb oxide by self-organizing anodization. *Chemistry* **2012**, *18*, 9521–9524.

9. Lee, K. Y.; Schmuki, P. Highly ordered nanoporous $Ta_2O_5$ formed by anodization of Ta at high temperatures in a glycerol/phosphate electrolyte. *Electrochem. Commun.* **2011**, *13*, 542–545.
10. Tsuchiya, H.; Macak, J. M.; Sieber, I.; Taveira, L.; Ghicov, A.; Sirotna, K.; Schmuki, P. Self-organized porous WO3 formed in NaF electrolytes. *Electrochem. Commun.* **2005**, *7*, 295–298.
11. Shin, H. C.; Dong, J.; Liu, M. L. Porous tin oxides prepared using an anodic oxidation process. *Adv. Mater.* **2004**, *16*, 237–240.
12. Berger, S.; Hahn, R.; Roy, P.; Schmuki, P. Self-organized $TiO_2$ nanotubes: Factors affecting their morphology and properties. *Phys. Status Solidi B* **2010**, *247*, 2424–2435.
13. Keller, F.; Hunter, M. S.; Robinson, D. L. Structural Features of Oxide Coatings on Aluminium. *J. Electrochem. Soc.* **1953**, *100*, 411–419.
14. O'Sullivan, J. P.; Wood, G. C. The Morphology and Mechanism of Formation of Porous Anodic Films on Aluminium. *Proc. R. Soc. A* **1970**, *317*, 511–543.
15. Masuda, H.; Satoh, M. Fabrication of gold nanodot array using anodic porous alumina as an evaporation mask. *Jpn. J. Appl. Phys., Part 2* **1996**, *35*, L126–L129.
16. Masuda, H.; Yamada, H.; Satoh, M.; Asoh, H.; Nakao, M.; Tamamura, T. Highly ordered nanochannel-array architecture in anodic alumina. *Appl. Phys. Lett.* **1997**, *71*, 2770–2772.
17. Cheng, C.; Ngan, A. H. W. Growth Sustainability of Nanopore Channels in Anodic Aluminum Oxide Guided with Prepatterns. *J. Phys. Chem. C* **2013**, *117*, 12183–12190.
18. Friedman, A. L.; Brittain, D.; Menon, L. Roles of pH and acid type in the anodic growth of porous alumina. *J. Chem. Phys.* **2007**, *127*, 154717.
19. Nielsch, K.; Choi, J.; Schwirn, K.; Wehrspohn, R. B.; Gösele, U. Self-ordering Regimes of Porous Alumina: The 10 Porosity Rule. *Nano Lett.* **2002**, *2*, 677–680.
20. Chen, W.; Wu, J. S.; Xia, X. H. Porous anodic alumina with continuously manipulated pore/cell size. *ACS Nano* **2008**, *2*, 959–965.
21. Ebihara, K.; Takahashi, H.; Nagayama, M. Structure and density of anodic oxide films formed on aluminum in oxalic acid solutions. *J. Met. Finish. Soc. Jpn.* **1983**, *34*, 548–553.
22. Li, F. Y.; Zhang, L.; Metzger, R. M. On the growth of highly ordered pores in anodized aluminum oxide. *Chem. Mater.* **1998**, *10*, 2470–2480.
23. Nagayama, M.; Tamura, K.; Takahashi, H. Dissolution of porous anodic oxide films on Al in $(COOH)_2$ solutions. *Corros. Sci.* **1970**, *10*, 617–627.
24. Macak, J. M.; Tsuchiya, H.; Ghicov, A.; Yasuda, K.; Hahn, R.; Bauer, S.; Schmuki, P. $TiO_2$ nanotubes: Self-organized electrochemical formation, properties and applications. *Curr. Opin. Solid State Mater. Sci.* **2007**, *11*, 3–18.
25. Ghicov, A.; Schmuki, P. Self-ordering electrochemistry: a review on growth and functionality of $TiO_2$ nanotubes and other self-aligned MO(x) structures. *Chem. Commun.* **2009**, 2791–2808.

26. Zwilling, V.; Darque-Ceretti, E.; Boutry-Forveille, A.; David, D.; Perrin, M. Y.; Aucouturier, M. Structure and physicochemistry of anodic oxide films on titanium and TA6V alloy. *Surf. Interface Anal.* **1999**, *27*, 629–637.
27. Ismail, S.; Ahmad, Z. A.; Berenov, A.; Lockman, Z. Effect of applied voltage and fluoride ion content on the formation of zirconia nanotube arrays by anodic oxidation of zirconium. *Corros. Sci.* **2011**, *53*, 1156–1164.
28. Tsuchiya, H.; Schmuki, P. Self-organized high aspect ratio porous hafnium oxide prepared by electrochemical anodization. *Electrochem. Commun.* **2005**, *7*, 49–52.
29. Rangaraju, R. R.; Panday, A.; Raja, K. S.; Misra, M. Nanostructured anodic iron oxide film as photoanode for water oxidation. *J. Phys. D: Appl. Phys.* **2009**, *42*, 135303.
30. Richter, C.; Wu, Z.; Panaitescu, E.; Willey, R. J.; Menon, L. Ultrahigh-aspect-ratio titania nanotubes. *Adv. Mater.* **2007**, *19*, 946–948.
31. Mohapatra, S. K.; Misra, M.; Mahajan, V. K.; Raja, K. S. Design of a Highly Efficient Photoelectrolytic Cell for Hydrogen Generation by Water Splitting: Application of $TiO_{2-x}C_x$ Nanotubes as a Photoanode and Pt/$TiO_2$ Nanotubes as a Cathode. *J. Phys. Chem. C* **2007**, *111*, 8677–8685.
32. Yahia, S. A. A.; Hamadou, L.; Kadri, A.; Benbrahim, N.; Sutter, E. M. M. Effect of Anodizing Potential on the Formation and EIS Characteristics of $TiO_2$ Nanotube Arrays. *J. Electrochem. Soc.* **2012**, *159*, K83–K92.
33. Macak, J. M.; Hildebrand, H.; Marten-Jahns, U.; Schmuki, P. Mechanistic aspects and growth of large diameter self-organized $TiO_2$ nanotubes. *J. Electroanal. Chem. Interfacial Electrochem.* **2008**, *621*, 254–266.
34. Paulose, M.; Prakasam, H. E.; Varghese, O. K.; Peng, L.; Popat, K. C.; Mor, G. K.; Desai, T. A.; Grimes, C. A. $TiO_2$ Nanotube Arrays of 1000 μm Length by Anodization of Titanium Foil: Phenol Red Diffusion. *J. Phys. Chem. C* **2007**, *111*, 14992–14997.
35. Meng, G.; Jung, Y. J.; Cao, A.; Vajtai, R.; Ajayan, P. M. Controlled fabrication of hierarchically branched nanopores, nanotubes, and nanowires. *Proc. Natl. Acad. Sci. U.S.A.* **2005**, *102*, 7074–7078.
36. Albu, S. P.; Kim, D.; Schmuki, P. Growth of aligned $TiO_2$ bamboo-type nanotubes and highly ordered nanolace. *Angew. Chem., Int. Ed.* **2008**, *47*, 1916–1919.
37. Oh, J.; Thompson, C. V. Abnormal Anodic Aluminum Oxide Formation in Confined Structures for Lateral Pore Arrays. *J. Electrochem. Soc.* **2011**, *158*, C71.
38. Lee, W.; Ji, R.; Gosele, U.; Nielsch, K. Fast fabrication of long-range ordered porous alumina membranes by hard anodization. *Nat. Mater.* **2006**, *5*, 741–747.
39. Losic, D.; Losic, D., Jr. Preparation of Porous Anodic Alumina with Periodically Perforated Pores. *Langmuir* **2009**, *25*, 5426–5431.
40. Våland, T.; Heusler, K. E. Reactions at the oxide-electrolyte interface of anodic oxide films on aluminum. *J. Electroanal. Chem. Interfacial Electrochem.* **1983**, *149*, 71–82.
41. Thompson, G. E. Porous anodic alumina: Fabrication, characterization and applications. *Thin Solid Films* **1997**, *297*, 192–201.

42. Davies, J. A.; Domeij, B.; Pringle, J. P. S.; Brown, F. The migration of metal and oxygen during anodic film formation. *J. Electrochem. Soc.* **1965**, *112*, 675–680.
43. Habazaki, H.; Shimizu, K.; Nagata, S.; Skeldon, P.; Thompson, G. E.; Wood, G. C. Ionic transport in amorphous anodic titania stabilised by incorporation of silicon species. *Corros. Sci.* **2002**, *44*, 1047–1055.
44. Lu, Q.; Skeldon, P.; Thompson, G. E.; Masheder, D.; Habazaki, H.; Shimizu, K. Transport numbers of metal and oxygen species in anodic tantala. *Corros. Sci.* **2004**, *46*, 2817–2824.
45. Khalil, N.; Leach, J. S. L. The anodic oxidation of valve metals—I. Determination of ionic transport numbers by α-spectrometry. *Electrochim. Acta* **1986**, *31*, 1279–1285.
46. Amsel, G.; Samuel, D. The mechanism of anodic oxidation. *J. Phys. Chem. Solids* **1962**, *23*, 1707–1718.
47. Pringle, J. P. S. Transport Numbers of Metal and Oxygen during the Anodic Oxidation of Tantalum. *J. Electrochem. Soc.* **1973**, *120*, 398–407.
48. Perrière, J.; Siejka, J.; Rigo, S. Investigation of ionic movements during anodic oxidation of superimposed metallic layers by the use of rutherford backscattering techniques and nuclear microanalysis. *Corros. Sci.* **1980**, *20*, 91–102.
49. Verwey, E. J. W. Electrolytic conduction of a solid insulator at high fields - The formation of the anodic oxide film on aluminium. *Physica* **1935**, *2*, 1059–1063.
50. Cabrera, N.; Mott, N. F. Theory of the oxidation of metals. *Rep. Prog. Phys.* **1949**, *12*, 163–184.
51. Dewald, J. F. A Theory of the Kinetics of Formation of Anode Films at High Fields. *J. Electrochem. Soc.* **1955**, *102*, 1–6.
52. Bernard, W. J.; Cook, J. W. The Growth of Barrier Oxide Films on Aluminum. *J. Electrochem. Soc.* **1959**, *106*, 643–646.
53. Harkness, A. C.; Young, L. High Resistance Anodic Oxide Films on Aluminium. *Can. J. Chem.* **1966**, *44*, 2409–2413.
54. DeWitt, S.; Thornton, K. Model for anodic film growth on aluminum with coupled bulk transport and interfacial reactions. *Langmuir* **2014**, *30*, 5314–5325.
55. Wang, M. H.; Hebert, K. R. Metal and oxygen ion transport during ionic conduction in amorphous anodic oxide films. *J. Electrochem. Soc.* **1999**, *146*, 3741–3749.
56. Fromhold, A. T. Nonsimultaneous Place Exchange: A Microscopic High Field Transport Mechanism in Solids. *J. Electrochem. Soc.* **1980**, *127*, 411–425.
57. Sato, N.; Cohen, M. The Kinetics of Anodic Oxidation of Iron in Neutral Solution. *J. Electrochem. Soc.* **1964**, *111*, 512–519.
58. Vermilyea, D. A. Stresses in Anodic Films. *J. Electrochem. Soc.* **1963**, *110*, 345–346.
59. Çapraz, Ö. Ö.; Hebert, K. R.; Shrotriya, P. In Situ Stress Measurement During Aluminum Anodizing Using Phase-Shifting Curvature Interferometry. *J. Electrochem. Soc.* **2013**, *160*, D501–D506.

60. Van Overmeere, Q.; Blaffart, F.; Proost, J. What controls the pore spacing in porous anodic oxides? *Electrochem. Commun.* **2010**, *12*, 1174–1176.
61. Çapraz, Ö. Ö.; Shrotriya, P.; Skeldon, P.; Thompson, G. E.; Hebert, K. R. Factors Controlling Stress Generation during the Initial Growth of Porous Anodic Aluminum Oxide. *Electrochim. Acta* **2015**, *159*, 16–22.
62. Garcia-Vergara, S. J.; Skeldon, P.; Thompson, G. E.; Habazaki, H. A flow model of porous anodic film growth on aluminium. *Electrochim. Acta* **2006**, *52*, 681–687.
63. Houser, J. E.; Hebert, K. R. The role of viscous flow of oxide in the growth of self-ordered porous anodic alumina films. *Nat. Mater.* **2009**, *8*, 415–420.
64. Fromhold, A. T.; Cook, E. L. Kinetics of Oxide Film Growth on Metal Crystals - Electronic and Ionic Diffusion in Large Surface-Charge and Space-Charge Fields. *Phys. Rev.* **1968**, *175*, 877–897.
65. Hickmott, T. W. Electrolyte effects on charge, polarization, and conduction in thin anodic AlO films. I. Initial charge and temperature-dependent polarization. *J. Appl. Phys.* **2007**, *102*, 93706.
66. Lambert, J.; Guthmann, C.; Ortega, C.; Saint-Jean, M. Permanent polarization and charge injection in thin anodic alumina layers studied by electrostatic force microscopy. *J. Appl. Phys.* **2002**, *91*, 9161–9169.
67. Hoar, T. P.; Mott, N. F. A Mechanism for the Formation of Porous Anodic Oxide Films on Aluminium. *J. Phys. Chem. Solids* **1959**, *9*, 97–99.
68. Parkhutik, V. P.; Shershulsky, V. I. Theoretical modelling of porous oxide growth on aluminium. *J. Phys. D: Appl. Phys.* **1992**, *25*, 1258–1263.
69. Thamida, S. K.; Chang, H. C. Nanoscale pore formation dynamics during aluminum anodization. *Chaos* **2002**, *12*, 240–251.
70. Cheng, C.; Ngan, A. H. W. Modelling and simulation of self-ordering in anodic porous alumina. *Electrochim. Acta* **2011**, *56*, 9998–10008.
71. Yasuda, K.; Macak, J. M.; Berger, S.; Ghicov, A.; Schmuki, P. Mechanistic Aspects of the Self-Organization Process for Oxide Nanotube Formation on Valve Metals. *J. Electrochem. Soc.* **2007**, *154*, C472–C478.
72. Singh, G. K.; Golovin, A. A.; Aranson, I. S.; Vinokur, V. M. Formation of nanoscale pore arrays during anodization of aluminum. *Europhys. Lett.* **2005**, *70*, 836–842.
73. Singh, G. K.; Golovin, A. A.; Aranson, I. S. Formation of self-organized nanoscale porous structures in anodic aluminum oxide. *Phys. Rev. B* **2006**, *73*, 205422.
74. Jessensky, O.; Muller, F.; Gosele, U. Self-organized formation of hexagonal pore arrays in anodic alumina. *Appl. Phys. Lett.* **1998**, *72*, 1173–1175.
75. Oh, J.; Thompson, C. V. The role of electric field in pore formation during aluminum anodization. *Electrochim. Acta* **2011**, *56*, 4044–4051.
76. Sulka, G. D.; Stroobants, S.; Moshchalkov, V. V.; Borghs, G.; Celis, J. P. Effect of tensile stress on growth of self-organized nanostructures on anodized aluminum. *J. Electrochem. Soc.* **2004**, *151*, B260–B264.
77. Stanton, L. G.; Golovin, A. A. Effect of Electrostriction on the Self-organization of Porous Nanostructures in Anodized Aluminum Oxide. *Math. Modell. Nat. Phenom.* **2008**, *3*, 73–91.

78. Gong, D.; Grimes, C. A.; Varghese, O. K.; Hu, W.; Singh, R. S.; Chen, Z.; Dickey, E. C. Titanium oxide nanotube arrays prepared by anodic oxidation. *J. Mater. Res.* **2011**, *16*, 3331–3334.
79. Houser, J. E.; Hebert, K. R. Modeling the Potential Distribution in Porous Anodic Alumina Films during Steady-State Growth. *J. Electrochem. Soc.* **2006**, *153*, B566–B573.
80. Stanton, L. G.; Golovin, A. A. Effect of ion migration on the self-assembly of porous nanostructures in anodic oxides. *Phys. Rev. B* **2009**, *79*, 035414.
81. Hebert, K. R.; Albu, S. P.; Paramasivam, I.; Schmuki, P. Morphological instability leading to formation of porous anodic oxide films. *Nat. Mater.* **2012**, *11*, 162–166.
82. DeWitt, S.; Thornton, K. Simulations of Anodic Nanopore Growth Using the Smoothed Boundary and Level Set Methods. In preparation, 2015.
83. Mathew, S.; Yella, A.; Gao, P.; Humphry-Baker, R.; Curchod, B. F.; Ashari-Astani, N.; Tavernelli, I.; Rothlisberger, U.; Nazeeruddin, M. K.; Gratzel, M. Dye-sensitized solar cells with 13 % efficiency achieved through the molecular engineering of porphyrin sensitizers. *Nat. Chem.* **2014**, *6*, 242–247.
84. Nelson, J. Continuous-time random-walk model of electron transport in nanocrystalline $TiO_2$ electrodes. *Phys. Rev. B* **1999**, *59*, 15374–15380.
85. Mor, G. K.; Shankar, K.; Paulose, M.; Varghese, O. K.; Grimes, C. A. Use of Highly-Ordered $TiO_{(2)}$ Nanotube Arrays in Dye-Sensitized Solar Cells. *Nano Lett.* **2006**, *6*, 215–218.
86. Mohammadpour, F.; Moradi, M.; Lee, K.; Cha, G.; So, S.; Kahnt, A.; Guldi, D. M.; Altomare, M.; Schmuki, P. Enhanced performance of dye-sensitized solar cells based on $TiO_2$ nanotube membranes using an optimized annealing profile. *Chem. Commun.* **2015**, *51*, 1631–1634.
87. Zhu, K.; Neale, N. R.; Miedaner, A.; Frank, A. J. Enhanced Charge-Collection Efficiencies and Light Scattering in Dye-Sensitized Solar Cells Using Oriented $TiO_2$ Nanotubes Arrays. *Nano Lett.* **2007**, *7*, 69–74.
88. Richter, C.; Schmuttenmaer, C. A. Exciton-like trap states limit electron mobility in $TiO_2$ nanotubes. *Nat. Nanotechnol.* **2010**, *5*, 769–772.
89. Kang, T. S.; Smith, A. P.; Taylor, B. E.; Durstock, M. F. Fabrication of Highly-Ordered TiO(2) Nanotube Arrays and Their Use in Dye-Sensitized Solar Cells. *Nano Lett.* **2009**, *9*, 601–606.
90. Jacobsson, T. J.; Fjallstrom, V.; Edoff, M.; Edvinsson, T. Sustainable solar hydrogen production: from photoelectrochemical cells to PV-electrolyzers and back again. *Energy Environ. Sci.* **2014**, *7*, 2056–2070.
91. Murphy, A. B.; Barnes, P. R. F.; Randeniya, L. K.; Plumb, I. C.; Grey, I. E.; Horne, M. D.; Glasscock, J. A. Efficiency of solar water splitting using semiconductor electrodes. *Int. J. Hydrogen Energy* **2006**, *31*, 1999–2017.
92. Mao, A.; Park, N. G.; Han, G. Y.; Park, J. H. Controlled growth of vertically oriented hematite/Pt composite nanorod arrays: use for photoelectrochemical water splitting. *Nanotechnology* **2011**, *22*, 175703.
93. Prakasam, H. E.; Varghese, O. K.; Paulose, M.; Mor, G. K.; Grimes, C. A. Synthesis and photoelectrochemical properties of nanoporous iron (III) oxide by potentiostatic anodization. *Nanotechnology* **2006**, *17*, 4285–4291.

94. Mushove, T.; Breault, T. M.; Thompson, L. T. Synthesis and Characterization of Hematite Nanotube Arrays for Photocatalysis. *Ind. Eng. Chem. Res.* **2015**, *54*, 4285–4292.
95. Mor, G. K.; Kim, S.; Paulose, M.; Varghese, O. K.; Shankar, K.; Basham, J.; Grimes, C. A. Visible to near-Infrared Light Harvesting in $TiO_2$ Nanotube Array-P3HT Based Heterojunction Solar Cells. *Nano Lett.* **2009**, *9*, 4250–4257.
96. Mor, G. K.; Shankar, K.; Paulose, M.; Varghese, O. K.; Grimes, C. A. High efficiency double heterojunction polymer photovoltaic cells using highly ordered $TiO_2$ nanotube arrays. *Appl. Phys. Lett.* **2007**, *91*, 152111.
97. Chen, D.; Zhao, W.; Russell, T. P. P3HT Nanopillars for Organic Photovoltaic Devices Nanoimprinted by AAO Templates. *ACS Nano* **2012**, *6*, 1479–1485.
98. Ortiz, G. F.; Hanzu, I.; Djenizian, T.; Lavela, P.; Tirado, J. L.; Knauth, P. Alternative Li-Ion Battery Electrode Based on Self-Organized Titania Nanotubes. *Chem. Mater.* **2009**, *21*, 63–67.
99. Wang, Y.; Liu, B.; Li, Q.; Cartmell, S.; Ferrara, S.; Deng, Z. D.; Xiao, J. Lithium and lithium ion batteries for applications in microelectronic devices: A review. *J. Power Sources* **2015**, *286*, 330–345.
100. Wagemaker, M.; Kearley, G. J.; Van Well, A. A.; Mutka, H.; Mulder, F. M. Multiple Li Positions Inside Oxygen Octahedra in Lithiated $TiO_2$ Anatase. *J. Am. Chem. Soc.* **2003**, *125*, 840–848.
101. Liu, D.; Xiao, P.; Zhang, Y.; Garcia, B. B.; Zhang, Q.; Guo, Q.; Champion, R.; Cao, G. $TiO_2$ Nanotube Arrays Annealed in $N_2$ for Efficient Lithium-Ion Intercalation. *J. Phys. Chem. C* **2008**, *112*, 11175–11180.
102. Xiong, H.; Yildirim, H.; Shevchenko, E. V.; Prakapenka, V. B.; Koo, B.; Slater, M. D.; Balasubramanian, M.; Sankaranarayanan, S. K. R. S.; Greeley, J. P.; Tepavcevic, S.; Dimitrijevic, N. M.; Podsiadlo, P.; Johnson, C. S.; Rajh, T. Self-Improving Anode for Lithium-Ion Batteries Based on Amorphous to Cubic Phase Transition in $TiO_2$ Nanotubes. *J. Phys. Chem. C* **2012**, *116*, 3181–3187.
103. Zhu, Q.; Hu, H.; Li, G.; Zhu, C.; Yu, Y. $TiO_2$ Nanotube Arrays Grafted with $MnO_2$ Nanosheets as High-Performance Anode for Lithium Ion Battery. *Electrochim. Acta* **2015**, *156*, 252–260.
104. Yoo, H. D.; Shterenberg, I.; Gofer, Y.; Gershinsky, G.; Pour, N.; Aurbach, D. Mg rechargeable batteries: an on-going challenge. *Energy Environ. Sci.* **2013**, *6*, 2245–2250.
105. Lamberti, A.; Garino, N.; Sacco, A.; Bianco, S.; Manfredi, D.; Gerbaldi, C. Vertically aligned $TiO_2$ nanotube array for high rate Li-based micro-battery anodes with improved durability. *Electrochim. Acta* **2013**, *102*, 233–239.
106. Wei, Z.; Liu, Z.; Jiang, R.; Bian, C.; Huang, T.; Yu, A. $TiO_2$ nanotube array film prepared by anodization as anode material for lithium ion batteries. *J. Solid State Electrochem.* **2009**, *14*, 1045–1050.
107. Madian, M.; Giebeler, L.; Klose, M.; Jaumann, T.; Uhlemann, M.; Gebert, A.; Oswald, S.; Ismail, N.; Eychmuller, A.; Eckert, J. Self-Organized $TiO_2$/CoO Nanotubes as Potential Anode Materials for Lithium Ion Batteries. *ACS Sustainable Chem. Eng.* **2015**, *3*, 909–919.

108. Ortiz, G. F.; Hanzu, I.; Lavela, P.; Tirado, J. L.; Knauth, P.; Djenizian, T. A novel architectured negative electrode based on titania nanotube and iron oxide nanowire composites for Li-ion microbatteries. *J. Mater. Chem.* **2010**, *20*, 4041–4046.
109. Wang, W.; Tian, M.; Abdulagatov, A.; George, S. M.; Lee, Y. C.; Yang, R. Three-Dimensional Ni/$TiO_2$ Nanowire Network for High Areal Capacity Lithium Ion Microbattery Applications. *Nano Lett.* **2012**, *12*, 655–660.
110. Cheah, S. K.; Perre, E.; Rooth, M.; Fondell, M.; Harsta, A.; Nyholm, L.; Boman, M.; Gustafsson, T.; Lu, J.; Simon, P.; Edstrom, K. Self-Supported Three-Dimensional Nanoelectrodes for Microbattery Applications. *Nano Lett.* **2009**, *9*, 3230–3233.
111. Xie, K.; Lu, Z.; Huang, H.; Lu, W.; Lai, Y.; Li, J.; Zhou, L.; Liu, Y. Iron supported C@$Fe_3O_4$ nanotube array: a new type of 3D anode with low-cost for high performance lithium-ion batteries. *J. Mater. Chem.* **2012**, *22*, 5560–5567.
112. Taberna, P. L.; Mitra, S.; Poizot, P.; Simon, P.; Tarascon, J. M. High rate capabilities $Fe_3O_4$-based Cu nano-architectured electrodes for lithium-ion battery applications. *Nat. Mater.* **2006**, *5*, 567–573.
113. Yang, Y.; Fan, X.; Casillas, G.; Peng, Z.; Ruan, G.; Wang, G.; Yacaman, M. J.; Tour, J. M. Three-Dimensional Nanoporous Fe(2)O(3)/Fe(3)C-Graphene Heterogeneous Thin Films for Lithium-Ion Batteries. *ACS Nano* **2014**, *8*, 3939–3946.
114. Cheng, H.; Lu, Z.; Ma, R.; Dong, Y.; Wang, H. E.; Xi, L.; Zheng, L.; Tsang, C. K.; Li, H.; Chung, C. Y.; Zapien, J. A.; Li, Y. Y. Rugated porous $Fe_3O_4$ thin films as stable binder-free anode materials for lithium ion batteries. *J. Mater. Chem.* **2012**, *22*, 22692–22698.
115. Kim, J. H.; Zhu, K.; Yan, Y.; Perkins, C. L.; Frank, A. J. Microstructure and Pseudocapacitive Properties of Electrodes Constructed of Oriented NiO-$TiO_2$ Nanotube Arrays. *Nano Lett.* **2010**, *10*, 4099–4104.
116. Hu, C. C.; Chang, K. H.; Lin, M. C.; Wu, Y. T. Design and Tailoring of the Nanotubular Arrayed Architecture of Hydrous $RuO_2$ for Next Generation Supercapacitors. *Nano Lett.* **2006**, *6*, 2690–2695.
117. Zhao, G.-Y.; Li, H.-L. Preparation of polyaniline nanowire arrayed electrodes for electrochemical supercapacitors. *Microporous Mesoporous Mater.* **2008**, *110*, 590–594.
118. Zhang, L. L.; Zhao, X. S. Carbon-based materials as supercapacitor electrodes. *Chem. Soc. Rev.* **2009**, *38*, 2520–2531.
119. Bo, G.; Xiaogang, Z.; Changzhou, Y.; Juan, L.; Long, Y. Amorphous $Ru_{1-y}Cr_yO_2$ loaded on $TiO_2$ nanotubes for electrochemical capacitors. *Electrochim. Acta* **2006**, *52*, 1028–1032.
120. Xu, J.; Gao, L.; Cao, J.; Wang, W.; Chen, Z. Preparation and electrochemical capacitance of cobalt oxide (Co3O4) nanotubes as supercapacitor material. *Electrochim. Acta* **2010**, *56*, 732–736.
121. Hu, X.; Jiang, J. L.; Zhang, F.; Ling, Z. Y. Fabrication and electrochemical properties of periodic pillared-layer nickel hybrid electrode. *Solid State Ionics* **2014**, *268*, 316–320.

122. Xing, H.; Long, J. J.; Hui, F. Z.; Zhiyuan, L. Fabrication and Electrochemical Performance of Porous Nickel-Based Hybrid Electrodes from Anodized Etched Aluminum. *J. Electrochem. Soc.* **2013**, *160*, A1425–A1429.
123. Xie, Y.; Huang, C.; Zhou, L.; Liu, Y.; Huang, H. Supercapacitor application of nickel oxide–titania nanocomposites. *Compos. Sci. Technol.* **2009**, *69*, 2108–2114.
124. Xia, H.; Feng, J.; Wang, H.; Lai, M. O.; Lu, L. $MnO_2$ nanotube and nanowire arrays by electrochemical deposition for supercapacitors. *J. Power Sources* **2010**, *195*, 4410–4413.
125. Lu, X.; Wang, G.; Zhai, T.; Yu, M.; Gan, J.; Tong, Y.; Li, Y. Hydrogenated $TiO_2$ Nanotube Arrays for Supercapacitors. *Nano Lett.* **2012**, *12*, 1690–1696.
126. Xie, K.; Li, J.; Lai, Y.; Zhang, Z.; Liu, Y.; Zhang, G.; Huang, H. Polyaniline nanowire array encapsulated in titania nanotubes as a superior electrode for supercapacitors. *Nanoscale* **2011**, *3*, 2202–2207.

## Chapter 3

# Nanoparticle Building Blocks as a Foundation for Advanced Thermoelectric Energy Generators

**D. M. Mott*** **and S. Maenosono**

**School of Materials Science, The Japan Advanced Institute of Science and Technology, 1-1 Asahidai, Nomi, Ishikawa, Japan 923-1292**
***E-mail: derrickm@jaist.ac.jp**

The field of thermoelectric materials has a long history spanning decades. After initial development phase, only incremental improvements were made to energy conversion materials in terms of their efficiency. Recently however, the integration of nanotechnology to thermoelectric materials has lead to a surge in research because of the development of new nano-based techniques for improving thermoelectric efficiency. The field is now advancing rapidly and there is a marked shift from bulkscale non-sustainable materials towards sustainable bottom-up produced nanoparticle materials. Herein, we present a summary of some of the key findings in the field that have led to our own studies and development from tellurium containing nanoparticles to sustainable sulfide nanomaterials. A summary of research advancements along with the challenges associated with the developing nanoparticle based thermoelectric technology will be discussed for nanoparticle systems including the classic tellurides as well as new sustainable materials such as copper sulfide and copper iron sulfide.

# Introduction

Technologically speaking, we have reached a crossroad in terms of global energy production and use. The realities of accelerated global warming coupled with the strictly controlled and limited supply of oil create an accelerating need for new energy sources. Nuclear energy has been a stopgap for many years, but with mounting nuclear waste, deteriorating public opinion and the same efficiency limits as coal or oil based energy production, is not enough to cover growing demand or to maintain safe environmental standards. The highly touted sustainable and renewable solar and wind technologies offer increasing efficiency with higher energy return as advancements are made, but it will still be some time before they are developed to a point where all of our energy demand can be met by them. In light of this global energy climate, a new strategy is needed that can quickly enhance the energy production techniques we are already using. The vast majority of electricity generation is based on the steam turbine, which offers an energy extraction efficiency of about 48% maximum. The basic premise relies on heating water (either by nuclear heat generation or by burning coal/oil, etc), which is limited by the Carnot heat cycle. One strategy to meet environmental and energy needs now is to enhance the efficiency of the steam turbine process by reclaiming energy from the otherwise wasted heat, which would lead to not only a global revolution in energy production, but would drastically lower greenhouse emissions. This is possible by using thermoelectric devices at the hot point where water emerges after the steam turbine (*1*). While they have received little mainstream attention over the years, the field of thermoelectric materials offers an exciting prospect for harnessing the otherwise lost heat energy not only in steam turbine technology, but also at any heat source such as in automobiles (*1*, *2*), for geothermal energy, or even heat producing catalytic reactions as well as other applications (*1*). In fact, even by using modest thermoelectric materials (ZT = 1 to 2) in these applications, the energy production efficiency can be increased by as much as 15% (*3*), representing a monumental increase in energy generation efficiency, making the technology highly important in meeting energy needs while minimizing material consumption and waste production.

Thermoelectric materials encompass a field where the Seebeck, Peltier and/or Thomson effect are harnessed to either create an electric current from a temperature gradient, or use an electric current to generate a temperature gradient (*4*). The efficiency of the phenomenon is defined by the thermoelectric figure of merit ($ZT=\sigma S^2T/k$), where $\sigma$ is electrical conductivity (S/m), S is the Seebeck coefficient ($\mu$V/K), T is temperature (K) and k is thermal conductivity (W/(m·K)) of the material, where Z is a dimensionless unit. Thermoelectric materials have been known of for quite a long time, along with their potential benefits, but have never found widespread use because of fundamental limitations in the energy conversion efficiency of even the best materials (*3*). The challenges associated with increasing the overall ZT value can be observed in the figure of merit equation where electrical conductivity must be increased while thermal conductivity suppressed. Physically these two parameters are challenging to individually control which places a hard maximum on the efficiency that can be achieved. Recently however, it was found that nanotechnology offers one route to suppressing the thermal conductivity while essentially maintaining the electrical conductivity of a material (*3, 5*). By imparting nano-scale crystal grain defects into the material, the thermal conductivity can be suppressed through scattering of the heat carrying phonon at these crystal grain interfaces, while the electrical conductivity is basically unimpeded because of the significantly smaller mean free path of the electron (*5, 6*). The finding has generated a resurgence in interest in thermoelectric materials and has led to the development of many more related techniques for enhancing the energy conversion efficiency of thermoelectric materials (*3, 7*).

Despite this progress however, there is still little understanding of how exactly the nanoscale structuring can be optimized for obtaining the best ZT value for a given material because there are few studies focusing on thermoelectric materials composed of nanoparticles with well-defined size, shape, composition and structure. Most studies of thermoelectric materials deal with top down techniques where a bulk material is broken down, oftentimes with ball-milling, then is compressed back into a large scale material. Enhanced techniques utilize spark plasma sintering to maximize electrical conductivity and many specialized methods of material preparation revolve around the approach (*3*). In contrast, very few studies utilize bottom up wet chemical nanoparticle synthesis as a production technique for the particles. While such synthetic approaches can provide high quality particles with uniform parameters, there are several drawbacks. When nanoparticles are used as building blocks for the construction of thermoelectric materials, many challenges are introduced which must be overcome in order for the material to exhibit good thermoelectric activity. First, any organic ligands must be removed from the nanoparticle surfaces, or replaced by something that is sufficiently electrically conductive (*3, 8*). Next, the natural high surface area of the particles makes the creation of a dense pellet, which is required for a thermoelectric material, a fundamental challenge (high pressure, high heat or other techniques are required that would destroy the nanoparticle characteristics) (*3*). Finally, most nanoparticle synthetic techniques only successfully produce a minute amount of sample (typically 100 mg or less), which is much less than is required to reliably measure characteristics such as thermal conductivity. In order for the role of nanotechnology to be more fully understood and utilized in thermoelectric materials, these challenges must be researched and overcome.

The collection of work discussed here seeks to cover some of the fundamental questions about the role of nanotechnology in developing new thermoelectric materials. Much information has been generated for both bulk materials and the nano-structured counterparts of many theoretically ideal materials such as bismuth telluride (*9, 10*), bismuth antimony telluride (*5, 11*), lead telluride (*2, 9*), etc. But, we should also consider some of the lesser known sustainable materials as well. The chalcogenide class of materials has proven highly interesting (of which tellurium is a member) as an alternative, particularly the selenium or sulfur based materials (*1, 12*). The sulfur based chalcogenides offer potentially high thermoelectric efficiency, and are inherently sustainable because of the high elemental abundance. A summary of techniques for producing high quality nanoparticles for thermoelectric materials will be covered. Important nanoscale parameters such as nanoparticle size, shape, structure and composition will be discussed as well as advancements in techniques for processing the nanoscale materials into larger bulk devices. Techniques for analyzing and optimizing the thermoelectric characteristics will be summarized.

## Tellurium Based Materials

The tellurides represent the most widely studied class of thermoelectric materials both at the bulk scale and for nanomaterials because of the highly beneficial properties of the element for thermoelectricity. Tellurium possesses an electronic structure that allows for optimum electrical conductivity in the overall material (*1*, *7*). Unfortunatley however, tellurlium is exceedingly rare in the earth, making all tellurium containing thermoelectric materials non-sustainable. Regardless, our understanding of how thermoelectric materials operate and how to refine the device components to achieve maximal energy conversion efficiency have been greatly aided by the study of tellurium based materials. They are an excellent place to begin for understanding the role that nanotechnology can play in thermoelectric technology.

### Synthetic Strategies for Tellurium Based Nanoparticles

Bismuth telluride ($Bi_2Te_3$) is perhaps the most iconic traditional thermoelectric material. The study of the basic material has a long history and is one of the few thermoelectric materials to be widely commercialized (in nich applications and more recently for portable refrigeration). The material also serves as a basis for the related alloys including bismuth antimony telluride (i.e. $Bi_{2-x}Sb_xTe_3$) where the incorporation of an additional element allows some control over the electrical conductivity parameters. In terms of bottom up development of thermoelectric type nanoparticles, it has received the most attention in terms of fundamental study of synthetic strategies that aim to control the nanoparticle size, shape, composition and structure. The resulting materials have also received the most attention in terms of delineating the thermoelectric characteristics.

One of the earliest and best known examples of bismuth telluride and bismuth antimony telluride nanoparticles synthesized using a bottom up technique with an aim to control the nanoparticle parameters was conducted by Clemens Burda (*11*). In the work, dodecanethiol coordinated bismuth (Bi) and antimony (Sb) precursors were used as well as trioctylphosphine Te. Benzyl ether was used as a solvent with a reaction temperature of 150 °C. The resulting particles possessed a uniform size and roughly spherical shape as demonstrated in the example TEM image inset to Figure 1. What is even more remarkable was the fine control that could be achieved over the resulting particle composition. Antimony content from X=0.2 to X=1.5 was achieved (in the formula $Bi_{2-X}Sb_XTe_3$), which is a relatively wide degree of control. The resulting particles were annealed at 380 °C and the Seebeck coefficient was measured. Most of the materials were n-type with a magnitude ranging from about 40 to 149 μV/K, while the single sample synthesized with the highest fraction of antimony showed p-type activity of 256.6 μV/K. Figure 1 shows a graph demonstrating the relationship between the particle composition and the resulting Seebeck coefficient. The analysis revealed that the sample with a composition of $Bi_{1.5}Sb_{0.5}Te_3$ possessed the highest magnitude of n-type activity (*11*).

The positive results obtained by Burda generated excitement in the field and encouraged us to pursue synthesis techniques with an eye on refining the particle composition and structure. In particular, we focused on fine control of bottom up wet chemical synthetic parameters such as type of coordinating ligand used, solvent, and heating profile, which were shown to be crucial in creating telluride based nanoparticles with controlled and increasingly complex composition and structure.

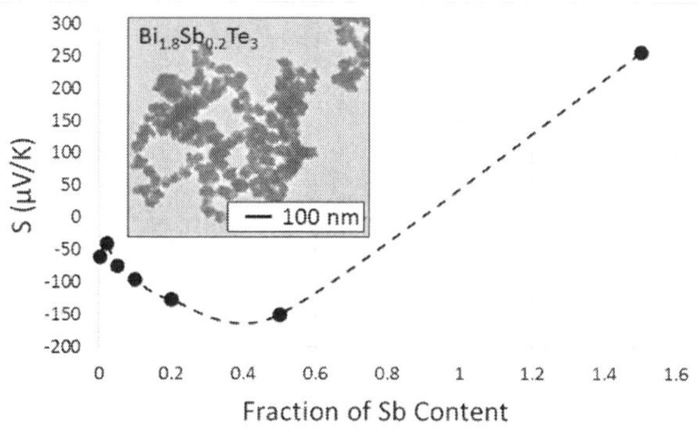


*Figure 1. Relationship between nanoparticle composition and resulting Seebeck coefficient for $Bi_{2-x}Sb_xTe_3$ materials created using a bottom up wet chemical approach. The inset shows an example TEM image of particles with a composition of $Bi_{1.8}Sb_{0.2}Te_3$. Reproduced with permission from reference (11). Copyright 2010, American Chemical Society.*

### *A Versatile Synthetic System for Producing Bismuth Antimony Telluride Nanoparticles with Various Structures, Shapes, and Composition*

A summary of our own work in this area gives an overview of some of the key synthetic accomplishments in creating bismuth telluride, antimony telluride and bismuth antimony telluride nanoparticles with complex and intriguing shape, structure and composition. By utilizing bismuth chloride, antimony chloride and tellurium chloride precursors, we created a versatile synthetic strategy for bismuth, antimony and tellurium containing nanoparticles. Three different organic protecting ligands (oleylamine, oleic acid and 1-decanethiol) were employed to control the resulting nanoparticle characteristics such as particle size, shape and structure. 1,2-hexadecanediol was used as a reducing agent and dioctylether served as a solvent (*13*). The essential results of our studies showed that by using a mixture of oleylamine and oleic acid ligands, very high aspect ratio nanowires were formed (Figure 2A). However, when 1-decanethiol was used, nanodiscs were observed (Figure 2D). Furthering the study, we also used a mixture of all three ligands and obtained particles that had a polyhedral flake-like shape (Figure 2F) (*13*). The results showed a stark contrast in the nanoparticle shapes that could be optained in the basic synthetic approach simply by changing the type of organic ligand used in the synthesis, providing a simple and straightforward way to tune these parameters for thermoelectric type nanoparticles. However, what became even more intriguing was the fact that the particles possessed different structures and compositions. For the nanowires synthesized with oleylamine and oleic acid, the very high resolution imaging technique scanning transmission electron microscopy with a high angle annular dark field detector (STEM-HAADF) revealed that part of the wires were composed of pure tellurium Figure 2B shows the STEM-HAADF image of a wire segment with atomic structure of tellurium),while other areas were composed of $Bi_2Te_3$. Figure 2C shows a STEM-HAADF image where a nanowire atomic structure transitions from pure tellurium to $Bi_2Te_3$ (*14*). Such structures had not been observed before and provide new avenues to engineering thermoelectric nanomaterials.

Similarly for the nanodiscs synthesized with 1-decanethiol, compositional and STEM-HAADF analysis (Figure 2E) showed that some areas of the disks were composed of $Bi_2Te_3$, while other areas were $Sb_2Te_3$ (*13, 14*). Interestingly when all three ligands were used, the nanoflakes appeared to be uniformly composed of $(Bi_{0.5}Sb_{0.5})_2Te_3$, the creation of which was one of our original goals in the research (*13*). Further study of the reaction condictions revealed the role that each type of precursor-ligand combination plays in the reaction (*15*). For the case of nanodiscs, where 1-decanethiol was used, we found that tellurium nanodiscs formed first, and then the bismuth or antimony was incorporated to the nanodisc structure by the catalytic decomposition of bismuth and antimony 1-decanethiol complexes, leading to the observed phase segregation shown in Figure 2G (*15*). Meanwhile for the nanowires, a slightly different phenomenon occurred. When oleylamine is used as an independent ligand, we found that pure tellurium nanowires form alongside individual bismuth/antimony particles. These small bismuth/antimony particles then undergo oriented attachment to the wire and become converted into areas of bismuth antimony telluride (*15*).

These results provided great insight into how to prepare tellurium containing thermoelectric nanoparticles with controllable characteristics, allowing the creation of nanoparticles with even more complex structure such as tellurium wires studded with $(BiSb)_2Te_3$ nanodiscs (Figure 2I). Part of the ongoing challenges with such materials though include scale up of the synthesis and processing into a large scale material for thermoelectric characterization. We attempted this by using the nanowire sample and found that we could reliably measure the Seebeck coefficient. The value was found to be 211 μV/K, which is reasonable and comparable to reference values (*13*). Determination of the materials thermal conductivity, which is required for evaluating the ZT value, was still elusive, partially because of the low mass of sample, poor mechanical stability and generally low electrical conductivity value.

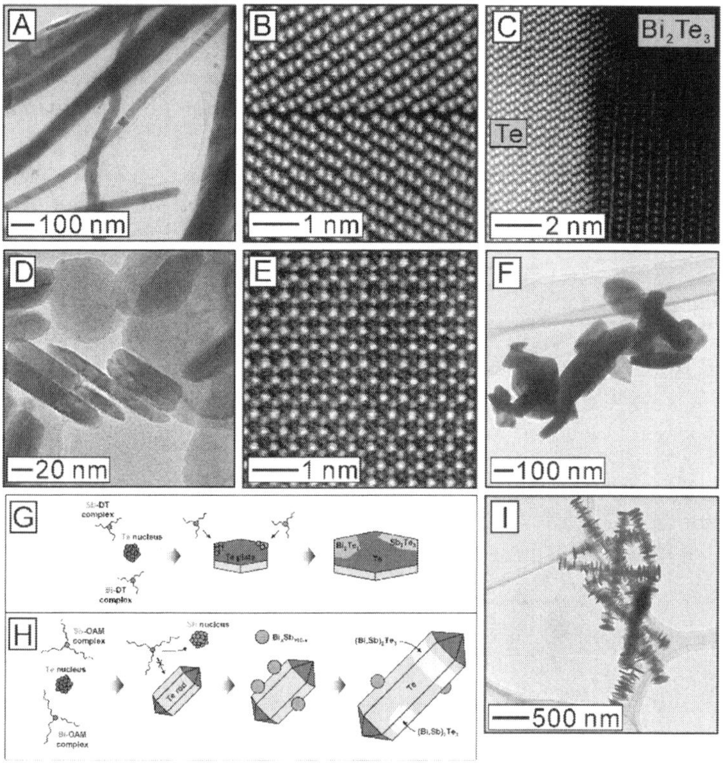


*Figure 2. TEM image of nanowires (A), STEM-HAADF image of tellurium segement (B) and transition from Te to $Bi_2Te_3$ (C), TEM image of nanodiscs (D), STEM-HAADF image of a nanodisc lying on face (E), TEM image of nanoflakes (F), scheme of synthetic formation pathway of nanodiscs (G), scheme of synthetic formation pathway of nanowires (H), TEM image of nanowires studded with nanodiscs (I). Reproduced with permission from references (13), (14) and (15). Copyright 2011, Wiley and Sons. Copyright 2011, American Chemical Society. Copyright 2011, Royal Society of Chemistry. (see color insert)*

## Alternatives to Tellurium Containing Thermoelectrics

One of the first alternatives we considered to tellurium containing thermoelectric materials was the intriguing zinc antimonide system. The bulk material had received some interest as an efficient thermoelectric material because of the inherently low thermal conductivity and moderate electrical conductivity (*1*), so we set about studying the synthesis and characterization of zinc animonide nanoparticles. The synthetic approach involved the use of both zinc and antimony chloride metallic precursors, di-octylether as a solvent, oleylamine as a particle protecting ligand and in this case the reducing agent was lithium triethylborohydride (*16*). The strong reducing agent was required to avoid oxidation of the forming particles and to ensure a complete reduction in the reaction. Figure 3 shows a representative transmission electron microscope (TEM) image, STEM-HAADF image, and energy dispersive spectroscopy (EDS) elemental maps for zinc and antimony (of the particle shown in the STEM-HAADF image), as well as an overlay map of the two elements. The resulting particles showed a unique composition/structural profile where the particle core is antimony rich and the particle periphery is zinc rich (*16*). Consequently the individual particles possess a unique structure profile comprising several different zinc antimonide phases including orthorhombic and hexagonal (and possibly rhombohedral antimony) phases (*16*). We measured the Seebeck value of the material by depositing a layer of the nanoparticles on a glass plate and simply applying gentle pressure, we then measured the Seebeck value using a four point probe method and found a value of 25 μV/K. The material is P-type, which is consistent with the corresponding bulk scale materials, but the magnitude of the value is depressed, which could be a result of compositional inhomogeneity in the sample. While the results were highly promising from a fundamental standpoint of eliminating tellurium from the system, we wished to further enhance the sustainability and safety of the system in light of the fact that antimony is toxic and is still a relatively rare element itself. The next step in our studies was to delve into the sulfide class of materials.

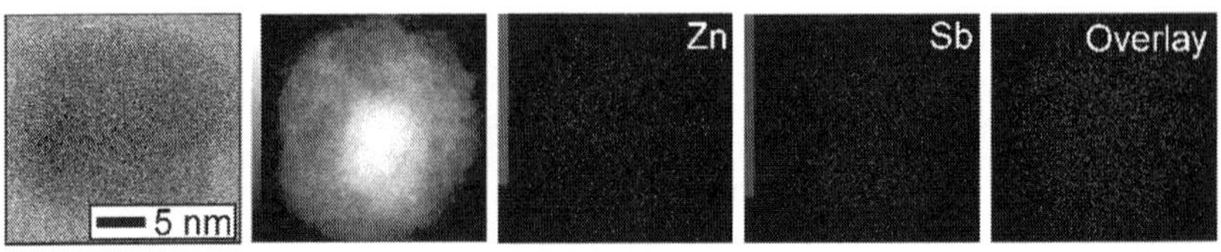


*Figure 3. HR-TEM image, STEM-HAADF image, EDS map of zinc, EDS map of antimony and an overlay of the zinc and antimony maps from left to right. The scale bar is the same for all images. (see color insert)*

## Sustainable Sulfide Based Materials

With some experience synthesizing the tellurium containing nanoparticles, we then turned our attention to more fundamentally sustainable materials. When considering thermoelectrics from the staindpoint of sustainability, the challenges associated with energy conversion efficiency become even more severe. Traditionally, the ZT value of non-tellurium containing materials was so low that they were completely excluded or dismissed as viable thermoelectrics. This has however recently changed with the discovery of nanoscale processing for thermoelectrics. Of particular interest are the selenides and the sulfides, which have provided some very intriguing characteristics for enhanced thermoelectric activity. One example is the so called "phonon glass electron crystal" concept for copper sulfides where a unique atomic structure allows electrons to propagate through the material with ease (as in a crystalline material) while the heat carrying phonon experiences low mobility (as in a glass) (*1*, *3*, *6*). Characteristics such as these guided our development of copper sulfide based nanoparticle materials for thermoelectrics.

## Copper Sulfide (Chalcocite) Thermoelectric Nanoparticles

One our our earliest attempts at a sustainable thermoelectric material was on the copper sulfide (chalcocite) system. Copper sulfide ($Cu_2S$) has a monoclinic crystal structure and the nanoparticles can be synthesized through a straightforward thermolysis approach. Our own nanoparticles were synthesized through a modified technique where copper acetate was dissolved in a small amount of methanol, then was injected to hot di-octylether and 1-dodecanethiol. The alkanethiol acts as a sulfur source for the creation of the copper sulfide through thermolysis and also as a nanoparticle surface ligand, assisting in control of particle size and shape (*17*, *18*). Figure 4 shows the basic characterization of the particles. The inset TEM image reveals uniformly shaped particles with a size of 7.2 ± 0.4 nm. The particles appear spherical in the TEM image, but actually possess a disk morphology. Nearly all of the particles lay face down on the TEM grid and they form well-ordered self assembled arrays to minimize surface energy. X-ray diffraction (XRD) analysis (left side of Figure 4) reveals the monoclinic crystal structure for these $Cu_2S$ nanoparticles, confirming the material identity. It is important to note that we cannot rule out the possibility that a copper deficient phase forms such as $Cu_{1.98}S$ (i.e. djurlite) because the small particle size causes the peaks in the XRD pattern to be greatly broadened. This makes fine assignment of the peak position more of a challenge than for a bulk material. In any case we are confident we have created predominantly chalcocite nanoparticle materials.

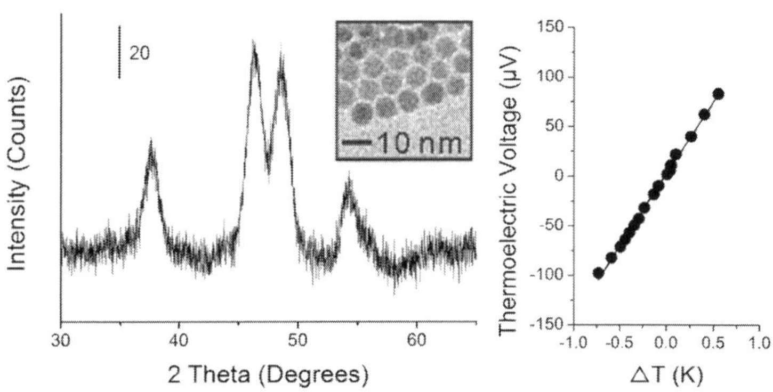


*Figure 4. XRD pattern (left side), TEM image (inset) and Seebeck characterization (right side) of $Cu_2S$ nanoparticles.*

Next, we set out to prepare this material for measuring the Seebeck properties, a first step to studying the full range of thermoelectric parameters. The challenges we faced were creating enough nanoparticle material for a reliable characterization, making a sufficiently dense material, and achieving acceptable electrical conductivity. We also wanted to avoid any harsh treatment techniques that might lead to loss of the particles nanoscale size. To collect enough nanoparticle sample, we opted for performing several individual nanoparticle synthesis and combining the resulting materials. This allowed us to bypass having to scale up the synthesis and re-evaluate the synthetic conditions required to obtain comparable nanoparticles to the low yield synthesis. Four individual synthesis were used to collect enough materal (about 120 mg of nanoparticle sample was sufficient for Seebeck measurement). Next, we settled on a three step treatment technique to create a dense nanoparticle pellet that exhibited good electrical conductivity. We first used ethylene diamine as a short chain linking molecule to remove the long chain (and poor electrical conductivity) 1-dodecanethiol molecules from the nanoparticle surfaces. Next, we pressed this material into a disk shape using a pellet press with a pressure of 25 Mpa for 3 minutes. A black glossy disk with a thickness of 2 mm and diameter of 10 mm resulted. Finally we thermally treated this disk at a temperature of 400 °C for 2 hours in a nitrogen gas flow. The thermal treatment step was the most destructive to the material where some particle sintering occurred, which was necessary to create better interparticle contacts and make the material electrically conductive. Subsequent XRD analysis showed that the material retained the chalcocite structure (oxidation and phase separation did not occur) with only a minimal grain size increase. The sample was then analyzed using a four point probe method to determine the Seebeck coefficient. The right side graph in Figure 4 shows the resulting data from the Seebeck analysis. The material is P-type, which we expected based on the reference data, however the Seebeck value was 146 μV/K which is significantly higher than for the bulk material (~10 μV/K) (*1*). Other reports on the Seebeck characteristics of chalcocite based nanoparticle materials also show elevated Seebeck value (up to 85 μV/K) (*18*), so the result is highly interesting in terms of the enhanced thermoelectric properties imparted by nanomaterials.

## Copper Iron Sulfide (Chalcopyrite) Thermoelectric Nanoparticles

After characterizing the $Cu_2S$ nanoparticles we set our sights on techniques for modifying or tuning the semiconducting characteristics. The most straightforward technique available to use that would give the greatest control was to add an additional element to the copper sulfide structure. We decided on incorporating iron to the basic copper sulfide nanoparticles because iron is abundant, non-toxic, and copper iron sulfide (chalcopyrite) has shown some interesting thermoelectric properties (*19*, *20*). We utilized a very similar synthetic protocol as compared to the copper sulfide nanoparticles. In this case we dissolved copper acetate and iron acetate precursors in a small amount of methanol and injected these simultaneously into hot octylether and 1-dodecanethiol. The metallic feeding ratio of copper and iron were kept the same to create $CuFeS_2$ (chalcopyrite) nanoparticles. In this case the iron can also undergo thermolysis, albeit at a higher temperature than for pure copper. In our synthetic analysis we also found that the presence of copper seems to provide a catalytic effect for the thermolysis of iron. The inset TEM image of Figure 5 shows that the resulting particles have a roughly polyhedral shape with an irregular size (*21*). The chalcopyrite nanoparticles are much larger than the pure chalcocite nanoparticles. XRD analysis shown on the left side of Figure 5 reveals the tetragonal structure of the material with no sign of phase impurities (*21*). The relatively narrow peak widths also indicate the larger crystalline size when compared to the copper sulfide nanoparticles.

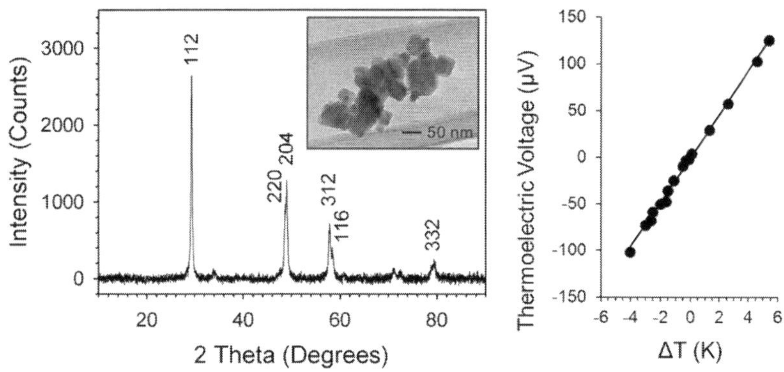


*Figure 5. XRD pattern (left side), TEM image (inset) and Seebeck characterization (right side) of $CuFeS_2$ nanoparticles. Reproduced with permission from reference (21). Copyright 2014, The Japan Society of Applied Physics.*

When preparing this sample for Seebeck characterization, we used an identical protocol as for the copper sulfide material. Several small scale nanoparticle synthesis results were collected and were treated in ethylene diamine to remove the long chain 1-dodecanethiol molecules from the surface of the nanoparticles. The material was then compressed into a pellet under 25 Mpa of pressure to give a disk with a diameter of 10 mm and thickness of 2 mm. The material became significantly more electrically conductive after being thermally treated at 400 °C under nitrogen atmosphere for 2 hours. XRD analysis at this point showed that the material structure and composition was retained (no oxidation or phase separation resulted) and the particle crystalline size underwent only a slight increase (*21*). We then used this sample to analyze the Seebeck properties of the material. The plot on the right side of Figure 5 shows the Seebeck analysis where P-type conductivity was measured with a value of 23.6 µV/K (*21*). The magnitude of the Seebeck coefficient for our material is lower than other comparable reference materials, but this is likely due to the high degree of dependence on doping or carrier concentration on the Seebeck value for this material (*19*, *20*). It is likely that this material has a relatively low population of charge carriers which could be enhanced through subtle changes in the material composition. The chalcopyrite material shows a modest Seebeck value, but has strong promise as a sustainable material that can easily be fabricated with nanoscale characteristics.

## Advanced Techniques for Processing Nanoparticles into Large Scale Materials for Thermoelectrics

As efforts continue towards refining nanoparticle materials and their processing from individual building blocks to large scale materials with good electrical conducitivity for thermoelectrics, some new techniques have emerged that are of significant note. The studies by Kanatzidis are particularly significant in light of the pursuit of methods to reliably remove the organic ligands that persist on the nanoparticle surface after wet chemical preparation (*8*). In a recent study, the Seebeck characteristics of both lead(II) sulfide, (PbS) and lead tellurium selenium ($PbTe_{0.2}Se_{0.8}$) core and PbS shell [$PbTe_{0.2}Se_{0.8}$@PbS ($PbTe_{0.2}Se_{0.8}$ nanoparticles coated by PbS] nanoparticles were studied both before removing oleic acid from the particle surface, and after treatment with various concentrations of hydrochloric acid. Treatment of the particles with dilute hydrochloric acid achieved two things, first the organic ligands were removed from the particle surface, but in addition the resulting material became doped with halide ions, beneficially impacting the resulting Seebeck values. The left hand side TEM image in Figure 6 the PbS nanoparticles where the organic ligands cause the particles to be well separated and ordered on the TEM grid. The accompanying graph shows a plot of the Seebeck coefficient at different temperatures for untreated particles as well as for materials prepared with increasing concentrations of hydrochloric acid (HCl). For the untreated particles, the Seebeck coefficient starts off as P-type, then sharply transitions to N-type with heating. In contrats, after treatment with HCl, the materials all show a relatively consistent N-type value approaching -200 μV/K (*8*). Similarly for the $PbTe_{0.2}Se_{0.8}$@PbS nanoparticle material, the particle surfaces are coated in organic ligands as observed in the TEM image of the particles (right hand TEM image in Figure 6). The Seebeck measurement of the untreated material shows a transition from P-type activity to N-type activity with increasing temperature while the HCl treated material was N-type at all temperatures with a maximum value of about -180 μV/K (*8*). While these materials still incorporate tellurium, they demonstrate an effective technique to not only remove the organic ligands from a nanoparticle system in preparation of thermoelectric characterization, but also offer an opportunity to manipulate the Seebeck coefficient through doping. Techniques such as these may serve as a basis for the continued development of nanoparticle based thermoelectric materials with enhanced characteristics and properties.

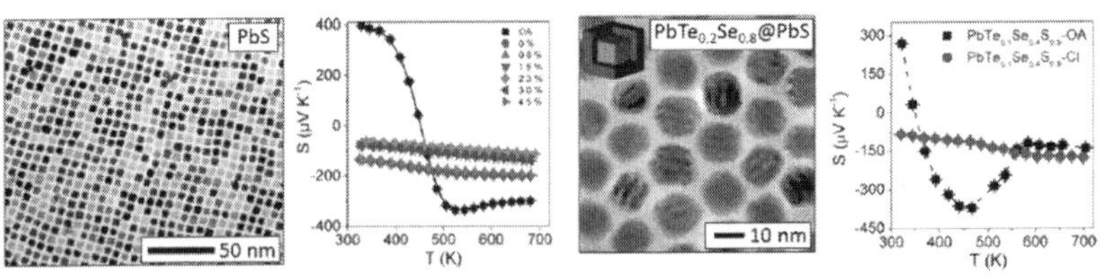


*Figure 6. TEM image of PbS nanoparticles, Seebeck measurements of PbS materials, TEM image of $PbTe_{0.2}Se_{0.8}$@PbS nanoparticles and Seebeck measurements of $PbTe_{0.1}Se_{0.4}S_{0.5}$ materials. Reproduced with permission from reference (8). Copyright 2015, American Chemical Society. (see color insert)*

## Conclusion and Outlook

In conculsion, the results that have been collected in this work serve as a representation of the current status in the field of nanotechnology driven thermoelectric materials today. There is increasing interest and compelling evidence that nanomaterials produced using bottom up techniques with a focus on control of the nanoparticle size, shape, structure and composition offer some of the best opportunities to enhance and control the thermoelectric parameters such as thermal conductivity, electrical conductivity and Seebeck value. Moreover, the gradual transition from tellurium containing materials to sustainable ones reflects the great need to produce thermoelectric devices for energy applications on a wide technological scale to meet growing energy needs and address environmental concerns. The transition to sustainable materials also signifies the overall advance in nanotechnology in enhancing the efficiency of traditionally non-ideal materials, opening the door to new discoveries in the field of thermoelectrics. The outlooks is promising, while there are still many challenges ahead in terms of developing strategies to scale up nanomaterial production, enhance interparticle contacts, and create robust thermoelectric devices from the nanoparticle building blocks, the time is right for a surge in sustainable thermoelectric device production and utilization. Applications such as automotives, steam turbine based energy production, and even solar produced heat can be greatly enhanced by using the materials that have already been developed. The energy, environmental and economic benefits are too good to pass up.

## References


1. Gonçalves, A. P.; Godart, C. New promising bulk thermoelectrics: intermetallics, pnictides and chalcogenides. *Eur. Phys. J. B* **2014**, *87*, 42.
2. Heremans, J. P.; Jovovic, V.; Toberer, E. S.; Saramat, A.; Kurosaki, K.; Charoenphakdee, A.; Yamanaka, S.; Snyder, G. J. Enhancement of thermoelectric efficiency in PbTe by distortion of the electronic density of states. *Science* **2008**, *321*, 554–557.
3. Kanatzidis, M. G. Nanostructured thermoelectics: the new paradigm. *Chem. Mater.* **2010**, *22*, 648–659.
4. Rowe, D. M. In *Thermoelectrics Handbook: Macro to Nano*; CRC Press: Boca Raton, Florida, 2006.
5. Poudel, B.; Hao, Q.; Ma, Y.; Lan, Y.; Minnich, A.; Yu, B.; Yan, X.; Wang, D.; Muto, A.; Vashaee, D.; Chen, X.; Liu, J.; Dresselhaus, M. S.; Chen, G.; Ren, Z. High-thermoelectric performance of nanostructured bismuth antimony telluride bulk alloys. *Science* **2008**, *320*, 634–638.
6. Liu, H.; Shi, X.; Xu, F.; Zhang, L.; Zhang, W.; Chen, L.; Li, Q.; Uher, C.; Day, T.; Snyder, G. J. Copper ion liquid-like thermoelectrics. *Nature Materials* **2012**, *11*, 422–425.
7. Zhao, L-D.; Hao, S.; Lo, S-H.; Wu, C-I.; Zhou, X.; Lee, Y.; Li, H.; Biswas, K.; Hogan, T. P.; Uher, C.; Wolverton, C.; Dravid, V. P.; Kanatzidis, M. G. High thermoelectric performance via hierarchichal compositionally alloyed nanostructures. *J. Am. Chem. Soc.* **2013**, *135*, 7364–7370.

8. Ibanez, M.; Korkosz, R. J.; Luo, Z.; Riba, P.; Cadavid, D.; Ortega, S.; Cabot, A.; Kanatzidis, M. G. Electron doping in bottom-up engineered thermoelectric nanomaterials through HCl-mediated ligand displacement. *J. Am. Chem. Soc.* **2015**, *137*, 4046–4049.
9. Kovalenko, M. V.; Spokoyny, B.; Lee, J-S.; Scheele, M.; Weber, A.; Perera, S.; Landry, D.; Talapin, D. V. Semiconductor nanocrystals functionalized with antimony telluride zintl ions for nanostructured thermoelectrics. *J. Am. Chem. Soc.* **2010**, *132*, 6686–6695.
10. Scheele, M.; Oeschler, N.; Meier, K.; Kornowski, A.; Klinke, C.; Weller, H. Synthesis and thermoelectric characterization of $Bi_2Te_3$ nanoparticles. *Adv. Funct. Mater.* **2009**, *19*, 3476–3483.
11. Zhao, Y.; Dyck, J. S.; Hernandez, B. M.; Burda, C. Enhancing thermoelectric performance through adjusting carrier concentration. *J. Am. Chem. Soc.* **2010**, *132*, 4982–4983.
12. Maignan, A.; Guilmeau, E.; Gascoin, F.; Breard, Y.; Hardy, V. Revisiting some chalcogenides for thermoelectricity. *Sci. Tech. Adv. Mater.* **2012**, *13*, 053003.
13. Mott, D. M.; Mai, N. T.; Thuy, N. T. B.; Maeda, Y.; Linh, T. P. T.; Koyano, M.; Maenosono, S. Bismuth, antimony and tellurium alloy nanoparticles with controllable shape and composition for efficient thermoelectric devices. *Phys. Status Solidi A* **2011**, *208*, 52–58.
14. Mott, D. M.; Mai, N. T.; Thuy, N. T. B.; Sakata, T.; Higashimine, K.; Koyano, M.; Maenosono, S. Elucidation of the complex structure of nanoparticles composed of bismuth, antimony, and tellurium using scanning transmission electron microscopy. *J. Phys. Chem. C* **2011**, *115*, 17334–17340.
15. Mai, N. T.; Mott, D. M.; Thuy, N. T. B.; Osaka, I.; Maenosono, S. Study on formation mechanism and ligand-directed architectural control of nanoparticles composed of Bi, Sb and Te: towards one-pot synthesis of ternary $(Bi,Sb)_2Te_3$ nanobuilding blocks. *RSC Adv.* **2011**, *1*, 1089–1098.
16. Mai, N. T.; Mott, D. M.; Higashimine, K.; Maenosono, S. One-pot chemical synthesis of zinc antimonide nanoparticles as building blocks for nanostructured thermoelectric materials. *Chem. Lett.* **2012**, *41*, 1529–1532.
17. Sigman, M. B.; Ghezelbash, A.; Hanrath, T.; Saunders, A. E.; Lee, F.; Korgel, B. A. Solventless synthesis of mondisperse Cu2S nanorods, nanodisks, and nanoplatelets. *J. Am. Chem. Soc.* **2003**, *125*, 16050–16057.
18. Ge, Z-H.; Zhang, B-P.; Chen, Y-X.; Yu, Z-X.; Liu, Y.; Li, J-F. Synthesis and transport property of Cu(1.8)S as a promising thermoelectric compound. *Chem. Commun.* **2011**, *47*, 12697–12699.
19. Li, J.; Tan, Q.; Li, J-F. Synthesis and property evaluation of $CuFeS_{2-X}$ as earth-abundant and environmentally-friendly thermoelectric materials. *J. Alloy Compd.* **2013**, *551*, 143–149.
20. Tsujii, N.; Mori, T.; Isoda, Y. Phase stability and thermoelectric properties of $CuFeS_2$-based magnetic semiconductor. *J. Electron. Mater.* **2014**, *43*, 2371–2375.
21. Verma, S.; Singh, M.; Ahuja, D.; Shimose, H.; Nishino, S.; Miyata, M.; Mott, D.; Koyano, M.; Maenosono, S. Chalcopyrite nanocomposite material for sustainable thermoelectrics. *Jpn. J. Appl. Phys.* **2014**, *53*, 120301.

# Section II

# Chapter 4

# Delafossite $CuAlO_2$ Nanoparticles with Electrocatalytic Activity toward Oxygen and Hydrogen Evolution Reactions

**Jahangeer Ahmed and Yuanbing Mao***

**Department of Chemistry, University of Texas – Pan American, Edinburg, Texas 78539, United States**
***E-mail: yuanbing.mao@utrgv.edu. Tel: +1-956-665-2417.**

As a type of novel electrocatalytic materials, delafossite $CuAlO_2$ nanoparticles (NPs) with an average particle size of ~ 20 nm were synthesized by a facile ultrasonic synthesis followed with annealing at 850°C in $N_2$ for the first time. Their phase purity, size and morphology were investigated by powder X-ray diffraction, Raman spectroscopy and electron microscopic studies. Their electrochemical performance studies demonstrated promising catalytic activity in the electrolysis of water to the hydrogen evolution reactions (HER) and oxygen evolution reactions (OER) compared to bulk $CuAlO_2$. More specifically, these $CuAlO_2$ nanoparticles showed 12 times higher current density to the OER than the bulk counterpart. Hence, being a low-cost semiconducting material, these delafossite $CuAlO_2$ nanoparticles with the size of ~ 20 nm could be useful to further study their other applications like solar cells, photochemical water splitting, proton exchange membrane water electrolysis etc.

## Introduction

To effectively address the fossil fuel depletion issue and serious environmental problems accompanying its combustion, new forms of energy that is clean, renewable, cheap, safe and viable are currently under intensive search. Hydrogen has received considerable attention recently as a next-generation energy carrier. Electrolysis of water with the assistance of catalysts provides hydrogen and oxygen which have a potential to provide sustainable energy via fuel cells. Hydrogen is used as a fuel in the fuel cells by converting the chemical energy into electrical energy and also used as a fuel in automobile and aerospace industries while oxygen is also useful in fuel cells to assist the power generation by enhancing the temperature of the reactors because of the completion of combustion reactions. However, efficient catalysts are still under intensive search for water splitting for good product selectivity from band alignment with chemical redox potentials, fast reaction kinetics from desirable efficiencies and without charge carrier recombination in addition to good stability, cost-effective and non-toxicity.

Delafossites with the general formula $A^IM^{III}O_2$ comprised of alternating layers of slightly distorted edge-shared $MO_6$ octahedra and two-dimensional close-packed A-cation planes forming linear O-$A^{1+}$-O "dumbbells", and the oxygens coordinated by 4 cations (1 $A^I$ & 3 $M^{III}$) are important p-type transparent conducting oxide materials (*1*) which exhibits a variety of applications including gas sensors (*2*), optical (*3–7*), electrical (*5, 8, 9*) and thermoelectric devices (*10*), dilute magnetic semiconductors (*11*), field emission displays (*12*), photo-catalytic water splitting (*13*), and solar cells applications (*14–17*). Delafossites have been progressively studied in the last decade , especially after $CuAlO_2$ was discovered as a good *p*-type transparent conducting oxide (TCO) with high conductivity, and then attracted attention for *p*-type dye-sensitized solar cells (DSSCs) due to their desirable low valence band (VB) edge position and high hole mobility, almost exclusively on $CuMO_2$ with M = Al, Ga or Cr. Copper based delafossites with copper at the A–site have a wide variety of interesting electronic properties (*1, 18*). Few examples of delafossites like $PtCoO_2$ (*19*), $PdCoO_2$ (*19*), $PdRhO_2$ (*19*), $CuRhO_2$ (*20, 21*) have also been reported in literature as the electro-catalysts for oxygen evolution reaction (OER). There is only one report in the literature regarding the electrocatalysis based on delafossite $CuAlO_2$ (*21*), and the electrochemical behavior of delafossite $CuAlO_2$ has not been studied extensively. Therefore, we are motivated to synthesize $CuAlO_2$ nanoparticles and study their electrochemical performance. Moreover, we demonstrated that these delafossite $CuAlO_2$ nanoparticles are promising catalysts in electrolysis of water to the hydrogen and oxygen evolutions.

High temperature (above 1000 °C) synthesis normally form micron–sized $CuAlO_2$ particles. Therefore, one of our aims was to synthesize $CuAlO_2$ nanoparticles in the nanometer region by lowering the synthesis temperature. In terms of the synthesis of delafossite $CuAlO_2$ nanoparticles, there are several used techniques including sol–gel (*3, 14, 22, 23*), hydrothermal (*15, 24, 25*), mechanical alloying (*9, 17*), pulsed laser deposition (*8*), sputtering (*12, 26*), and electro-spinning (*7*) methods. Nanocrystalline $CuAlO_2$ particles were also synthesized from a biological process at low temperature (50 °C) (*6*) and by using boehmite nanorods (γ-AlOOH) and copper (I) acetate as the precursor at high temperature (1150 °C) (*27*). At the meantime, bulk delafossite $CuAlO_2$ could be easily synthesized by solid state method at high temperature range from 1100 °C to 1200 °C (*27–29*).

To the best of our knowledge, there is no report in literature based on the ultrasonic synthesis of $CuAlO_2$ nanoparticles. Herein, we report a facile synthesis of $CuAlO_2$ nanoparticles (~20 nm) using the ultrasonic process followed by the annealing in $N_2$ atmosphere at 850 °C for 48 h. It is noteworthy that the ultrasonic process provides the smallest $CuAlO_2$ nanoparticles compared with reported data on the literature based on our best knowledge. The smallest $CuAlO_2$ nanoparticles previously reported were in the range from 35 nm to 50 nm by using chemical routes (*3*, *14*, *27*). Furthermore, we carefully investigate the electrochemical performance of $CuAlO_2$ nanoparticles to the OER and HER by electrolysis of water using 0.5 M KOH and Ag/AgCl as the electrolyte and reference electrode, respectively. The structural and morphological characterizations of the nanopowders were investigated by powder X-ray diffraction (PXRD), Raman spectroscopy, field emission scanning electron microscopy (FESEM), transmission electron microscopy (TEM), and energy dispersive x–ray studies.

## Experimental

### Materials and Synthesis Methods

$Cu(NO_3)_2{\bullet}3H_2O$ (Sigma Aldrich, 99 %), $Al(NO_3)_3{\bullet}3H_2O$ (Sigma Aldrich, 98 %), and $NaHCO_3$ (Sigma Aldrich, 99.7 %) were used in the ultrasonic synthesis of the $CuAlO_2$ nanoparticles. The ultrasonic process involved the mixing of parent materials (0.1 M $Cu(NO_3)_2{\bullet}3H_2O$ and 0.1 M $Al(NO_3)_3{\bullet}3H_2O$) followed by the titration with 0. 5 M $NaHCO_3$ aqueous solution drop by drop at 25 °C under the ultrasonic wave (operating at 117 V with the frequency of 60 kHz) and kept for 3 h. The light green precipitates were centrifuged followed by washing with deionized water for three times. The resulting light green colored powders were then dried at 60 °C for 6 h and used as a precursor in the synthesis of grey colored $CuAlO_2$ nanoparticles at 850 °C for 48 h in $N_2$ (99.999 %). Bulk counterpart, i.e. $CuAlO_2$ micron sized particles, were also synthesized via solid state route to compare the electrochemical performance with the $CuAlO_2$ nanoparticles. More specifically, $Al_2O_3$ (Sigma Aldrich, 98%) and CuO nanopowders (size ~50 nm, Sigma Aldrich) were taken at 1:2 molar ratio and ground together for half an hour in an agate mortar pestle followed by firing at 1150 °C for 36 h in $N_2$ atmosphere.

### Materials Characterization

The resulting products of both cuprous aluminate delafossite [$CuAlO_2$] nanoparticles and bulk counterpart were initially characterized by powder X-ray diffraction (PXRD) spectroscopy carried out on a Bruker D8 Advanced diffractometer with Ni-filtered Cu-Kα radiation (λ = 1.5406 Å) in the 2θ range from 10° to 60° with the scan rate of 0.02°/second on a zero background sample holder. Raman scattering experiments were carried out by a Bruker SENTERRA RAMAN microscope with a 785 nm $Ar^+$ laser as the excited light and an objective of 20X of optical microscope. The laser beam was focused to a spot size of 5 μm with the power of 10 mW. Both the particle size and morphology of the $CuAlO_2$ nanoparticles were investigated by the ZEISS SIGMA VP FESEM and ZEISS LEO 900 TEM operated at 5 kV and 80 kV, respectively. The TEM samples were prepared by the dispersion of 2 mg of the nanoparticles in 10 ml of ethanol under the ultra-sonication for 5 min and a drop of suspension was placed on to the surface of porous carbon film supported by a copper grid followed by drying in air. Thermo-gravimetric analysis (TGA) of the nanomaterials was carried out using a Perkin–Elmer system with a heating rate of 10 °C/min in oxygen atmosphere.

### Electrochemical Measurements

Electrolysis of water for gas ($O_2$ and $H_2$) evolutions was carried out with a computer controlled potentiostat/galvanostat (Metrohm Autolab B.V.) at room temperature. The electrochemical work station contains three electrodes, i.e. Pt wire, Ag/AgCl, and glassy carbon electrodes as the counter electrode, reference electrode and working electrode, respectively. The working electrodes were prepared by pasting $CuAlO_2$ powders with carbon black and polyvinylidene fluoride (PVDF, weight fractions of $CuAlO_2$:carbon black:PVDF of 75:15:10 w/w) on to the surface of glassy carbon electrode followed by drying at 65 °C in vacuum oven. The prepared electrodes were used as the working electrodes in the electrochemical measurements, including cyclic voltammetry (CV) and chronoamperometry, for oxygen evolving reaction (OER) and hydrogen evolution reactions (HER). Cyclic voltammetry was carried at the scan rate of 5, 10, 25, 50 and 100 mV/s versus Ag/AgCl electrode for the OER and from 0 to -1.2 V versus the Ag/AgCl electrode for the HER in alkaline electrolyte (0.5 M KOH). Chronoamperometric measurements were carried out for 200 s at 0.60 V and -1.2 V versus Ag/AgCl for OER and HER, respectively, in 0.5 M KOH electrolyte. The nitrogen gas was passed through the electrolyte for 10 minutes prior to the electrochemical measurements to remove the dissolved air from the electrolyte toward the OER and HER. Also electrochemical measurements in $O_2$–saturated 0.5 M potassium (I) hydroxide (KOH) were run for OER. All measurements to study the OER and HER were conducted with freshly prepared electrodes and electrolytes. The current density of the $CuAlO_2$ electrocatalysts was calculated using the electro-active surface area of electrode which is calculated by using the Randles–Sevcik equation (*30–32*).

## Results and Discussion

### Materials Characterization

Based on the powder X-ray diffraction pattern, the as-synthesized precursor after the ultrasonic process could be analyzed on the basis of crystalline phases of basic copper carbonate ($Cu_2(OH)_2CO_3$) and aluminum (III) hydroxide ($Al(OH)_3$) (Figure 1a). The annealing of this precursor in $N_2$ atmosphere at 850 °C for

48 h leaded to the formation of grey monophasic copper aluminate ($CuAlO_2$) nanoparticles (Figure 1b). Figure 1c shows the PXRD pattern of $CuAlO_2$ bulk particles obtained by solid state route, which cannot be synthesized at or below 1000 °C, either in air or inert atmosphere (*28, 29*). All the reflections (003, 006, 101, 012, 104, 009, 107 and 018) could be indexed on the basis of pure phase of rhombohedral structure with space group of R3-m (JCPDS file # 01-075-2356).

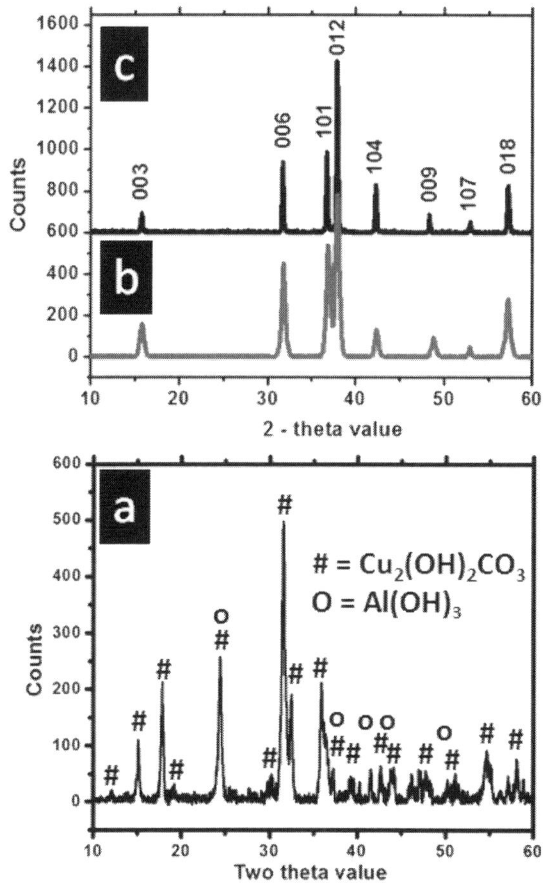


*Figure 1. PXRD patterns of (a) the as-synthesized precursor by ultrasonic process at room temperature, (b) the $CuAlO_2$ nanoparticles obtained subsequently at 850 °C for 48h in $N_2$, and (c) bulk $CuAlO_2$ particles synthesized by solid state route at 1150 °C for 36 h.*

Figure 2 shows the FESEM and TEM micrographs of the synthesized $CuAlO_2$ nanoparticles and the bulk counterpart. The FESEM and TEM studies (Figures 2a and 2b) showed the formation of spherical shaped nanoparticles with an average particles size of ~20 nm. By considering the reported particle size of $CuAlO_2$ nanoparticles (*3, 14, 27*), what presented here obtained by the ultrasonic process followed by firing at 850 °C has the smallest size. Specifically, $CuAlO_2$ nanoparticles reported recently on the literature with the smallest average size (35 nm) were prepared by an acid catalyzed sol–gel method followed by the heat treatment in sealed silica tube under a controlled partial pressure (p) of oxygen ($pO_2$) of $10^{-5}$ atm (*14*). Figure 2c shows the FESEM image of bulk $CuAlO_2$ with micron–sized plates-like particles (~2 μm) which were synthesized by the solid state route at 1150°C. From Brunauer, Emmett and Teller (BET) surface area measurements, the specific surface area of the $CuAlO_2$ nanoparticles and $CuAlO_2$ bulk particles were found to be 1.15 and 0.025 $m^2/g$, respectively. As expected, the specific BET surface area increases with decreasing the particle size of these $CuAlO_2$ particles. As reported in the literature, delafossite $CuAlO_2$ nanoparticles are an important transparent semiconducting oxides for the light to electricity conversion in dye-sensitized solar cells (DSSCs). The conversion efficiencies depend not only on the nature of semiconducting materials and dye molecules but also on semiconducting particle size with uniform distribution (*14–17*). Therefore, the synthesized $CuAlO_2$ nanoparticles here with the smallest particle size of ~20 nm could be an ideal choice for further studies of their various applications, including the electrocatalytic performance studied here.

The rhombohedral crystal structure of the $CuAlO_2$ samples were also characterized using Raman spectroscopy. Figure 3 shows Raman spectra of the rhombohedral $CuAlO_2$ nanoparticles which were collected by employing the normal backscattering geometries. The Raman active modes ($E_g$ and $A_{1g}$ modes) of rhombohedral $CuAlO_2$ with R3-m space group could be summarized according to the following equation (irreducible representation Γ): $= E_g + A_{1g}$, where the $A_{1g}$ mode has atomic vibrations in the direction of c-axis along the O–Cu–O bonds and the $E_g$ mode describes the atomic vibrations perpendicular to c-axis (*33–35*). In the Raman spectrum of the $CuAlO_2$ nanoparticles, only two sharp bands were observed at 415 and 760 $cm^{-1}$, which could be assigned to Raman active $E_g$ and $A_{1g}$ modes, respectively, of the rhombohedral crystal structure of delafossite $CuAlO_2$. Its $A_{1g}$ Raman active mode is slightly shifted when compared to the bulk $CuAlO_2$ particles, which appears at 764 $cm^{-1}$. As reported on pressure and temperature dependent Raman studies in the literature, the frequencies of Raman active modes ($E_g$ and $A_{1g}$) of delafossite $CuAlO_2$ nanoparticles can be compared with those of bulk $CuAlO_2$, of which the phonon modes of $E_g$ and $A_{1g}$ were reported at 418 and 767 $cm^{-1}$, respectively (*33, 35*). Even though the specific reason needs further studies, the shift in the phonon modes from the $CuAlO_2$ nanoparticles could be explained on the basis of thermal expansion, lattice expansion, anharmonic phonon-phonon interaction, and polaronic effects with pressure and temperature (*33–35*).

*Figure 2. (a) FESEM and (b) TEM images of the $CuAlO_2$ nanoparticles. (c) FESEM image of the bulk $CuAlO_2$ particles.*

Thermal studies of these $CuAlO_2$ nanoparticles could be useful to the research of their stability and in the preparation of their thin films at high temperature for various applications. Figure 4a shows the TGA plot of the $CuAlO_2$ nanoparticles under the flow of oxygen gas with the heating rate of 5 °C/min from 30 to 900 °C. A very small amount of weight loss occurs up to ~250 °C which could be due to the loss of adsorbed moisture by the nanoparticles. Subsequently, we observe that the $CuAlO_2$ nanoparticles demonstrate no obvious weight change up to 675 °C in $O_2$ atmosphere. Thermal studies of $CuAlO_2$ nanoparticles up to this temperature (~675°C) confirm the absence of $O_2$ deficiency in this material (*14*). Hereafter, the weight increases and that could be attributed the beginning of decomposition of the $CuAlO_2$ nanoparticles to CuO and $CuAl_2O_4$ (*14*). Powder X-ray diffraction studies of the residual material confirm the complete decomposition of the $CuAlO_2$ nanoparticles at 900 °C in $O_2$ (Figure 4b). Our results also agree with other previous reports (*14*, *28*, *29*).

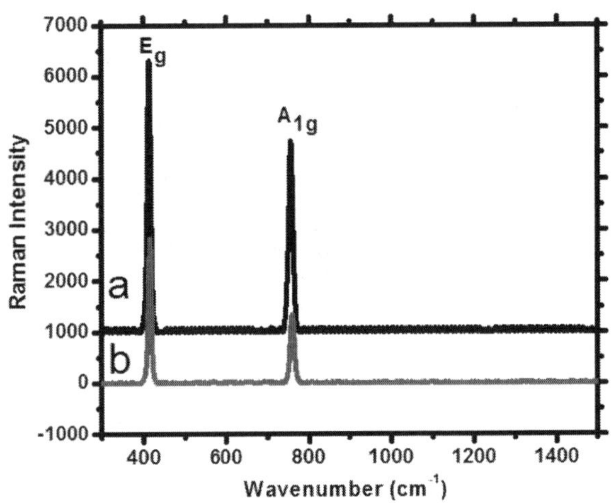


*Figure 3. Raman spectra of (a) the $CuAlO_2$ nanoparticles and (b) the bulk $CuAlO_2$ particles.*

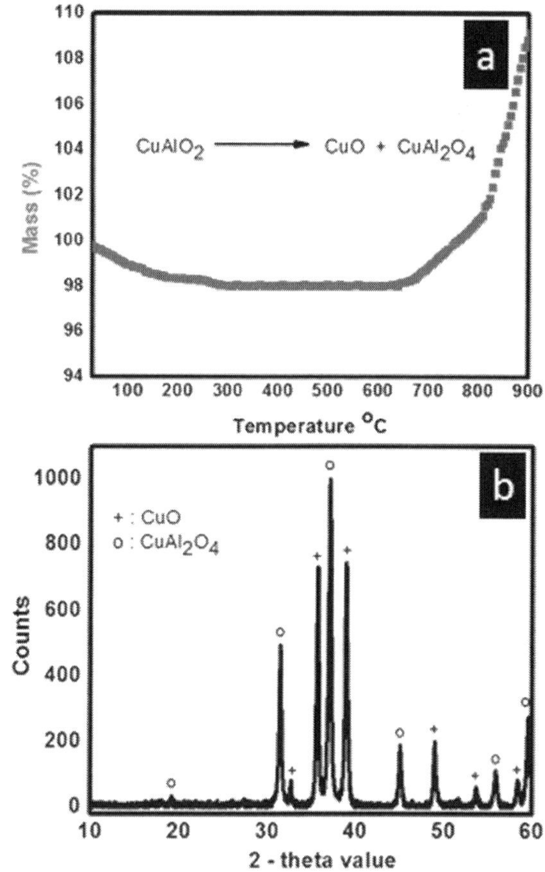


*Figure 4. (a) TGA study of the $CuAlO_2$ nanoparticles in oxygen. (b) PXRD patterns of the product obtained after the TGA study in oxygen.*

## Electrocatalytic Performance

The electro-chemical measurements (cyclic voltammetry, CV) and chrono-amperometric, CA)) were carried out for the HER and OER using 0.5 M KOH electrolyte and Ag/AgCl reference electrode by passing $N_2$ gas in the electrolyte for 5 min before starting the electrolysis of water. Figure 5a and 5b show the CV curves of the $CuAlO_2$ nanoparticle and bulk particles for HER from −1.2 to 0.4 V versus Ag/AgCl while figure 6c and 6d show their CV curves for OER from −0.1 to 1.0 V versus Ag/AgCl at the scan rates of 5, 10, 25, 50, 100 mV/s. Cyclic voltammograms (Figure 5a and 5b) show a redox peak around −0.20 V which could be assigned to the $Cu^{+}$ to $Cu^{2+}$ conversion as reported in the literature (*36*, *37*). From the CV studies at various scan rates, i.e. from 5 to 100 mV/s, the $CuAlO_2$ nanoparticles generate slightly higher current than the bulk $CuAlO_2$ particles for the HER. From Randles–Sevcik equation (*30–32*), the current density of the $CuAlO_2$ nanoparticles was found to be slightly higher (64 mA/cm²) than that of the bulk $CuAlO_2$ particles (61 mA/cm²) calculated by using the electrode electroactive surface area of 0.0786 cm². The current density could be a combination of both faradaic and non-faradaic processes and a function of the surface area and the morphology of the electrode materials. In the faradaic process, the charge transferred across the electrode–electrolyte which is governed by Faraday's law. Herein, electrons transferred from electrocatalysts to water molecules to generate hydroxide ion and hydrogen according to the following reaction: $2H_2O + 2e^- \rightarrow H_2 + 2OH^-$. Noticeably, there is no report in literature on HER by the electrolysis of water using delafossite $CuAlO_2$ as an electrocatalyst while copper based materials including intermetallic (Cu-Co) (*37*), $Cu_2O$ (*38*), Cu (*36*) were used as electrocatalysts in hydrogen generation. As shown here, delafossite $CuAlO_2$ shows an enhancement in the generation of hydrogen from the electrolysis of water compared to the other copper based materials. Figure 5c and 5d show the CV curves of $CuAlO_2$ nanoparticles and bulk $CuAlO_2$ particles, respectively, for OER at the various scan rates from 5 to 100 mV/s in 0.5 M KOH versus Ag/AgCl.

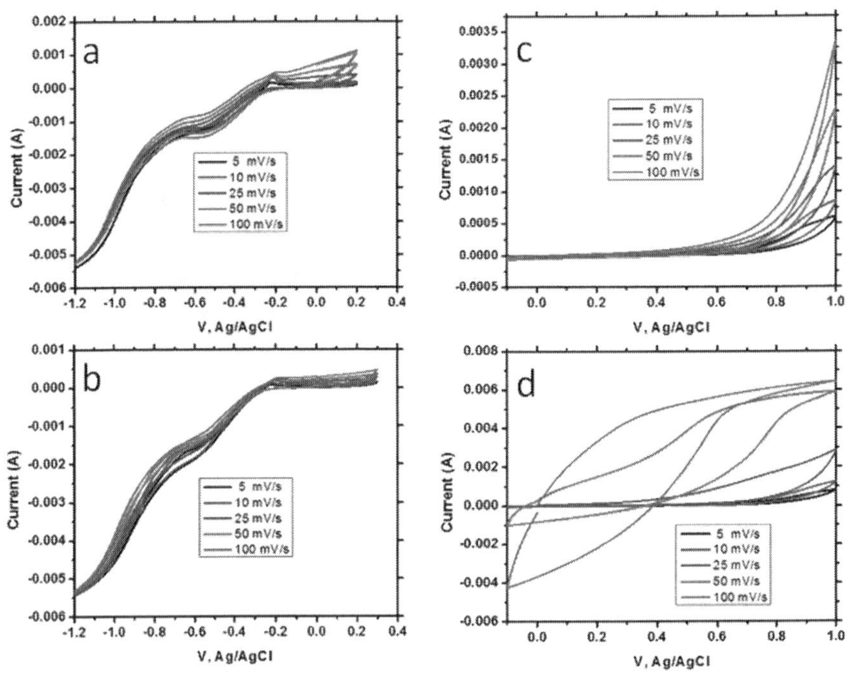


*Figure 5. Cyclic voltammetry of (a) the $CuAlO_2$ nanoparticles and (b) the bulk $CuAlO_2$ particles for HER at various scan rates, and (c) the $CuAlO_2$ nanoparticles and (d) the bulk $CuAlO_2$ particles for OER at various scan rates.*

Chronoamperometry (CA) is a potentiostatic true quantitative measurement which is used to investigate the electrocatalytic performance of the materials at the fixed potential. The charge transfer processes were also studied by CA experiments using these $CuAlO_2$ samples as electrocatalysts for HER and

OER (Figure 6). Insets of Figures 6a and 6b reveal the schematic diagrams of electrochemical cell and cell reactions occur at the surface of working electrodes to the HER and OER from the electrolysis of water. Figure 6a shows the CA experiments of the $CuAlO_2$ electrocatalysts for the HER at the fixed potential of –1.2 V for 200 seconds. In our experiment, the electrochemical reaction stops when the potential is switched off and hence current drops drastically. From the CA measurements, the nanocrystalline $CuAlO_2$ particles generate slightly higher current than the bulk $CuAlO_2$ particles (Figure 6a). The cathodic current generated is directly proportional to the rate of hydrogen evolution. The current densities were found to be 67 mA/cm$^2$ and 64 mA/cm$^2$ for the nanosized and bulk $CuAlO_2$ particles, respectively, which are consistent with the current density observed by CV measurements in Figure 6. A slightly higher current density clearly indicates that the $CuAlO_2$ nanoparticles are showing better performance than the bulk $CuAlO_2$ particles in the electrolysis of water to HER. This is noteworthy that the delafossite $CuAlO_2$ particles have been showing a promising electrocatalytic behavior to the HER. Figure 6b shows the CA measurements of the nanosized and bulk $CuAlO_2$ particles for the OER at the fixed potential of 0.60 V for 200 seconds. By comparing the catalytic activity of these electrodes in CA, we clearly observe that the nanosized $CuAlO_2$ particles generate higher current than the bulk $CuAlO_2$ particles up to 200 seconds. The current densities for the nanosized $CuAlO_2$ nanoparticles (14 mA/cm$^2$) and bulk $CuAlO_2$ particles (1.13 mA/cm$^2$) were calculated by using the electroactive surface area of the electrode. The electroactive surface area of the electrode (nanosized particles = 0.053 cm$^2$, bulk particles = 0.133 cm$^2$) for OER was calculated by using Randles–Sevcik equation. In the present study, we found that the nanosized $CuAlO_2$ particles show much better electrocatalytic activity (12 times) than the bulk particles toward the OER.

We have also investigated the catalytic efficiencies of the delafossite $CuAlO_2$ samples in $O_2$–saturated electrolyte (0.5M KOH) versus Ag/AgCl to OER (Figure 7). These data show an enhancement with consistency of current generation at the various scan rates (from 10–100 mV/s). Figure 7a and 7b show the CV curves of $CuAlO_2$ nanoparticles and bulk particles. From the CV studies, we observe that the OER starts beyond the potential of 0.5 V in both cases and the current generated is consistent with the $CuAlO_2$ nanoparticles in the $O_2$–saturated electrolyte versus Ag/AgCl compared to the bulk materials. $CuAlO_2$ nanoparticles generate higher current than the bulk particles in $O_2$–saturated 0.5M KOH compared to the $N_2$–saturated 0.5 M KOH to the OER which is also consistent with the CA studies (Figure 7c). The electroactive surface area of the nanoparticles (0.114 cm$^2$) and bulk particles (0.046 cm$^2$) of $CuAlO_2$ was used to calculate the current density. The observed current density of the $CuAlO_2$ nanoparticles (17.5 mA/cm$^2$) to OER was found to be higher than the bulk particles (10.8 mA/cm$^2$) in $O_2$–saturated 0.5 M KOH electrolyte. Therefore, the present work demonstrated that the $CuAlO_2$ nanoparticles are promising electrocatalysts to the HER and OER in alkaline electrolyte versus Ag/AgCl and show enhanced catalytic activity compared to the bulk counterpart.

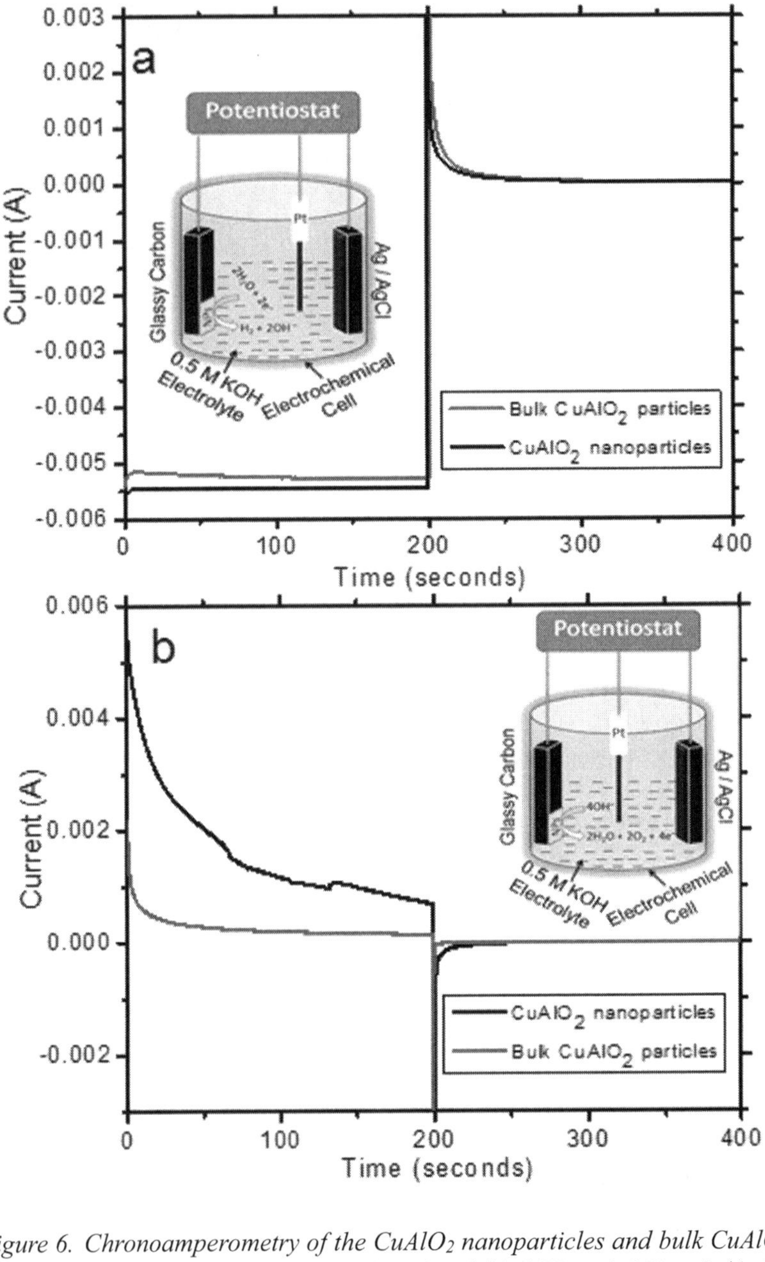


*Figure 6. Chronoamperometry of the $CuAlO_2$ nanoparticles and bulk $CuAlO_2$ particles for (a) HER at -1.2 V vs Ag/AgCl and (b) OER at 0.6 V vs Ag/AgCl. Insets of Figures 6a and 6b represent the schematic diagrams of HER and OER, respectively.*

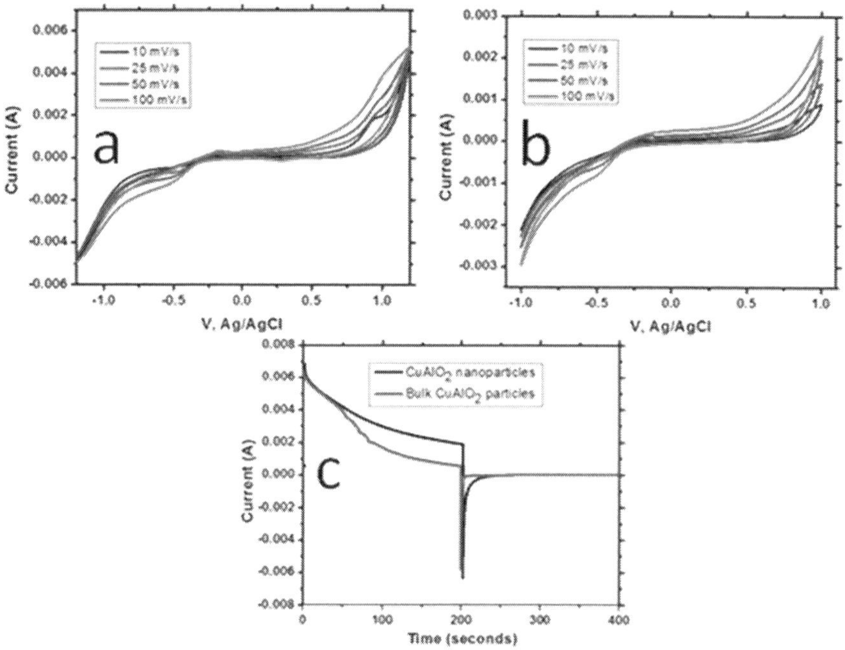


*Figure 7. Cyclic voltammetry of (a) the $CuAlO_2$ nanoparticles and (b) the bulk $CuAlO_2$ particles for OER at various scan rates in $O_2$–saturated 0.5 M KOH. (c) Chronoamperometry of the $CuAlO_2$ nanoparticles and bulk particles for OER in $O_2$–saturated 0.5 M KOH at 0.6 V versus Ag/AgCl.*

## Conclusions

Herein, we report the synthesis of $CuAlO_2$ nanoparticles with average size of ~20 nm for the first time via an ultrasonic process followed with annealing at 850 °C in $N_2$. The phase purity of the $CuAlO_2$ nanoparticles was investigated by powder X-ray diffraction and Raman spectroscopic studies while their morphology was studied by SEM and TEM. More importantly, it was found that the $CuAlO_2$ nanoparticles show higher electrocatalytic activity than the micron-sized particles toward the hydrogen and oxygen evolution reactions in 0.5 M KOH electrolytes versus Ag/AgCl. Therefore, these results confirmed that delafossite $CuAlO_2$ can serve as promising electrocatalyst for water splitting and will draw intensive research interest on their other applications like solar cells, photochemical water splitting, proton exchange membrane water electrolysis etc.

## Acknowledgments

The authors thank the support from the American Chemical Society – Petroleum Research Fund #51497, the Air Force Office of Scientific Research (award # FA9550-12-1-0159), and the National Science Foundation under DMR grant # 0934157 (PREM-UTPA/UMN-Science and Engineering of Polymeric and

Nanoparticle-based Materials for Electronic and Structural Applications). The authors also thank to Dr. M. Pokhrel for taking SEM images, Ms. T. Olmedo for taking TEM images, and Ms. J. Cruz for performing BET measurements.

## References


1. Kawazoe, H.; Yasukawa, M.; Hyodo, H.; Kurita, M.; Yanagi, H.; Hosono, H. P-type electrical conduction in transparent thin films of $CuAlO_2$. *Nature* **1997**, *389* (6654), 939–942.
2. Zheng, X. G.; Taniguchi, K.; Takahashi, A.; Liu, Y.; Xu, C. N. Room temperature sensing of ozone by transparent p-type semiconductor $CuAlO_2$. *Appl. Phys. Lett.* **2004**, *85* (10), 1728–1729.
3. Mo, R.; Liu, Y. Synthesis and properties of delafossite $CuAlO_2$ nanowires. *J. Sol-Gel Sci. Technol.* **2011**, *57* (1), 16–19.
4. Banerjee, A. N.; Chattopadhyay, K. K. Size-dependent optical properties of sputter-deposited nanocrystalline p-type transparent $CuAlO_2$ thin films. *J. Appl. Phys.* **2005**, *97* (8), 084308.
5. Li, G.; Zhu, X.; Lei, H.; Jiang, H.; Song, W.; Yang, Z.; Dai, J.; Sun, Y.; Pan, X.; Dai, S. Preparation and characterization of $CuAlO_2$ transparent thin films prepared by chemical solution deposition method. *J. Sol-Gel Sci. Technol.* **2010**, *53* (3), 641–646.
6. Ahmad, A.; Jagadale, T.; Dhas, V.; Khan, S.; Patil, S.; Pasricha, R.; Ravi, V.; Ogale, S. Fungus-based synthesis of chemically difficult-to-synthesize multifunctional nanoparticles of $CuAlO_2$. *Adv. Mater.* **2007**, *19* (20), 3295–3299.
7. Zhao, S.; Li, M.; Liu, X.; Han, G. Synthesis of $CuAlO_2$ nanofibrous mats by electrospinning. *Mater. Chem. Phys.* **2009**, *116* (2–3), 615–618.
8. Yanagi, H.; Inoue, S.-i.; Ueda, K.; Kawazoe, H.; Hosono, H.; Hamada, N. Electronic structure and optoelectronic properties of transparent p-type conducting $CuAlO_2$. *J. Appl. Phys.* **2000**, *88* (7), 4159–4163.
9. Prakash, T.; Prasad, K. P.; Kavitha, R.; Ramasamy, S.; Murty, B. S. Dielectric relaxation studies of nanocrystalline $CuAlO_2$ using modulus formalism. *J. Appl. Phys.* **2007**, *102* (10), 104104.
10. Koumoto, K.; Koduka, H.; Seo, W.-S. Thermoelectric properties of single crystal $CuAlO_2$ with a layered structure. *J. Mater. Chem.* **2001**, *11* (2), 251–252.
11. Kizaki, H.; Sato, K.; Yanase, A.; Katayama-Yoshida, H. Ab initio calculations of $CuAlO_2$-based dilute magnetic semiconductor. *Phys. B* **2006**, *376–377*, 812–815.
12. Arghya Narayan, B.; Sang, W. J. Low-macroscopic field emission properties of wide bandgap copper aluminium oxide nanoparticles for low-power panel applications. *Nanotechnology* **2011**, *22* (36), 365705.
13. Smith, J. R.; Van Steenkiste, T. H.; Wang, X.-G. Thermal photocatalytic generation of $H_2$ over $CuAlO_2$ nanoparticle catalysts in $H_2O$. *Phys. Rev. B* **2009**, *79* (4), 041403.

14. Ahmed, J.; Blakely, C. K.; Prakash, J.; Bruno, S. R.; Yu, M.; Wu, Y.; Poltavets, V. V. Scalable synthesis of delafossite $CuAlO_2$ nanoparticles for p-type dye-sensitized solar cells applications. *J. Alloys Compd.* **2014**, *591*, 275–279.
15. Bandara, J.; Yasomanee, J. P. P-type oxide semiconductors as hole collectors in dye-sensitized solid-state solar cells. *Semicond. Sci. Technol.* **2007**, *22* (2), 20.
16. Nattestad, A.; Zhang, X.; Bach, U.; Cheng, Y.-B. Dye-sensitized $CuAlO_2$ photocathodes for tandem solar cell applications. *J. Photon. Energy* **2011**, *1* (1), 011103.
17. Prakash, T.; Ramasamy, S. Dye Sensitized Solar Cells Using Nanocrystalline $CuAlO_2$-CuSCN Embedded Poly(vinyl carbazole) Composites as Solid Hole-Transporter. *Sci. Adv. Mater.* **2012**, *4* (1), 29–34.
18. Herraiz-Cardona, I.; Fabregat-Santiago, F.; Renaud, A.; Julián-López, B.; Odobel, F.; Cario, L.; Jobic, S.; Giménez, S. Hole conductivity and acceptor density of p-type $CuGaO_2$ nanoparticles determined by impedance spectroscopy: The effect of Mg doping. *Electrochim. Acta* **2013**, *113*, 570–574.
19. Carcia, P. F.; Shannon, R. D.; Bierstedt, P. E.; Flippen, R. B. $O_2$ electrocatalysis on thin film metallic oxide electrodes with the delafossite structure. *J. Electrochem. Soc.* **1980**, *127* (9), 1974–1978.
20. Hinogami, R.; Toyoda, K.; Aizawa, M.; Kawasaki, T.; Gyoten, H. Copper delafossite anode for water electrolysis. *ECS Trans.* **2013**, *58* (2), 27–31.
21. Hinogami, R.; Toyoda, K.; Aizawa, M.; Yoshii, S.; Kawasaki, T.; Gyoten, H. Active copper delafossite anode for oxygen evolution reaction. *Electrochem. Commun.* **2013**, *35* (0), 142–145.
22. Ghosh, C. K.; Popuri, S. R.; Mahesh, T. U.; Chattopadhyay, K. K. Preparation of nanocrystalline $CuAlO_2$ through sol–gel route. *J. Sol-Gel Sci. Technol.* **2009**, *52* (1), 75–81.
23. Deng, Z.; Zhu, X.; Tao, R.; Dong, W.; Fang, X. Synthesis of $CuAlO_2$ ceramics using sol-gel. *Mater. Lett.* **2007**, *61* (3), 686–689.
24. Xiong, D.; Zeng, X.; Zhang, W.; Wang, H.; Zhao, X.; Chen, W.; Cheng, Y.-B. Synthesis and characterization of $CuAlO_2$ and $AgAlO_2$ delafossite oxides through low-temperature hydrothermal methods. *Inorg. Chem.* **2014**, *53* (8), 4106–4116.
25. Sheets, W. C.; Mugnier, E.; Barnabé, A.; Marks, T. J.; Poeppelmeier, K. R. Hydrothermal synthesis of delafossite-type oxides. *Chem. Mater.* **2005**, *18* (1), 7–20.
26. Banerjee, A. N.; Maity, R.; Chattopadhyay, K. K. Preparation of p-type transparent conducting $CuAlO_2$ thin films by reactive DC sputtering. *Mater. Lett.* **2004**, *58* (1–2), 10–13.
27. Thu, T. V.; Thanh, P. D.; Suekuni, K.; Hai, N. H.; Mott, D.; Koyano, M.; Maenosono, S. Synthesis of delafossite $CuAlO_2$ p-type semiconductor with a nanoparticle-based Cu(I) acetate-loaded boehmite precursor. *Mater. Res. Bull.* **2011**, *46* (11), 1819–1827.

28. Kumekawa, Y.; Hirai, M.; Kobayashi, Y.; Endoh, S.; Oikawa, E.; Hashimoto, T. Evaluation of thermodynamic and kinetic stability of $CuAlO_2$ and $CuGaO_2$. *J. Therm. Anal. Calorim.* **2010**, *99* (1), 57–63.
29. Amrute, A. P.; Łodziana, Z.; Mondelli, C.; Krumeich, F.; Pérez-Ramírez, J. Solid-state chemistry of cuprous delafossites: Synthesis and stability aspects. *Chem. Mater.* **2013**, *25* (21), 4423–4435.
30. Taurino, I.; Carrara, S.; Giorcelli, M.; Tagliaferro, A.; De Micheli, G. Comparison of two different carbon nanotube-based surfaces with respect to potassium ferricyanide electrochemistry. *Surf. Sci.* **2012**, *606* (3–4), 156–160.
31. Min-Jung, S.; Dong-Hwa, Y.; Joon-Hyung, J.; Nam-Ki, M.; Suk-In, H. Comparison of effective working electrode areas on planar and porous silicon substrates for cholesterol biosensor. *Jap. J. Appl. Phys.* **2006**, *45* (9R), 7197.
32. Aljabali, A. A. A.; Barclay, J. E.; Butt, J. N.; Lomonossoff, G. P.; Evans, D. J. Redox-active ferrocene-modified Cowpea mosaic virus nanoparticles. *Dalton Trans.* **2010**, *39* (32), 7569–7574.
33. Singh, M. K.; Dussan, S.; Sharma, G. L.; Katiyar, R. S. Raman scattering measurements of phonon anharmonicity in $CuAlO_2$ thin films. *J. Appl. Phys.* **2008**, *104* (11).
34. Ingram, B. J.; Mason, T. O.; Asahi, R.; Park, K. T.; Freeman, A. J. Electronic structure and small polaron hole transport of copper aluminate. *Phys. Rev. B* **2001**, *64* (15), 155114.
35. Pellicer-Porres, J.; Martínez-García, D.; Segura, A.; Rodríguez-Hernández, P.; Muñoz, A.; Chervin, J. C.; Garro, N.; Kim, D. Pressure and temperature dependence of the lattice dynamics of $CuAlO_2$ investigated by Raman scattering experiments and ab initio calculations. *Phys. Rev. B* **2006**, *74* (18), 184301.
36. Ahmed, J.; Trinh, P.; Mugweru, A. M.; Ganguli, A. K. Self-assembly of copper nanoparticles (cubes, rods and spherical nanostructures): Significant role of morphology on hydrogen and oxygen evolution efficiencies. *Solid State Sci.* **2011**, *13* (5), 855–861.
37. Ahmed, J.; Ganguly, A.; Saha, S.; Gupta, G.; Trinh, P.; Mugweru, A. M.; Lofland, S. E.; Ramanujachary, K. V.; Ganguli, A. K. Enhanced electrocatalytic activity of copper–cobalt nanostructures. *J. Phys. Chem. C* **2011**, *115* (30), 14526–14533.
38. Kumar, B.; Saha, S.; Ganguly, A.; Ganguli, A. K. A facile low temperature (350 °C) synthesis of $Cu_2O$ nanoparticles and their electrocatalytic and photocatalytic properties. *RSC Adv.* **2014**, *4* (23), 12043–12049.

## Chapter 5

# Hybrid Electrolytes

**Manuel Maréchal,[*,1] Christel Laberty-Robert,[2] and Sébastien Livi[3]**

**[1]CNRS, INAC-SPRAM, F-38000 Grenoble, France; Univ. Grenoble Alpes, INAC-SPRAM, F-38000 Grenoble, France; CEA, INAC-SPRAM, F-38000 Grenoble, France**
**[2]Sorbonne Universités, UPMC Univ. Paris 06, CNRS-UMR 7574, Collège de France, Laboratoire de Chimie de la Matière Condensée de Paris, 11 place Marcelin Berthelot, F-75005 Paris, France**
**[3]Université de Lyon, F-69003, Lyon, France; INSA Lyon, F-69621, Villeurbanne, France; CNRS, UMR 5223, Ingénierie des Matériaux Polymères**
**[*]E-mail: manuel.marechal@cea.fr**

The need for clean, safe and efficient energy is a major societal issue that involves the development of new technologies for storage and energy production. The electrolytic media are at the center of the electrochemical devices like fuel cells, dye–sensitized solar cells, also called Grätzel, or super-capacitors or lithium-ion batteries. The finding of new electrolytes is rather challenging as it must fit certain constraints such as wide electrochemical windows, sufficient ionic conductivities, excellent thermal stability and sufficient mechanical properties. Many studies have converged towards the study of hybrid electrolytes consisting of organic and inorganic phases, or polymer-ionic liquid mixtures, to obtain solid electrolytes and to fulfill all these constraints. Controlling the structure at the nanoscale and understanding their properties at various length and space scales are crucial for their optimization. Recent and different ways were proposed to design hybrid electrolytes opening new perspectives in the energy field. The genesis and the presentation of these developments are presented through three main applications: the fuel cells, the organic photovoltaic cells and the energy storage.

# Introduction

The consumption of energy is continuously increasing and one of the answer is to improve the energy devices. Then the need to develop new advanced materials for the production and storage of energy in safe, efficient and environmentally friendly conditions is prominent. Electrolytes, at the center of the various devices, follow an incremental evolution together with breakthroughs.

Hybrid electrolytes have been widely investigated for various energy devices and fall into three broad types: (A) Hybrid membranes for Proton Exchange Membranes for Fuel Cells (PEMFCs) especially for temperatures up to 120 degrees Celsius (°C), (B) Hybrid ionic conductor xerogels for organic photovoltaic devices, and (C) Hybrid polymer/ionic liquid membranes for use in the energy storage. All these hybrid electrolytes have recently experienced new advances by controlling their structure at various length scales from macro- to nano-scales.

# A. Hybrid Polymer Membranes for Proton Exchange Membranes for Fuel Cells

The hybrid electrolytic polymer membranes have been extensively studied for fuel cells (*1–4*). A scheme of fuel devices equipped with polymer membrane is presented in the figure 1. These membranes can be obtained by different processing routes after mixing the polymers and the inorganic components. The main processes are the casting (*5, 6*), the spray-coating (*7*), the extrusion (*8*) and more recently the electrospinning (*6*).

### a. Hybrid Polymer Membranes with Inorganic Fillers

This family can be divided into several subfamilies depending on whether the polymer phase is an ionic conductor and whether the inorganic or hybrid phase contains ionic groups. One of the main families is constituted of membranes with a polymer electrolyte such as Nafion®-like perfluoroalkylsulfonic acid polymers (benchmark membranes) (*9–15*), and sulfonated polyaromatic polymers (*16, 17*). These polymers include a second phase, (summarized in table 1) either an inorganic filler like mineral (laponite (*11, 16, 18*), montmorillonite (*12–14*), zinc oxide ($ZrO_2$) (*19*), Silica ($SiO_2$) (*20*), calcium hydrophosphonate (*10*), titania ($TiO_2$) (*13*)) or a heteropolyacid (hydrophosphoric acid (*21*), orthophosphoric acid (*22, 23*), phosphotungstic acid (*19, 24–28*), phosphomolybdic acid (*25, 26, 29*), silicotungstic acid (*25, 27, 30*), phosphoantinomic acid (*31*), zirconium phosphate sulfophenyl phosphate (*32*), zirconium phosphate (*15, 27*), mesoporous acid silicate (*16*)).

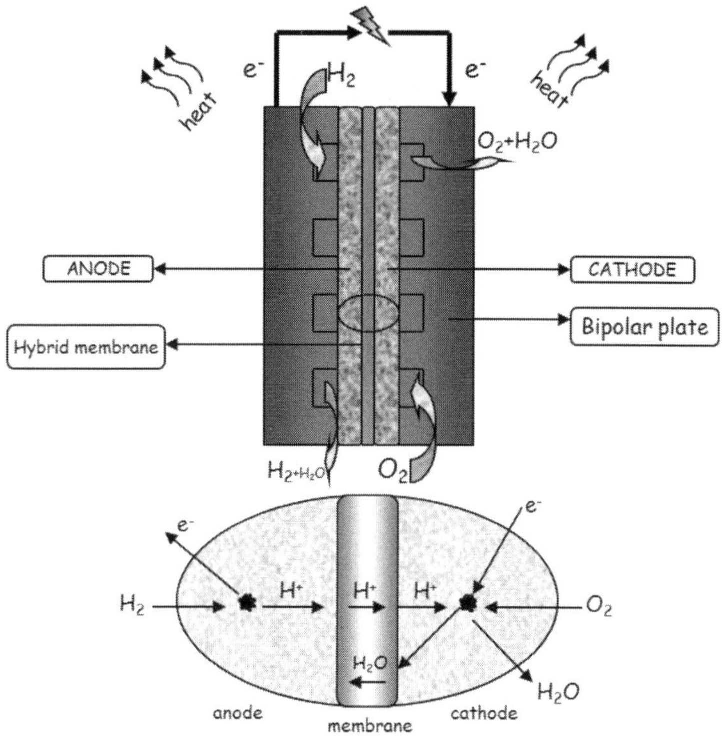


*Figure 1. Synoptic layout of a fuel cell with a proton exchange membrane and, in the enlargement, at both electrode/membrane interfaces, the catalytic redox reactions.*

The inorganic or hybrid fillers can also be embedded in non-ionic polymers, such as partially or perfluorinated polymers like Poly(Vinylidene Fluoride) (PVdF), Poly(Vinylidene Fluoride-co-HexaFluoroPropene) (P(VdF-HFP), and so forth (*36*, *37*). The obtained membranes are known to be chemically and electrochemically stable, with respect to the potential window of the applications (*38*).

The overall objective of the hybrid approach is to decouple or couple the properties of the obtained membranes depending on the applications. For fuel cells, the properties include ionic conductivity, mechanical strength and/or water retention at elevated temperatures (*15*).

**Table 1. Mineral fillers and heteropolyacids used in the hybrid membranes for fuel cells**

| *Mineral fillers* | *Formulae or composition* |
|---|---|
| Laponite (*18*, *33*)<br>(Southern Clay Co.) | $(Na_{0,7}{}^{+}[(Si_8Mg_{5,5}Li_{0,3})\text{-}O_{0,2}(OH)_4)]^{0,7-})$ |
| MMT (*12*)<br>MontMorillonite ($Na^+$-MMT) (*18*) | $(Na, Ca)_{0.3}(Al, Mg)_2Si_4O_{10}(OH)_2 \cdot nH_2O$ |
| m-MMT (*12*)<br>modified MontMorillonite<br>(dodecylamine-exchange montmorillonite) | as above |
| Zirconium isopropoxide (*19*)<br>(Alfa Co. Ltd) | $[Zr(OCH(CH_3)_2)_4\text{-}(CH_3)_2CHOH]$ |
| Heteropolyacids | |
| hydrophosphoric acid (*21*) | $H_3PO_2$ |
| Orthophosphoric acid (*22*, *23*) | $H_3PO_4$ |
| 12-PhosphoTungstic Acid (PTA or PWA) (*19*, *24–27*) | $(H_3(P(W_3O_{10})_4) \cdot nH_2O)$ |
| Phosphomolybdic Acid (PMA or $PMo_{12}$) (*25*, *26*, *29*) | $(12(MoO_3 \cdot 2H_2O) \cdot H_3PO_4 . 4H_2O)$ |
| SilicoTungstic Acid (STA or SiWA) (*25*, *27*, *30*) | $(H_4(Si(W_3O_{10})_4) \cdot nH_2O)$ |
| Phosphatoantimonic acid ($H_3$) (*31*, *33*, *34*) | $(HSb\ (PO_4)_2 \cdot nH_2O)$ |
| Zirconium Phosphate Sulfophenyl Phosphonate (ZrPSPhP1.5) (*32*) | $(Zr(HPO_4) \cdot (O_3PC_6H_4SO_3H)_{1.5})$ |
| Zirconium Phosphate (ZrP) (*27*, *35*) | $(Zr(HPO_4)_2\ nH_2O)$ |

*Proton Conductivity*

The addition of mineral fillers (layered silicates or metallic oxides) or heteropolyacids allows improving conductivities at low relative humidity (Figure 2). Most of the fillers/heteropolyacids has a substantial hydrophilicity and/or significant acidity, which explains the high water uptake. The addition of inorganic fillers may decrease (*12*) or increase (*31*) conductivities while improving other properties. For heteropolyacids, there is often an optimal loading (*27*), around 7 weight-percent (wt.%), for example, for sulfonated polysulfones (SPS) /phosphotoantimonic acid (H3) (*31*, *34*, *39*). The size and the shape of the fillers also have its importance (*34*) since the different properties are highly dependent on the structuration of the fillers at the nanoscale and on the structuration of the organic components vs. inorganic ones at the various length scales.

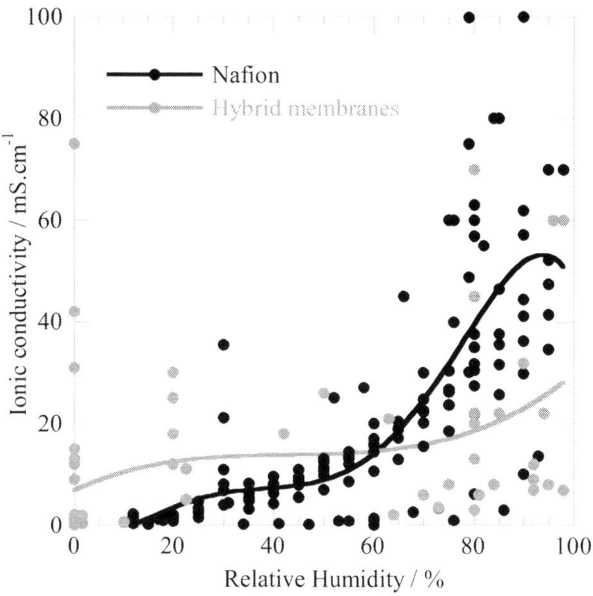


*Figure 2. Ionic conductivity as a function of relative humidity* ($T \in [\![0\,,120°C]\!]$ *for each relative humidity). The curves are the weighted average conductivities at each relative humidity between the different temperatures (i) for Nafion®, and (ii) of the conductivities of the hybrid membranes previously cited.*

The hydrophilic nature of these fillers should increase water uptakes and ionic conductivities in the final obtained composite membranes. The assumptions are questionable (*40*) since the literature reported contradictory results on the role played by the inorganic fillers (*13–15*). It can be noticed that the improvement of the proton conductivity was low the last 20 years (*4*, *14–16*) by using hybrid membranes in the field of low/medium-temperature fuel cells. In the hybrid membranes, the interplay between the fillers and the polymer electrolytes is not clearly and definitely elucidated (*9*, *10*, *16*): (i) The water contained in the particles cannot easily be shared with the rest of the polymer electrolyte phase, and (ii) the conductivity of the inorganic/hybrid phase is in general at least one order lower than the one of the electrolytic organic phases. Nevertheless some composite membranes have worthwhile properties in the application uses (*12*, *15*).

### *Mechanical Properties*

Improving effects of the mechanical properties at the local scale and the localization of the crystalline zones (*10*) are invoked to explain this paradox. The ionic transport has been shown to be intimately linked with the local mechanical properties (*41*) and the multiscale structure (*42*, *43*).

The addition of heteropolyacids (Table 2) can reinforce the membrane, either with ionic charges (heteropolyacids) or not (*33*). This addition can also weaken the membrane in the case of weak interactions. The system with sodiated forms of heteropolyacids (*25*) shows that, if the interaction is not strong enough, the network is weakened.

**Table 2. Mechanical properties of hybrid membranes**

| *Hybrid membrane* | *Ultimate tensile strength/MPa (σ)* | *Young Modulus/ MPa (E)* | *Elongation at break/% (ε)* |
|---|---|---|---|
| Nafion (23°C, 50%RH) (*44*) | 23/28 | 288/281 | 252/311 |
| NaSTA/Nafion®117 (*25*)<br>NaPMA/Nafion®117<br>NaPTA/Nafion®117 (wet) | 14<br>8<br>3 | | 45<br>70<br>170 |
| SPS 1.1 mol $H^{+}\cdot kg^{-1}$ 3.5/7.1/12 wt.% of $H_3$ (*39*) | | 158/88/59 | 6.4/5.4/1.9 |
| SPS 3/25 wt.% of $H_3$ (*39*) | | 370/340 | 48/60 |

*Degradation Temperatures*

The effect of heteropolyacids, which globally has high degradation temperatures, is not systematically studied (Table 3). In general, the obtained composites do not exhibit very different degradation temperatures from those of associated polymers. This observation seems logical since the degradation is always conditioned by the least stable component. For NaSTA/Nafion, the degradation temperature is lower than that of Nafion alone. In the case of the composite made of phosphotungstic acid/SBPS, the addition of the mineral acid only reduces the degradation temperature of the low degrees of sulfonation (-44°C between 20 and 60 mol.% of sulfonation). This fact means that the acidic functional groups are more or less protected as the degree of sulfonation, the loading, the level of interaction (from repulsion to ionic crosslinking), and acid or alkaline forms of the heteropolyacid (*45*). The ionic interactions of phosphotungstic acid/SBPS is such that the degradation temperature increases with the sulfonation rate instead of decreasing as for the pure ionomer(+64°C between 20 and 60 mol.% of sulfonation). The composites formed of non-ionic polymers and heteropolyacids have high degradation temperatures (>300°C). The combination of components having high degradation temperatures explains this observation. If all other characteristics are combined, they would be relevant candidates for "medium temperature" PEMFCs.

**Table 3. Degradation temperatures at 5 wt.% of weight loss of hybrid membranes**

| *Hybrid ionomers/fillers* | *Degradation temperature/°C (Inert atmosphere 10°C·min$^{-1}$)* |
|---|---|
| Nafion/MMT(3/5/7 wt.%) (*12*)<br>Nafion/m-MMT(3/5/7 wt.%) | 340/330/330<br>340/340/350 |
| Sulfonated PolyEtherEtherKetone/MMT(10 wt.%) (*18*)<br>Sulfonated PolyEtherEtherKetone /Laponite(10 wt.%) | 320<br>320 |
| Hybrid polymers/mineral fillers/acids | |
| $Al_2O_3$ (50 wt.%) + $H_2SO_4$ (6 mol·L$^{-1}$) + Poly(VinylideneFluoride)/15 mol.%HexaFluoroPropylene (PVdF-HFP) (*46*) | 180 |
| Zr/Polydimethylsiloxane/$H_3PO_4$ (*19*) | 300 |
| Hybrid ionomers/heteropolyacids | |
| NaSTA/ Nafion® (*25*) | 300 |
| phosphotungstic acid (*24*)<br>phosphotungstic acid/Sulfonated bisphenolsulfone (SPBS) 20/40/60 (6/4) | T>800<br>347/359/411 |
| Polymer/heteropolyacids | |
| phosphotungstic acid/bisphenolsulfone (*24*) | 313 |

## b. Polymer Membranes with Hybrid Particles

Many associated properties still remain elusive, such as the sizes, the distribution and the structure of the inorganic or hybrid particles, the interactions and the interplay of structure/property relationships between these subsystems. The structural studies of such hybrid membranes are very rare, notably due to the presence of two subsystems. Despite this complexity, the investigation of multiscale materials using direct and reciprocal spaces allows reaching a thorough structural description (*47*, *48*). Bulk and surface information are obtained in a large range of characteristic length scales ranging from the nanometer to angstrom scale. In the reciprocal space, the size, shape and interface effects can be discriminated between the different subsystems due to the large gaps between the related characteristic correlation lengths. The distribution and the size of the hybrid particles are notably investigated (*47–49*).

A recent study of an original composite membrane composed of conducting polystyrenesulfonic acid-grafted silica particles (Figure 3) into a PVdF-HFP matrix (*37*) was used to address and clarify these questions (*49*).

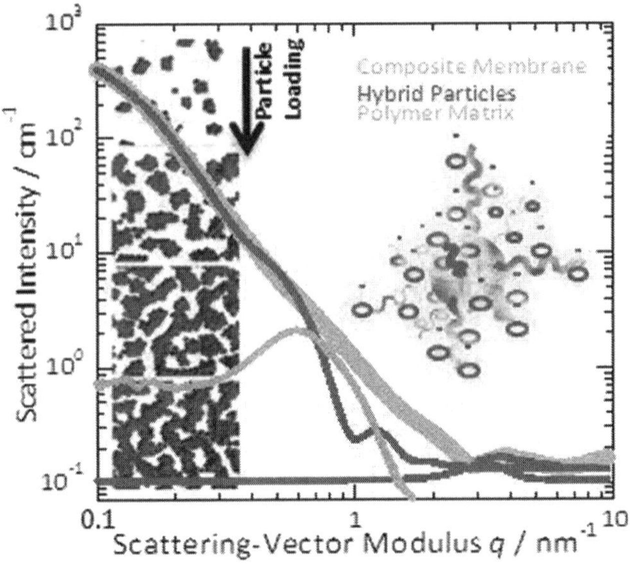


*Figure 3. SAXS absolute intensity as a function of the scattering-vector modulus q for (i) a swelled* ***composite membrane*** *with 50 wt.% of fillers at room temperature, (ii) a simulation of the form factor due to the* ***hybrid particles*** *with core-shell spheres having shell thicknesses equal to 70 nm distributed according a Schulz distribution (σ=0.5) and a core radius equal to 60 nm as well as the contribution of the polyelectrolyte chains, and (iii) the experimental contribution of the* ***polymer matrix****. (Reproduced with permission from reference (47). Copyright [2015][the Royal Society of Chemistry]).*

The structural variation of the various phases constituting the composite membranes i.e. the aggregates of particles, the polymer crystalline phase and their interactions, as a function of the amount of hybrid particles, were also described. The SAXS study, presented in the figure 3, shows the deconvolution of these different contributions. An incremental approach allowed to simulate the form factor of the hybrid particles with relevant parameters, to isolate the contribution of the polymer matrix, and to describe the polyelectrolyte chains of the hybrid particles. These results stress the possibility of a dichotomous interpretation leading to an overall quantitative and qualitative description of the nanostructure.

### c. Nanostructured Polymer Membranes Coupling Electrospinning and Sol-Gel Chemistry

Understanding the relationship, between the nanostructure and the properties, is also highlighted by recent studies on hybrid membranes including silica sol-gel chemistry with precursors having sulfonic acid functions and PVdF-HFP (*4, 50–52*). In fact, it has been possible to locate and to control the inorganic phase (size, structure) (*6, 7, 52, 53*), even at the nanoscale (*6*). The figure 4 shows the Small Angle Neutron Scattering (SANS) spectra (the absolute intensity as a function of the scattering-vector modulus q in $nm^{-1}$) of membranes obtained by electrospinning coupled to the sol-gel chemistry (*54, 55*) compared with the membrane having the same composition obtained by casting. These results clearly demonstrate the ability of the hybrid strategy to control the multiscale structure of these hybrid membranes particularly at the nanoscale. Thus, it is possible to modify the organization of ionic/hydrophilic domains (large q), of the semi-crystalline phase of the polymer phase (intermediate q), and at longer distances (small q), which describes the characteristic correlation lengths of the fibers.

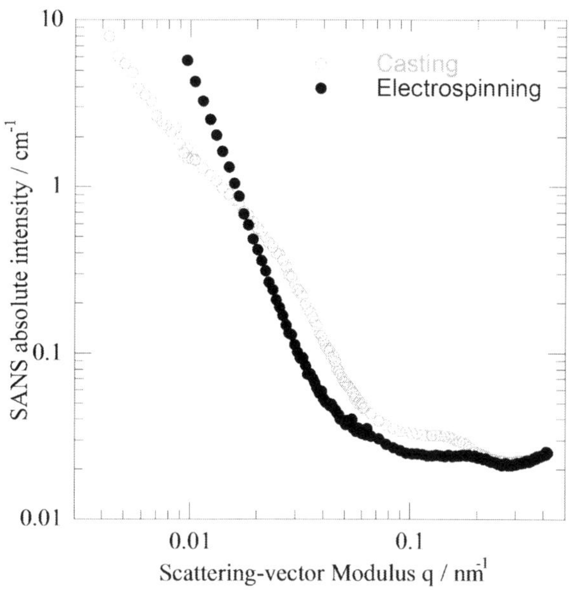


*Figure 4. Scattered intensity versus scattering-vector modulus q of electrospun and casting membranes with 55 wt.% of inorganic phase.*

The control of the ionic inorganic hybrid phase has significantly enabled to improve the ionic conductivity up to 80°C (*56*) with values in the same order of magnitude as Nafion membranes. This innovative coupling has achieved notable mechanical properties (*56*). Beyond the macroscopic properties, this family has deepened the understanding of proton transport in an ionic multiscale host matrix. Indeed the original nanostructure is made of fibers with cylindrical hybrid "mille-feuille" structure (*6, 43, 56*). This system allowed to increase knowledge in the field of multiscale transport in such membranes. By various techniques, the dynamics of the proton have been then described from the nano- to the macro-scales (*43, 57*).

The hybrid membranes for fuel cells have been widely studied in the past 50 years. Renewed interest has also been held to increase the operation temperature up to 120 or even 150°C. But these membranes are struggling to replace the benchmark perfluorinated membranes. Hybrid membranes are evolved from polymer blended with inorganic materials to nanostructured hybrid membranes. This complexity, both with the choice of components, such as hybrid particles, and with different process and chemistry options, shows the importance of the obtained multiscale structures to improve the performance of electrolytes. The control of these structures at different scales including the nanoscale seems to be a key to their evolution towards an application. Indeed recent studies present results distinguished from those of the previous families by coupling sol-gel chemistry and electrospinning as elaboration process (*6*, *43*, *56*).

## B. Electrolytic Xerogels for Organic Photovoltaic Cells

Among the technologies available to the global energy mix, photovoltaic conversion appears as a key player. This renewable and inexhaustible energy is mainly based on silicon technology. Its interesting performances helped the photovoltaic industrialization. However, the manufacturing cost of silicon photovoltaic cells remains a real drawback to its development. Described for the first time in the seventies, organic photovoltaic materials are an interesting alternative to inorganic materials. Specifically, the new flexible photovoltaic devices with photosensitized dyes have great potential due to their low cost, low weight, flexibility and ability to large scale implementation. Recent studies point to their high potential energy payback (*58*). However, to become a true complement to the inorganic technology, several challenges must be overcome, particularly in relation to the conversion efficiency, which is still extremely low. Indeed, the best yields reported in the literature for organic cells are nearly 11-13% in the laboratory against 15-25% for silicon cells.

### a. Organic Solar cells

Dye–sensitized solar cells DSSCs made of a mesoporous network of interconnected titania ($TiO_2$) nanocrystals are an interesting alternative to conventional inorganic photovoltaic devices because of their high–efficiency and low–cost (*59*). But the presence of liquid electrolyte causes often sealing problems that impact the long term operation of the devices and raises safety issues (*60*). Therefore, different strategies to solidify the electrolyte have been explored (*61*). The most studied ones are based on hole conductors (HTM) such as inorganic materials, conjugated polymers or small doped molecules (*62*). Such cells are reaching performances of 7.2% (*63*) but are still limited by their naturally low electronic conductivity, by the insufficient penetration of solid electrolytes in the nanoporous electrodes (*64*) which can reduce the cell performances by 10 to 20% (*65*). Alternatively, the ionic conductor may play a role of solid electrolyte to obtain an efficiency up to 8.1% (*66*). An original approach is the use of quasi-solid electrolytes like gels. But electrolyte gels do not achieve the objectives of security and stability required by the technology for a sustainable industrial development. Indeed organic solvents are trapped in the matrix and have a high risk of instability at high temperatures, due to the increase of their volatilities, and flammability (*67*). Moreover, these liquid electrolytes can "corrode" electrodes and dyes.

In Grätzel cells, the electrolyte contains a redox couple which carries electrons from the cathode to the molecules of the photosensitized dye. The most promising redox couple discovered to date is the iodide/triiodide ($I^-$ / $I_3^-$) (*68*). Ionic liquids (ILs), which are organic salts having a melting temperature below 100°C, have key advantages over organic solvents: thermal and chemical stability, low volatility, non-flammability, and wide electrochemical window (*69*). Among the various existing ionic liquids, imidazolium salts have been extensively tested as electrolytes for Grätzel cells. However, pure imidazolium iodide/triiodide ($I^-/I_3^-$) exhibit high viscosity, which often yield DSSCs with poor performance. Therefore, the search for alternative electrolytes remains a strategic issue for practical applications. Recently, eutectic ternary mixtures of iodide ionic liquids as electrolyte, combined with nano–crystalline $TiO_2$ particles as photo–electrode, lead to DSSCs with efficiency close to 8% under spectra (of 1 sun, 1000 W/m$^2$ at 25°C perpendicular to the cell plane) [AM 1.5] illuminations (*70*). However, the long–term stabilities of such DSSCs have not been reported.

### b. Ionogels

The development of ionogels, solid or semi-solid electrolytes, containing ionic liquids is an original way to overcome the limitations of certain electrolytes or to meet new specifications (*71*). The ionogels are composed of an ionic liquid and an organic network, constraining the liquid within a solid system, to provide a physical barrier between the electrodes and thereby prevent any short circuits. Accordingly, electrolytic ionogels should be safer and more suitable for the design of flexible devices. But some articles also disclose the use of electrolytes based on ionic liquids, including silica nanoparticles or networks to improve the performance of the obtained systems. Cells containing ionogels as electrolytes reached a yield of 7% in a Grätzel cell, and ionogels based on protic ionic liquids have shown significant transport properties (*72*).

An alternative method for producing ionogels relies on the synthesis, in one step by sol-gel chemistry, of a silica network immobilizing in-situ ionic liquids (*73*, *74*). But the ionic liquids are only trapped in the silica network and are not bound to the silica network leading to problems of leakage and therefore long-term stability. A modification of these strategies with silica nanoparticles, preformed and assembled with ionic liquids modified with silane groups, has been proposed (*75*). The control of the "chemistry" allows obtaining powders in which the silica nanoparticles are connected to each other via imidazolium functions.

### c. Mono-Ionic Xerogels

An original and innovative route is to obtain an ionogel by a simple elaboration in one step, while (i) maintaining the main properties of gels trapping ionic liquids, and while (ii) mitigating their prohibitive functional drawbacks. Indeed, the approach is to co-condense a silica precursor with a functionalized ionic liquid with a trialkoxysilane function in acidic conditions to form a silica network wherein the cation or the anion of the ionic liquid is bonded to the silica network via covalent bonds (Figure 5). This approach is well positioned relative to technological constraints of safety and efficiency, and uses the advantages of ionic liquids while avoiding the usual drawbacks during operation of a cell.

In particular, this approach is interesting because the synthesis of the ionogels occurs directly into the DSCCs (see figure 6) and, as a consequence, a better control of the electrolyte/electrode interfaces may be achieved, enhancing the performances of the DSSCs (*76*). More precisely, the DSSCs are filled with the liquid electrolyte that becomes gel with time. Thus, the electrode/electrolyte interfaces will be enhanced and then the reaction with the hole and the iodide will be facilitated, avoiding electron/home ($e^-$/hole) recombination. The presence of covalent bonds between silica and ionic liquids may avoid leaks or volatilization of ionic liquids and prevent electrolyte losses. These conditions are more favorable for long term performances. Furthermore this cation immobilization strategy may have the same effect as its absorption on the semiconductor (*77*), i.e. to induce the oxidation of iodide ions via a fast mechanism.

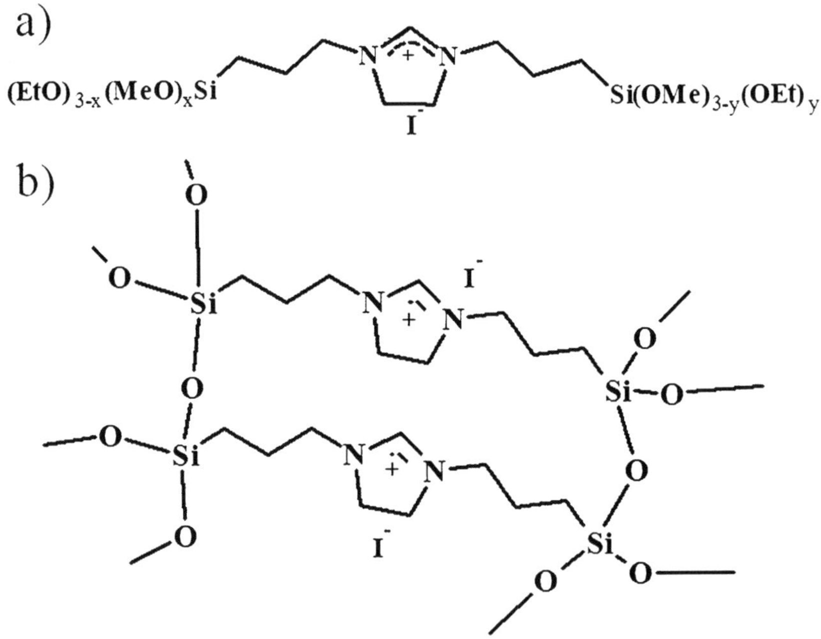


*Figure 5. a) Example of an ionic liquid precursor: 1,3-di(3-trimethoxysilylpropyl)imidazolium iodide, and b) schematic of the organization of the imidazolium groups into the silica network.*

This concept of new materials immediately removes some critical limitations such as the penetration of semiconductor pores, volatility and exudation. The wide selection of precursors of the sol-gel chemistry, the nature (lengths, ionic groups, and functionality) of ionic liquid precursors and additives will create a model family and will be a relevant lever to evaluate and optimize the performance of the obtained gels. To adjust the concentrations, volumes and steric flexibility of ionic functions will indeed have a decisive influence on the transport properties and the structure of the ionic liquid xerogels. As it was ascertained, it is possible to obtain ionic mobilities identical to those obtained in the corresponding pure ionic liquids (*78–80*). Varying the concentration of diiodine directly alters the ratio of ion pairs (*81*) and therefore the conduction mechanisms. Various parameters impact the diffusion coefficient of $I_3^-$.

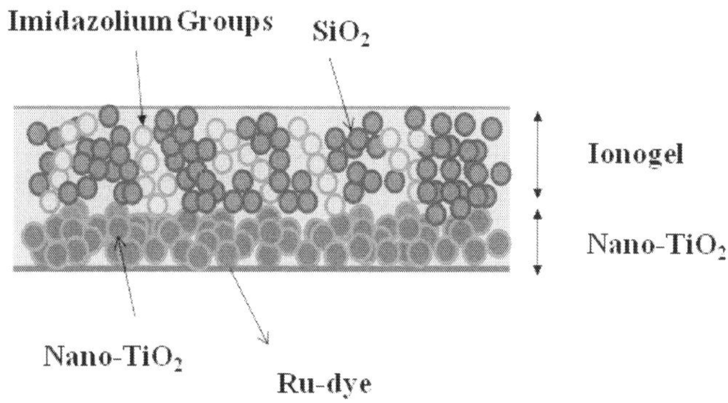


*Figure 6. Schematic representation of the DSSC using an in-situ reaction of the ionic liquid precursor with silica precursor. (Reproduced with permission from reference (78). Copyright [2013] [the Centre National de la Recherche Scientifique (CNRS) and the Royal Society of Chemistry])*

**d. Description of the Nanostructure by Small and Wide Angle X−rays Scattering( SWAXS)**

The ionic liquid subunits influence the nanostructure of the ionogels as observed by SWAXS performed on ionogels with various wt.% of the ionic-liquid monomer (Figure 7).

All the xerogels exhibit a broad peak at $q \sim 16.5$ nm$^{-1}$ that is characteristic of the silica tetrahedrons within the amorphous silica network. For xerogels containing more than 60% in weight of the ionic-liquid monomer, a peak is also observed at $q = 4.4$ nm$^{-1}$. This distance is related to the length of the ionic liquid subunits, and is very close to the lengths found in ionic silica nanoparticle networks including ligands with one imidazolium chloride unit (*82*) as well as in the bulk liquid-like ionic liquid (*83*). A shoulder is also observed *ca.* 9 nm$^{-1}$. It corresponds to a correlation length of 0.7 nm, close to the one reported in literature for π stacking interactions (*84–86*). Accordingly, this distance was attributed to the correlation length between the aromatic units in between the ionic liquid backbones. Additionally, the curves profile at low q is highly dependent on the quantity of ionic liquid precursor in the ionogels. For ionogel with more than 60% of the ionic liquid, the flat profile at low q indicates that no aggregate and no domain, up to a size of 100 nm, are present in the xerogels. The system appears homogeneous at this scale. Conversely, the intensity increases toward low q values for xerogels containing less ionic-liquid precursor. This would correspond to the existence of small cross−linked domains in the 20 nm range. The broad peak between 0.8 and 0.9 nm$^{-1}$ indicates a correlation length of ~ 8 nm for ionogels containing up to 25 wt.% of ionic-liquid monomers. Xerogels with 30 and 40 wt.% of ionic-liquid monomer exhibit an upturn at low q. Correlation length of about 1 nm was obtained. The observed results clearly show that the presence of functionalized ionic liquid dramatically influence the nanostructure of the resulting ionogels.

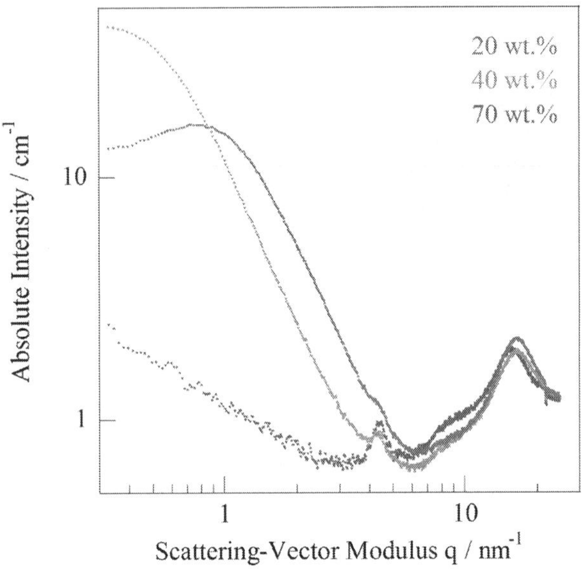


*Figure 7. SWAXS absolute intensity as a function of the scattering-vector modulus q for different weight content of the ionic-liquid monomer. (Reproduced with permission from reference . Copyright [2013] [the Centre National de la Recherche Scientifique (CNRS) and the Royal Society of Chemistry])*

The knowledge gained on the ionic transport, multi-scale structure and their correlation in these mono-ionic ionogels allow developments in other applications such as in energy storage systems or fuel cells operating to over 100°C. Indeed, an understanding of mechanisms of gelation with ionic-liquid precursors and the involvement of the structure on ion transport will support the choice of the precursors according to the intended applications and theoretical developments of the ion dynamics in an ionic confined medium. To take few examples, modulate or change the $I_2$ concentration in a Grätzel cell has a direct implication on the role of the Grotthuss mechanism. In a fuel cell, the choice of the mobile cation is essential for the electrode reactions. In general, the expected benefits go beyond Grätzel cells, since this approach by immobilization is also relevant in fuel cells (*87*) and batteries.

The history of electrolytes for DSSCs showed the possibility to lift the various scientific and technological challenges. These latest results, including solid electrolytes without liquid phases, especially make possible to solve a large part of these obstacles. This evolution is possible through the structuring of the host matrix in a wide range of spatial and temporal scales in these inorganic/organic hybrid environments, which are ionic and confined. Furthermore, to control the nanostructure of these systems allows an improvement of their properties. Indeed this new generation of electrolytes for application in future Grätzel cells permits to correlate the ion transport properties and the diffusion of species. The use of ionic liquid precursors, regularly cited for use in "green chemistry" (*88*), co-condensed by sol-gel chemistry, called "soft chemistry" (*54*), also allows to propose a class of electrolytes using little energy resources and widespread molecules having lower toxicity than usual solvents (*89*, *90*).

## C. Polymer/Ionic Liquid Electrolytes for Energy Storage

In the last few years, academic and industry research have had to overcome many challenges in the fields of environment, production, storage and conversion of energy but also in the reduction of costs in the transportation. Thus, polymer materials scientists have focused on the design and development of new functional polymer materials with unprecedented physical properties via the introduction of ionomers, nanoparticles or block copolymers (*91–93*). Recently, the use of ionic liquids (ILs) as functional additives of thermosets as well as thermoplastics has opened new horizons in polymer science (*94–98*). In fact, ILs have been investigated as building blocks of polymer matrices and as ionic conducting agents of polymer electrolytes (*99*, *100*).

### a. Gel-Like Electrolytes

For several years, "gel" electrolytes based on the combination of polar organic solvents and electrolyte salts have been widely studied in the fields of ionic devices, sensors and secondary batteries (*101*, *102*). However, their volatility and their toxicity have led researchers to a new alternative through the use of ILs known for their excellent thermal stability, their low vapor pressure and their high ionic conductivity (*103*, *104*). Thus, different routes to design polymer electrolytes have been described: 1) impregnation of gel electrolytes with hydrophilic or hydrophobic ILs (*71*) and 2) inclusion of polymers in ILs (*105*, *106*).

Many polymer matrices have been also investigated such as poly(acrylonitrile) (PAN), poly(ethylene oxide) (PEO), poly(methylmethacrylate) (PMMA) and fluorinated polymers (polytetrafluoroethylene (PTFE) and poly(vinylidene fluoride) (PVDF)) where good ionic conductivities, high mechanical strength and transparency are obtained (*107*, *108*). Very recently, research has focused on the development of solid or quasi-solid polymer electrolytes requiring a good confinement of the organic salts in the "host" polymer matrix as well as the high thermal-mechanical properties (*109–111*). Thus, different authors have investigated the influence of the chemical nature (pyridinium, phosphonium, imidazolium) of the ILs on the structuration and the final properties of thermosets based on epoxy networks (*112–115*). They have highlighted that the decomposition of pyridinium or imidazolium salts induces the formation of imidazole and pyridine which act as catalyst of the homopolymerization of epoxy networks (*112*, *113*). In addition, the use of a small amount of ILs in epoxy-amine networks led to an increase of the storage modulus combined with a plasticizer effect by a decrease of the glass transition temperature ($T_g$). Then , Livi and coworkers highlighted that phosphonium ionic liquids associated with phosphinate, dicyanamide or phosphate counter anions, could be used as curing agents of epoxy prepolymer composed of diglycidylether of bisphenol A (DGEBA). Thus, nanostructured thermosets based on epoxy-IL mixtures have been designed with glass transition temperature included between 80 °C and 140 °C coupled with excellent mechanical performances (*116–118*). Nevertheless, improving the ionic conductivity of these new systems remains a real challenge.

**b. Nanostructured Polymer/Ionic Liquid Electrolytes**

Another way was to develop nanostructured polymer/ionic liquid electrolytes with enhanced mechanical, thermal and barrier properties. Thus, ionic liquids were used in small amounts as building blocks in fluorinated matrices (*99*). The influence of the chemical nature of the counter anion as well as the cation on the morphology and on the final properties of polymer matrices has been also investigated (*99*). For the first time, pyridinium, imidazolium and phosphonium ILs have been synthesized and used as functional additives in PTFE matrix. The authors have demonstrated that imidazolium and pyridinium salts led to the phase separation characterized by the presence of ionic clusters (nm to µm) whereas phosphonium ionic liquid combined with iodide anion induced a "spider web" morphology (Figure 8).

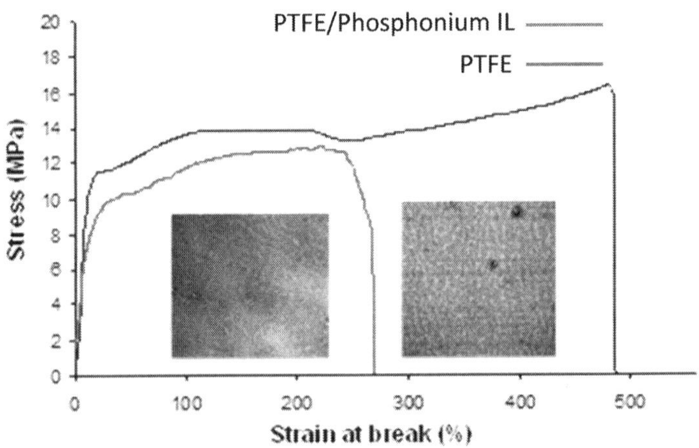


*Figure 8. Mechanical curves and TEM micrographs of PTFE without and with phosphonium ionic liquids.*

The behavior of the ionic liquid is very similar to ionomers where interactions between the polymer matrix and the cation-anion pairs are the key parameter of the phase separation inducing by the formation of multiplet structures (*119*, *120*). Moreover, the mechanical performances were significantly improved with increases in the Young's modulus (E) of 160 % coupled with an exceptional increase of the strain at break (+ 190 %). Very recently, Yang et al. have also demonstrated that phosphonium ILs are excellent nanostructuration agents on a semicrystalline copolymer denoted Poly(Vinylidene Fluoride-ChloroTriFluoroEthylene [P(VdF-CTFE)] (*121*). In addition, the authors have highlighted that the addition of few percent of ILs induced more polar β-phase and more homogeneous distribution morphology as well as an excellent compromise between stiffness and strain to failure (*121*). These early results showed the feasibility of creating efficient polymer electrolytes based on epoxy networks or fluorinated matrices. Nevertheless, the challenge to overcome is to introduce the electrolyte salts in these membranes without changing the final properties of the materials obtained.

In conclusion, these different approaches based on endless ionic liquids combinations open up new horizons in the development of high-performance polymer electrolytes. However, a lot of work is required to obtain good mechanical strength and excellent ionic conductivity.

## Conclusion and Brief Summary

Hybrid electrolytes with improved properties were obtained for three energy devices (Fuel Cells, Organic Photovoltaic and energy storage). These new family of electrolytes, in their respective applications, tend to lessen the disadvantages of liquid phases present in the conventional electrolytes while improving their properties. These evolutions were possible thanks to an elucidation of the multiscale complex structure of the hybrid electrolytes and then the control of their nano-structuration. Beyond these improvements, this overall comprehension gives an opportunity to better understand various related properties including the mechanical properties and the transport mechanisms.

## References

1. Jones, D. J.; Rozière, J. Advances in the development of inorganic–organic membranes for fuel cell applications. *Adv. Polym. Sci.* **2008**, *215*, 219–264.
2. Alberti, G.; Casciola, M. Composite membranes for medium-temperature PEM fuel cells. *Annu. Rev. Mater. Res.* **2003**, *33*, 129–154.
3. Zhang, H. W.; Shen, P. K. Recent development of polymer electrolyte membranes for fuel cells. *Chem. Rev.* **2012**, *112* (5), 2780–2832.
4. Laberty-Robert, C.; Valle, K.; Pereira, F.; Sanchez, C. Design and properties of functional hybrid organic-inorganic membranes for fuel cells. *Chem. Soc. Rev.* **2011**, *40* (2), 961–1005.
5. Belieres, J. P.; Maréchal, M.; Saunier, J.; Alloin, F.; Sanchez, J. Y. Swollen polymethacrylonitrile urethane networks for lithium batteries. *J. Electrochem. Soc.* **2003**, *150* (1), A14–A20.
6. Maneeratana, V.; Bass, J. D.; Azais, T.; Patissier, A.; Valle, K.; Maréchal, M.; Gebel, G.; Laberty-Robert, C.; Sanchez, C. Fractal inorganic-organic interfaces in hybrid membranes for efficient proton transport. *Adv. Funct. Mater.* **2013**, *23* (22), 2872–2880.

7. Sel, O.; Azais, T.; Maréchal, M.; Gebel, G.; Laberty-Robert, C.; Sanchez, C. Sulfonic and phosphonic acid and bifunctional organic-inorganic hybrid membranes and their proton conduction properties. *Chem. - Asian J.* **2011**, *6* (11), 2992–3000.
8. Sanchez, J. Y.; Chabert, F.; Iojoiu, C.; Salomon, J.; El Kissi, N.; Piffard, Y.; Marechal, M.; Galiano, H.; Mercier, R. Extrusion: An environmentally friendly process for PEMFC membrane elaboration. *Electrochim. Acta* **2007**, *53* (4), 1584–1595.
9. Ramani, V.; Kunz, H. R.; Fenton, J. M. Effect of particle size reduction on the conductivity of Nafion/phosphotungstic acid composite membranes. *J. Membr. Sci.* **2005**, *266* (1-2), 110–114.
10. Park, Y. S.; Yamazaki, Y. Low water/methanol permeable Nafion/CHP organic-inorganic composite membrane with high crystallinity. *Eur. Polym. J.* **2006**, *42* (2), 375–387.
11. Bebin, P.; Caravanier, M.; Galiano, H. Nafion/clay-$SO_3H$ membrane for proton exchange membrane fuel cell application. *J. Membr. Sci.* **2006**, *278* (1-2), 35–42.
12. Jung, D. H.; Cho, S. Y.; Peck, D. H.; Shin, D. R.; Kim, J. S. Preparation and performance of a Nafion/montmorillonite nanocomposite membrane for direct methanol fuel cell. *J. Power Sources* **2003**, *118* (1-2), 205–211.
13. Mura, F.; Silva, R. F.; Pozio, A. Study on the conductivity of recast Nafion/montmorillonite and Nafion/$TiO_2$ composite membranes. *Electrochim. Acta* **2007**, *52* (19), 5824–5828.
14. Lee, W.; Kim, H.; Kim, T. K.; Chang, H. Nafion based organic/inorganic composite membrane for air-breathing direct methanol fuel cells. *J. Membr. Sci.* **2007**, *292* (1-2), 29–34.
15. Yang, C.; Srinivasan, S.; Bocarsly, A. B.; Tulyani, S.; Benziger, J. B. A comparison of physical properties and fuel cell performance of Nafion and zirconium phosphate/Nafion composite membranes. *J. Membr. Sci.* **2004**, *237* (1-2), 145–161.
16. Karthikeyan, C. S.; Nunes, S. P.; Prado, L. A. S. A.; Ponce, M. L.; Silva, H.; Ruffmann, B.; Schulte, K. Polymer nanocomposite membranes for DMFC application. *J. Membr. Sci.* **2005**, *254* (1-2), 139–146.
17. Zhang, Y.; Zhang, H. M.; Zhai, Y. F.; Zhu, X. B.; Bi, C. Investigation of self-humidifying membranes based on sulfonated poly(ether ether ketone) hybrid with sulfated zirconia supported Pt catalyst for fuel cell applications. *J. Power Sources* **2007**, *168* (2), 323–329.
18. Chang, J. H.; Park, J. H.; Park, G. G.; Kim, C. S.; Park, O. O. Proton-conducting composite membranes derived from sulfonated hydrocarbon and inorganic materials. *J. Power Sources* **2003**, *124* (1), 18–25.
19. Kim, J. D.; Honma, I. Proton conducting polydimethylsiloxane/zirconium oxide hybrid membranes added with phosphotungstic acid. *Electrochim. Acta* **2003**, *48* (24), 3633–3638.
20. Zoppi, R. A.; deCastro, C. R.; Yoshida, I. V. P.; Nunes, S. P. Hybrids of $SiO_2$ and poly(amide-6-b-ethylene oxide). *Polymer* **1997**, *38* (23), 5705–5712.

21. Vargas, R. A.; Zapata, V. H.; Matallana, E.; Vargas, M. A. More thermal studies on the PVOH/$H_3PO_2$/$H_2O$ solid proton conductor gels. *Electrochim. Acta* **2001**, *46* (10-11), 1699–1702.
22. Li, S. W.; Liu, M. L. Synthesis and conductivity of proton-electrolyte membranes based on hybrid inorganic-organic copolymers. *Electrochim. Acta* **2003**, *48* (28), 4271–4276.
23. Tadanaga, K.; Yoshida, H.; Matsuda, A.; Minami, T.; Tatsumisago, M. Inorganic-organic hybrid films from 3-glycidoxypropyltrimethoxysilane and orthophosphoric acid for medium temperature fuel cells. *Electrochem. Commun.* **2003**, *5* (8), 644–646.
24. Kim, Y. S.; Wang, F.; Hickner, M.; Zawodzinski, T. A.; McGrath, J. E. Fabrication and characterization of heteropolyacid ($H_3PW_{12}O_{40}$)/directly polymerized sulfonated poly(arylene ether sulfone) copolymer composite membranes for higher temperature fuel cell applications. *J. Membr. Sci.* **2003**, *212* (1-2), 263–282.
25. Tazi, B.; Savadogo, O. Effect of various heteropolyacids (HPAs) on the characteristics of Nafion/HPAS membranes and their $H_2$ /$O_2$ polymer electrolyte fuel cell parameters. *J. New Mater. Electrochem. Syst.* **2001**, *4* (3), 187–196.
26. Zaidi, S. M. J.; Mikhailenko, S. D.; Robertson, G. P.; Guiver, M. D.; Kaliaguine, S. Proton conducting composite membranes from polyether ether ketone and heteropolyacids for fuel cell applications. *J. Membr. Sci.* **2000**, *173* (1), 17–34.
27. He, R. H.; Li, Q. F.; Xiao, G.; Bjerrum, N. J. Proton conductivity of phosphoric acid doped polybenzimidazole and its composites with inorganic proton conductors. *J. Membr. Sci.* **2003**, *226* (1-2), 169–184.
28. Park, M. W.; Yang, J. C.; Han, H. S.; Shul, Y. G.; Lee, T. H.; Cho, Y. I. Heteropolyacid ($H_3PW_{12}O_{40}$) incorporated solid polymer electrolyte for PEMFC. *Denki Kagaku* **1996**, *64* (6), 743–748.
29. Asensio, J. A.; Borros, S.; Gomez-Romero, P. Enhanced conductivity in polyanion-containing polybenzimidazoles. Improved materials for proton-exchange membranes and PEM fuel cells. *Electrochem. Commun.* **2003**, *5* (11), 967–972.
30. Tazi, B.; Savadogo, O. Parameters of PEM fuel-cells based on new membranes fabricated from Nafion, silicotungstic acid and thiophene. *Electrochim. Acta* **2000**, *45* (25-26), 4329–4339.
31. Baradie, B.; Poinsignon, C.; Sanchez, J. Y.; Piffard, Y.; Vitter, G.; Bestaoui, N.; Foscallo, D.; Denoyelle, A.; Delabouglise, D.; Vaujany, M. Thermostable ionomeric filled membrane for $H_2$/$O_2$ fuel cell. *J. Power Sources* **1998**, *74* (1), 8–16.
32. Bonnet, B.; Jones, D. J.; Roziere, J.; Tchicaya, L.; Alberti, G.; Casciola, M.; Massinelli, L.; Bauer, B.; Peraio, A.; Ramunni, E. Hybrid organic-inorganic membranes for a medium temperature fuel cell. *J. New Mater. Electrochem. Syst.* **2000**, *3* (2), 87–92.
33. Poinsignon, C.; Amodio, I.; Foscallo, D.; Sanchez, J.-Y. Low cost filled thermostable ionomer membrane for PEMFC. *Mater. Res. Soc. Symp. Proc.* **1999**, *548*, 307–312.

34. Genova-Dimitrova, P.; Baradie, B.; Foscallo, D.; Poinsignon, C.; Sanchez, J.-Y. Ionomeric membranes for proton exchange membrane fuel cell (PEMFC): sulfonated polysulfone associated with phophatoantimonic acid. *J. Membr. Sci.* **2001**, *185*, 59–71.
35. Yoshiji, K. EP 04110352, 1992.
36. Di Noto, V.; Piga, M.; Giffin, G. A.; Negro, E.; Furlan, C.; Vezzu, K. New nanocomposite hybrid inorganic-organic proton-conducting membranes based on functionalized silica and PTFE. *ChemSusChem* **2012**, *5* (9), 1758–1766.
37. Niepceron, F.; Lafitte, B.; Galiano, H.; Bigarre, J.; Nicol, E.; Tassin, J. F. Composite fuel cell membranes based on an inert polymer matrix and proton-conducting hybrid silica particles. *J. Membr. Sci.* **2009**, *338* (1-2), 100–110.
38. Maréchal, M.; Wessel, R.; Diard, J. P.; Guindet, J.; Sanchez, J. Y. Study of PEMFC ionomers through model molecules mimicking the ionomer repeat units. *Electrochim. Acta* **2007**, *52* (28), 7953–7963.
39. Poinsignon, C.; Sanchez, J.-Y.; Piffard, Y.; Vitter, G.; Baradie, B.; Denoyelle, A. EP 97401613.1, 1997.
40. Kreuer, K. D.; Paddison, S. J.; Spohr, E.; Schuster, M. Transport in proton conductors for fuel-cell applications: Simulations, elementary reactions, and phenomenology. *Chem. Rev.* **2004**, *104* (10), 4637–4678.
41. Zapata, P.; Mountz, D.; Meredith, J. C. High-throughput characterization of novel PVDF/acrylic polyelectrolyte semi-interpenetrated network proton exchange membranes. *Macromolecules* **2010**, *43* (18), 7625–7636.
42. Lyonnard, S.; Gebel, G. Neutrons for fuel cell membranes: Structure, sorption and transport properties. *Eur. Phys. J.-Spec. Top.* **2012**, *213* (1), 195–211.
43. Dos Santos, L.; Maréchal, M.; Guillermo, A.; Lyonnard, S.; Moldovan, S.; Ersen, O.; Sel, O.; Perrot, H.; Laberty-Robert, C. Proton transport in hybrid organic-inorganic membrane: An illuminating paradox. *Adv. Funct. Mater.* **2015**, submitted.
44. DuPont, *Nafion® Membranes NR-111*, NR-112. 2001; pp 1–5.
45. Sanchez, J.-Y.; Iojoiu, C.; Mercier, R.; Maréchal, M.; El Kissi, N.; Galiano, H.; Chabert, F. PCT WO 000830, 2006.
46. Navarra, M. A.; Materazzi, S.; Panero, S.; Scrosati, B. PVDF-based membranes for DMFC applications. *J. Electrochem. Soc.* **2003**, *150* (11), A1528–1532.
47. Truffier-Boutry, D.; De Geyer, A.; Guetaz, L.; Diat, O.; Gebel, G. Structural study of zirconium phosphate-nafion hybrid membranes for high-temperature proton exchange membrane fuel cell applications. *Macromolecules* **2007**, *40* (23), 8259–8264.
48. Rubatat, L.; Gebel, G.; Diat, O. Fibrillar structure of Nafion: Matching Fourier and real space studies of corresponding films and solutions. *Macromolecules* **2004**, *37* (20), 7772–7783.
49. Maréchal, M.; Niepceron, F.; Gebel, G.; Mendil-Jakani, H.; Galiano, H. Inside the structure of a nanocomposite electrolyte membrane: how hybrid particles get along with the polymer matrix. *Nanoscale* **2015**, *7* (7), 3077–3087.

50. Hogarth, W. H. J.; da Costa, J. C. D.; Drennan, J.; Lu, G. Q. Proton conductivity of mesoporous sol-gel zirconium phosphates for fuel cell applications. *J. Mater. Chem.* **2005**, *15* (7), 754–758.
51. Sel, O.; Soules, A.; Ameduri, B.; Boutevin, B.; Laberty-Robert, C.; Gebel, G.; Sanchez, C. Original fuel-cell membranes from crosslinked terpolymers via a "sol-gel" strategy. *Adv. Funct. Mater.* **2010**, *20* (7), 1090–1098.
52. Valle, K.; Belleville, P.; Pereira, F.; Sanchez, C. Hierarchically structured transparent hybrid membranes by in situ growth of mesostructured organosilica in host polymer. *Nat. Mater.* **2006**, *5* (2), 107–11.
53. Sanchez, C.; Boissiere, C.; Cassaignon, S.; Chaneac, C.; Durupthy, O.; Faustini, M.; Grosso, D.; Laberty-Robert, C.; Nicole, L.; Portehault, D.; Ribot, F.; Rozes, L.; Sassoye, C. Molecular engineering of functional inorganic and hybrid materials. *Chem. Mater.* **2014**, *26* (1), 221–238.
54. Nicole, L.; Laberty-Robert, C.; Rozes, L.; Sanchez, C. Hybrid materials science: A promised land for the integrative design of multifunctional materials. *Nanoscale* **2014**, *6* (12), 6267–6292.
55. Sanchez, C.; Rozes, L.; Ribot, F.; Laberty-Robert, C.; Grosso, D.; Sassoye, C.; Boissiere, C.; Nicole, L. "Chimie douce": A land of opportunities for the designed construction of functional inorganic and hybrid organic-inorganic nanomaterials. *C. R. Chim.* **2010**, *13* (1-2), 3–39.
56. Dos Santos, L.; Rose, S.; Sel, O.; Maréchal, M.; Perrot, H.; Laberty-Robert, C. Electrospinning a Versatile Tool for Designing Hybrid Proton Conducting Membrane. 2015, submitted.
57. Dos Santos, L.; Laberty-Robert, C.; Maréchal, M.; Perrot, H.; Sel, O. Proton diffusion coefficient in electrospun hybrid membranes by electrochemical impedance spectroscopy. *Langmuir* **2015**, in press.
58. Espinosa, N.; Hosel, M.; Angmo, D.; Krebs, F. C. Solar cells with one-day energy payback for the factories of the future. *Energy Environ. Sci.* **2012**, *5* (1), 5117–5132.
59. Hagfeldt, A.; Boschloo, G.; Sun, L. C.; Kloo, L.; Pettersson, H. Dye-sensitized solar cells. *Chem. Rev.* **2010**, *110* (11), 6595–6663.
60. Grätzel, M. Corrigendum to "Conversion of sunlight to electric power by nanocrystalline dye-sensitized solar cells" [vol 164, pg 3, 2004]. *J. Photochem. Photobiol., A* **2004**, *168* (3), 235–235.
61. Li, B.; Wang, L. D.; Kang, B. N.; Wang, P.; Qiu, Y. Review of recent progress in solid-state dye-sensitized solar cells. *Sol. Energy Mater. Sol. Cells* **2006**, *90* (5), 549–573.
62. Hsu, C. Y.; Chen, Y. C.; Lin, R. Y. Y.; Ho, K. C.; Lin, J. T. Solid-state dye-sensitized solar cells based on spirofluorene (spiro-OMeTAD) and arylamines as hole transporting materials. *Phys. Chem. Chem. Phys.* **2012**, *14* (41), 14099–14109.
63. Burschka, J.; Dualeh, A.; Kessler, F.; Baranoff, E.; Cevey-Ha, N. L.; Yi, C. Y.; Nazeeruddin, M. K.; Gratzel, M. Tris(2-(1H-pyrazol-1-yl)pyridine)cobalt(III) as p-type dopant for organic semiconductors and its application in highly efficient solid-state dye-sensitized solar cells. *J. Am. Chem. Soc.* **2011**, *133* (45), 18042–18045.

64. Yang, L.; Cappel, U. B.; Unger, E. L.; Karlsson, M.; Karlsson, K. M.; Gabrielsson, E.; Sun, L. C.; Boschloo, G.; Hagfeldt, A.; Johansson, E. M. J. Comparing spiro-OMeTAD and P3HT hole conductors in efficient solid state dye-sensitized solar cells. *Phys. Chem. Chem. Phys.* **2012**, *14* (2), 779–789.
65. Planells, M.; Abate, A.; Hollman, D. J.; Stranks, S. D.; Bharti, V.; Gaur, J.; Mohanty, D.; Chand, S.; Snaith, H. J.; Robertson, N. Diacetylene bridged triphenylamines as hole transport materials for solid state dye sensitized solar cells. *J. Mater. Chem. A* **2013**, *1* (23), 6949–6960.
66. Komiya, R.; Han, L. Y.; Yamanaka, R.; Islam, A.; Mitate, T. Highly efficient quasi-solid state dye-sensitized solar cell with ion conducting polymer electrolyte. *J. Photochem. Photobiol., A* **2004**, *164* (1-3), 123–127.
67. Kay, A.; Grätzel, M. Low cost photovoltaic modules based on dye sensitized nanocrystalline titanium dioxide and carbon powder. *Sol. Energy Mater. Sol. Cells* **1996**, *44* (1), 99–117.
68. Oregan, B.; Grätzel, M. A low-cost, high-efficiency solar-cell based on dye-sensitized colloidal $TiO_2$ films. *Nature* **1991**, *353* (6346), 737–740.
69. Armand, M.; Endres, F.; MacFarlane, D. R.; Ohno, H.; Scrosati, B. Ionic-liquid materials for the electrochemical challenges of the future. *Nat. Mater.* **2009**, *8* (8), 621–629.
70. Bai, Y.; Cao, Y. M.; Zhang, J.; Wang, M.; Li, R. Z.; Wang, P.; Zakeeruddin, S. M.; Gratzel, M. High-performance dye-sensitized solar cells based on solvent-free electrolytes produced from eutectic melts. *Nat. Mater.* **2008**, *7* (8), 626–630.
71. Neouze, M. A.; Le Bideau, J.; Leroux, F.; Vioux, A. A route to heat resistant solid membranes with performances of liquid electrolytes. *Chem. Commun.* **2005** (8), 1082–1084.
72. Delahaye, E.; Gobel, R.; Lobbicke, R.; Guillot, R.; Sieber, C.; Taubert, A. Silica ionogels for proton transport. *J. Mater. Chem.* **2012**, *22* (33), 17140–17146.
73. Sharp, K. G.; Scherer, G. W. Interaction of formic acid with the silica gel network. *J. Sol-Gel Sci. Technol.* **1997**, *8* (1-3), 165–171.
74. Dai, S.; Ju, Y. H.; Gao, H. J.; Lin, J. S.; Pennycook, S. J.; Barnes, C. E. Preparation of silica aerogel using ionic liquids as solvents. *Chem. Commun.* **2000** (3), 243–244.
75. Litschauer, M.; Neouze, M. A. Nanoparticles connected through an ionic liquid-like network. *J. Mater. Chem.* **2008**, *18* (6), 640–646.
76. Nogueira, A. F.; Longo, C.; De Paoli, M. A. Polymers in dye sensitized solar cells: Overview and perspectives. *Coord. Chem. Rev.* **2004**, *248* (13-14), 1455–1468.
77. Pelet, S.; Moser, J. E.; Grätzel, M. Cooperative effect of adsorbed cations and iodide on the interception of back electron transfer in the dye sensitization of nanocrystalline $TiO_2$. *J. Phys. Chem. B* **2000**, *104* (8), 1791–1795.
78. Goebel, R.; Hesemann, P.; Weber, J.; Moller, E.; Friedrich, A.; Beuermann, S.; Taubert, A. Surprisingly high, bulk liquid-like mobility of silica-confined ionic liquids. *Phys. Chem. Chem. Phys.* **2009**, *11* (19), 3653–3662.

79. Lunstroot, K.; Driesen, K.; Nockemann, P.; Van Hecke, K.; Van Meervelt, L.; Gorller-Walrand, C.; Binnemans, K.; Bellayer, S.; Viau, L.; Le Bideau, J.; Vioux, A. Lanthanide-doped luminescent ionogels. *Dalton Trans.* **2009** (2), 298–306.
80. Fontaine, O.; Touidjine, A.; Maréchal, M.; Bonhomme, C.; Ribot, C.; Geffroy, B.; Jousselme, B.; Sanchez, C.; Laberty−Robert, C. One-pot route to class II hybrid ionogel electrolytes for DSSC. *New J. Chem.* **2014**, *38*, 2008–2015.
81. Call, F.; Stolwijk, N. A. Impact of I-2 additions on iodide transport in polymer electrolytes for dye-sensitized solar cells: Reduced pair formation versus a Grotthuss-like mechanism. *J. Phys. Chem. Lett.* **2010**, *1* (14), 2088–2093.
82. Martinelli, A.; Maréchal, M.; Ostlund, A.; Cambedouzou, J. Insights into the interplay between molecular structure and diffusional motion in 1-alkyl-3-methylimidazolium ionic liquids: A combined PFG NMR and X-ray scattering study. *Phys. Chem. Chem. Phys.* **2013**, *15* (15), 5510–5517.
83. Clement, S.; Tizit, A.; Desbief, S.; Mehdi, A.; De Winter, J.; Gerbaux, P.; Lazzaroni, R.; Boury, B. Synthesis and characterisation of pi-conjugated polymer/silica hybrids containing regioregular ionic polythiophenes. *J. Mater. Chem.* **2011**, *21* (8), 2733–2739.
84. Pichon, B. P.; Scampini, S.; Bied, C.; Moreau, J. J. E.; Man, M. W. C. The influence of arylene and alkylene units on the structuring of urea-based bridged silsesquioxanes. *Eur. J. Inorg. Chem.* **2012** (32), 5312–5322.
85. Fernandes, M.; Nobre, S. S.; Xu, Q.; Carcel, C.; Cachia, J. N.; Cattoen, X.; Sousa, J. M.; Ferreira, R. A. S.; Carlos, L. D.; Santilli, C. V.; Man, M. W. C.; Bermudez, V. d. Z. Self-structuring of lamellar bridged silsesquioxanes with long side spacers. *J. Phys. Chem. B* **2011**, *115* (37), 10877–10891.
86. Litschauer, M.; Peterlik, H.; Neouze, M.-A. Nanoparticles/ionic linkers of different lengths: Short-range order evidenced by small-angle x-ray scattering. *J. Phys. Chem. C* **2009**, *113* (16), 6547–6552.
87. Lee, S. Y.; Ogawa, A.; Kanno, M.; Nakamoto, H.; Yasuda, T.; Watanabe, M. Nonhumidified intermediate temperature fuel cells using protic ionic liquids. *J. Am. Chem. Soc.* **2010**, *132* (28), 9764–9773.
88. Anastas, P.; Eghbali, N. Green chemistry: Principles and practice. *Chem. Soc. Rev.* **2010**, *39* (1), 301–312.
89. Zhao, D. B.; Liao, Y. C.; Zhang, Z. D. Toxicity of ionic liquids. *Clean: Soil, Air, Water* **2007**, *35* (1), 42–48.
90. Ranke, J.; Molter, K.; Stock, F.; Bottin-Weber, U.; Poczobutt, J.; Hoffmann, J.; Ondruschka, B.; Filser, J.; Jastorff, B. Biological effects of imidazolium ionic liquids with varying chain lengths in acute *Vibrio fischeri* and WST-1 cell viability assays. *Ecotoxicol. Environ. Saf.* **2004**, *58* (3), 396–404.
91. Maiez-Tribut, S.; Pascault, J. P.; Soule, E. R.; Borrajo, J.; Williams, R. J. J. Nanostructured epoxies based on the self-assembly of block copolymers: A new miscible block that can be tailored to different epoxy formulations. *Macromolecules* **2007**, *40* (4), 1268–1273.

92. Vermogen, A.; Masenelli-Varlot, K.; Seguela, R.; Duchet-Rumeau, J.; Boucard, S.; Prele, P. Evaluation of the structure and dispersion in polymer-layered silicate nanocomposites. *Macromolecules* **2005**, *38* (23), 9661–9669.
93. Krishnamoorti, R.; Vaia, R. A.; Giannelis, E. P. Structure and dynamics of polymer-layered silicate nanocomposites. *Chem. Mater.* **1996**, *8* (8), 1728–1734.
94. Livi, S.; Duchet-Rumeau, J.; Gerard, J.-F. Nanostructuration of ionic liquids in fluorinated matrix: Influence on the mechanical properties. *Polymer* **2011**, *52* (7), 1523–1531.
95. Yousfi, M.; Livi, S.; Duchet-Rumeau, J. Ionic liquids: A new way for the compatibilization of thermoplastic blends. *Chem. Eng. J* **2014**, *255*, 513–524.
96. Livi, S.; Duchet-Rumeau, J.; Gerard, J.-F.; Thi Nhan, P. Polymers and ionic liquids: A successful wedding. *Macromol. Chem. Phys.* **2015**, *216* (4), 359–368.
97. Livi, S.; Gérard, J.-F.; Duchet-Rumeau, J., Ionic Liquids as Polymer Additives. In *Applications of Ionic Liquids in Polymer Science and Technology*; Mecerreyes, D., Ed.; Springer: Berlin, Heidelberg: 2015; pp 1–21.
98. Shaplov, A. S.; Marcilla, R.; Mecerreyes, D. Recent advances in innovative polymer electrolytes based on poly(ionic liquid)s. *Electrochim. Acta* **2015**, *175*, 18–34.
99. Livi, S.; Gerard, J.-F.; Duchet-Rumeau, J. Ionic liquids: Structuration agents in a fluorinated matrix. *Chem. Commun.* **2011**, *47* (12), 3589–3591.
100. Gray, F. M.; Maccallum, J. R.; Vincent, C. A.; Giles, J. R. M. Novel polymer electrolytes based on A-B-A block copolymers. *Macromolecules* **1988**, *21* (2), 392–397.
101. Hapiot, P.; Lagrost, C. Electrochemical reactivity in room-temperature ionic liquids. *Chem. Rev.* **2008**, *108* (7), 2238–2264.
102. Southall, J. P.; Hubbard, H. V. S. A.; Johnston, S. F.; Rogers, V.; Davies, G. R.; McIntyre, J. E.; Ward, I. M. Ionic conductivity and viscosity correlations in liquid electrolytes for incorporation into PVDF gel electrolytes. *Solid State Ionics* **1996**, *85* (1–4), 51–60.
103. Kim, K. S.; Choi, S.; Demberelnyamba, D.; Lee, H.; Oh, J.; Lee, B. B.; Mun, S. J. Ionic liquids based on N-alkyl-N-methylmorpholinium salts as potential electrolytes. *Chem. Commun.* **2004** (7), 828–829.
104. Simone, P. M.; Lodge, T. P. Phase behavior and ionic conductivity of concentrated solutions of polystyrene-poly(ethylene oxide) diblock copolymers in an ionic liquid. *ACS Appl. Mater. Interfaces* **2009**, *1* (12), 2812–2820.
105. Ohno, H. Molten salt type polymer electrolytes. *Electrochim. Acta* **2001**, *46* (10-11), 1407–1411.
106. Noda, A.; Watanabe, M. Highly conductive polymer electrolytes prepared by in situ polymerization of vinyl monomers in room temperature molten salts. *Electrochim. Acta* **2000**, *45* (8–9), 1265–1270.

107. Susan, M. A.; Kaneko, T.; Noda, A.; Watanabe, M. Ion gels prepared by in situ radical polymerization of vinyl monomers in an ionic liquid and their characterization as polymer electrolytes. *J. Am. Chem. Soc.* **2005**, *127* (13), 4976–4983.
108. Singh, B.; Sekhon, S. S. Polymer electrolytes based on room temperature ionic liquid: 2,3-dimethyl-1-octylimidazolium triflate. *J. Phys. Chem. B* **2005**, *109* (34), 16539–16543.
109. Maka, H.; Spychaj, T. Epoxy resin crosslinked with conventional and deep eutectic ionic liquids. *Polimery* **2012**, *57* (6), 456–462.
110. Maka, H.; Spychaj, T.; Pilawka, R. Epoxy resin/ionic liquid systems: The influence of imidazolium cation size and anion type on reactivity and thermomechanical properties. *Ind. Eng. Chem. Res.* **2012**, *51* (14), 5197–5206.
111. Kowalczyk, K.; Spychaj, T. Ionic liquids as convenient latent hardeners of epoxy resins: Rapid comminucations. *Polimery* **2003**, *48* (11-12), 833–835.
112. Soares, B. G.; Livi, S.; Duchet-Rumeau, J.; Gerard, J.-F. Synthesis and characterization of epoxy/MCDEA networks modified with imidazolium-based ionic liquids. *Macromol. Mater. Eng.* **2011**, *296* (9), 826–834.
113. Guenther Soares, B.; Livi, S.; Duchet-Rumeau, J.; Gerard, J. F. Preparation of epoxy/MCDEA networks modified with ionic liquids. *Polymer* **2012**, *53* (1), 60–66.
114. Soares, B. G.; Silva, A. A.; Livi, S.; Duchet-Rumeau, J.; Gerard, J.-F. New epoxy/jeffamine networks modified with ionic liquids. *J. Appl. Polym. Sci.* **2014**, *131* (3).
115. Soares, B. G.; Silva, A. A.; Pereira, J.; Livi, S. Preparation of epoxy/jeffamine networks modified with phosphonium based ionic liquids. *Macromol. Mater. Eng.* **2015**, *300* (3), 312–319.
116. Thi Khan Ly, N.; Livi, S.; Pruvost, S.; Soares, B. G.; Duchet-Rumeau, J. Ionic liquids as reactive additives for the preparation and modification of epoxy networks. *J. Polym. Sci., Part A: Polym. Chem.* **2014**, *52* (24), 3463–3471.
117. Silva, A. A.; Livi, S.; Netto, D. B.; Soares, B. G.; Duchet, J.; Gérard, J.-F. New epoxy systems based on ionic liquid. *Polymer* **2013**, *54* (8), 2123–2129.
118. Livi, S.; Silva, A. A.; Thimont, Y.; Thi Khanh Ly, N.; Soares, B. G.; Gerard, J.-F.; Duchet-Rumeau, J. Nanostructured thermosets from ionic liquid building block-epoxy prepolymer mixtures. *RSC Adv.* **2014**, *4* (53), 28099–28106.
119. Nyrkova, I. A.; Khokhlov, A. R.; Doi, M. Microdomains in block-copolymers and multiplets in ionomers - Parallels in behavior. *Macromolecules* **1993**, *26* (14), 3601–3610.
120. Eisenberg, A.; Hird, B.; Moore, R. B. A new multiplet-cluster model for the morphology of ramdom ionomers. *Macromolecules* **1990**, *23* (18), 4098–4107.
121. Yang, J.; Pruvost, S.; Livi, S.; Duchet-Rumeau, J. Understanding of versatile and tunable nanostructuration of ionic liquids on fluorinated copolymer. *Macromolecules* **2015**, *48*, 4581–4590.

## Chapter 6

# Modification of Nafion Membranes: Tailoring Properties for Function

**Johna Leddy**[*]

**Department of Chemistry, University of Iowa, Iowa City, Iowa 52240, United States**
**[*]E-mail: johna-leddy@uiowa.edu**

The perfluorinated cation exchange polymer Nafion® finds application in many electrochemical technologies and sensors because Nafion is mechanically robust and sustains high cation flux. On hydration, Nafion segregates into fluorocarbon domains and hydrated domains. Mechanical properties are largely set by the fluorocarbon components and ionic conductivity is restricted to the hydrated domains where dielectric constants are sufficient to support ion formation. Here, a range of Nafion properties are reviewed. Water content and intercalated ions strongly impact Nafion properties. These impacts are qualitatively considered in terms of electrochemical potential and activity effects. Means to alter the properties of Nafion are outlined based on qualitative thermodynamics and the length scale of the modification. Modifications to Nafion membranes in fuel cells are suggested.

Nafion®, a mechanically robust, biphasic, ion conducting polymer, exhibits high selectivity for cations and sustains high cation flux. These properties lead to widespread use of Nafion as a separator in electrochemical systems and as a sensor platform. Nafion was first developed for use in chloralkali processes (*1*). In electrochemical energy systems such as fuel cells, Nafion serves both as a membrane separator and as a catalyst layer additive that balances ion flux with electron, water, and gas flux (*1–3*) and increases concentrations of gaseous fuels and oxidants about the catalyst. In sensors, Nafion provides a stable ion conducting matrix that selects for cationic mediators, indicators, and analytes. Transition metal complexes such as luminescent $Ru(bpy)_3^{2+}$ are common cations (*4*). Nafion is used in organic synthesis as a superacid catalyst (*1*).

Nafion is available as commercially processed, insoluble, mechanically tough cation exchange membranes in several equivalent weights and thicknesses. For example, Nafion 117, a common electrochemical separator has an equivalent weight of 1100 (1100 g of Nafion per mole of sulfonate exchange sites) and thickness of 7 mil (180 μm). Commercial suspensions in aqueous alcohols typically contain 5 wt % Nafion. Suspensions are formed by extraction of membranes in water alcohol mixtures at high temperature and pressure (*1*). Superacid catalysis uses insoluble Nafion microbeads.

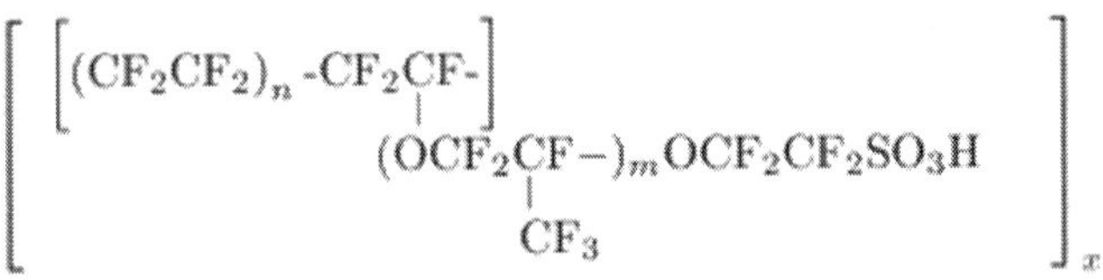

*Scheme 1. Chemical Structure of Nafion where m is usually 1; 5 ≤ n ≤ 7. Nafion is the succinct name for the material also known as ethanesulfonyl fluoride, 2-[1-[difluoro-[(trifluoroethenyl)oxy]methyl] -1,2,2,2-tetrafluoro-ethoxy]-1,1,2,2,-tetrafluoro-, with tetrafluoroethylene.*

Walther Grot at DuPont (*1*) first synthesized Nafion. Traditionally, hydrolysis of the precursor sulfuryl fluoride polymer forms the insoluble perfluorosulfonate polymer, but more recently Nafion membranes are cast from suspension (*1*).

The chemical structure of Nafion is shown in Scheme 1. The sulfonate anions $SO_3$ pendant on the polymer side chains serve as the cation exchange sites. Once hydrated, Nafion forms a biphasic matrix where the hydrophilic sulfonates extract into the water domains that segregate from the highly hydrophobic fluorocarbon backbone. The segregated aqueous and fluorocarbon phases establish nanostructured domains. Because sulfonates are anions, Nafion is highly selective for cations. With appropriate nanostructure, as in Nafion membranes, Nafion is superselective so that Nafion in contact with electrolyte exchanges cations but largely excludes anions even when the contacting electrolyte is at ionic strength higher than in the hydrated domains of Nafion. Cations concentrate into the hydrated domains to provide charge balance of the sulfonates. Because electroneutrality is well maintained and cation transport in the hydrated domain is facile, Nafion sustains high cation flux, on the order of an amp per $cm^2$ for well-hydrated membranes. Nafion properties are highly dependent on the water content.

Many and varied ion exchange polymers have been synthesized and evaluated since Nafion became available, but Nafion remains the most commonly used separator in electrochemical systems. Nafion is prized for its physical properties of mechanical robustness and high ionic conductivity. Several excellent compendiums and reviews of Nafion and its properties have appeared (*1, 5–7*).

The Chapter is not intended as an exhaustive review. Papers on Nafion properties of common interest are reported.

This Chapter inspects the relationship between the physical properties of Nafion and its nanoenvironment. From the electrostatics and water content that dominate the physical properties of Nafion, a semiquantitative sketch of the hydrated domain in biphasic Nafion is made. Means to tailor Nafion properties by manipulation of nanostructure and microstructure are outlined with possible impacts on properties such as conductivity and mechanical robustness noted.

## 1. Nanostructure and Some Known Physical Parameters for Nafion

It is well established that Nafion is nanostructured and that nanostructure is a dominant factor that sets Nafion properties. Common physical parameters measured for Nafion include density, equivalent weight, and dielectric constant. Density and equivalent weight mark water content and ion concentrations in the water domains. Density and so water content varies with extracted cation. Ions and water content establish chemical activity in Nafion that in turn set properties of conductance and flux of cations and water. Nanostructure and physical parameters help map the environment in Nafion. In this Section, a compendium of commonly used Nafion properties is summarized.

### 1.1. Nanostructure

Most models of Nafion biphasic structure adopt a characteristic dimension of 5 nm for well-hydrated membranes. The first reports of nanostructure by Gierke and coworkers (*8*, *9*) derived from small angle x-ray scattering measurements on $Ag^+$ exchanged Nafion, where 5 nm hydrated domains interconnect with 1 nm channels. Recent measurements by Allen and coworkers were fit by interconnected hydrated domains of 5 nm characteristic length (*10*). Interconnected hydrated domains are embedded in the fluorocarbon phase. Sulfonates available for ion exchange are ionized and therefore in the hydrated domain, where the dielectric constant is higher than the fluorocarbon phase. A volume of $(5\ nm)^3$ holds ~4200 water molecules. Biphasic Nafion segregates into fluorocarbon and hydrated domains. Nanostructure of the hydrated domains exhibits regular but not uniform dimensions. Such structures establish spatially varying charges and concentrations that lead to gradients to enhance transport.

### 1.2. Density: $\rho$

In biphasic Nafion, nanodomains of fluorocarbon and water exist, where fluorocarbon and water densities differ substantially. As an example, polytetrafluoroethylene (PTFE, Teflon®) has a density of 2.15 $g/cm^3$ and water density is 0.9982 $g/cm^3$ at 25 °C. Nafion density varies with membrane preparation and exchanged cation. Density characterizes Nafion water, a pivotal parameter in conductance and mechanical robustness.

The densities of Nafion 117 membranes and films formed by recasting Nafion 1100 suspension were determined by the hydrostatic weighing method and reported in two papers (*11, 12*). Films cast from suspension and membranes received in the proton form were first soaked in either water or concentrated nitric acid. For Nafion 117 membranes, the reported densities are 1.83 ± 0.10 and 1.77 ± 0.10 g/cm$^3$ for the water pretreated membranes. For the strong acid treated membranes, 1.76 ± 0.14 g/cm$^3$ was found (*11*). The 117 membrane densities were similar with water and acid pretreatment. For films cast from Nafion suspension, the water pretreated films had densities of 1.65 ± 0.02 and 1.40 ± 0.15 g/cm$^3$ and for the strong acid treated films, 1.77 ± 0.03 and 1.67 ± 0.14 g/cm$^3$. Densities for the membranes and films are similar with water treated films perhaps slightly lower. Lower density correlates with higher water content, and so typically higher ionic conductivity but lesser mechanical strength.

### *1.2.1. Electrostatics and Density*

Densities were also measured when cations are exchanged into Nafion membranes and films after pretreatment with water or acid (*11, 12*). Cations varied in radius $r_+$ and charge z: proton $H^+$, trimethylammonium $TMA^+$, ferrous $Fe^{2+}$, tris-bipyridal ruthenium $\mathrm{Ru(bpy)_3^{2+}}$, and hexaamine ruthenium $Ru(NH_3)_6^{3+}$. Correlation between density and electrostatic potential energy was found. For electrostatic potential energy E (J/mol) calculated for ions in a crystal,

$$E = \frac{z_1 z_2 e^2 \bar{A} N_0}{4\pi\epsilon\epsilon_0 r} \quad (1)$$

$z_1$ = -1 for sulfonate and $z_2$ is the charge of the extracted cation; e is charge on an electron, 1.602 ×10$^{-19}$ C; the Madelung constant $\bar{A} \cong 0.874\ (|z_1| + (|z_2|)$; and the separation distance r is the average of radius of the cation and sulfonate as r = 0.5 ($r_-$ + $r_+$) where the $r_-$ is estimated as 0.35 nm. $N_0$ is Avogadro number, $\varepsilon_0$ is permittivity of free space, 8.854 × 10 $^{14}$ C/Vcm, and ε is the dielectric constant, taken as 20 (*13*). Plots of measured density $\rho_{meas}$ with E are shown in Figure 1 a, b for Nafion 117 membranes where cations were exchanged after either water (a) or acid (b) pretreatment. In Figure 1 c, d, analogous plots are shown for ~ 200 μm thick Nafion films cast from suspension (*11*). Regression data, reported in Table 1 for Nafion 117 water ($\rho_{meas}^{117,water}$) and acid ($\rho_{meas}^{117,acid}$) pretreated and for Nafion 1100 films water ($\rho_{meas}^{1100,water}$) and acid ($\rho_{meas}^{1100,acid}$) pretreated, are of good linearity. Intercepts, which describe the density when electrostatic interactions are minimized, are statistically distinct and vary over the range between 1.4 and 1.95 g/cm$^3$. For acid pretreated Nafion 1100 films, density ($\rho_{meas}^{1100,acid}$) is fairly invariant for all but the $H^+$ exchanged membrane, so the slope can be taken as zero and average density of 1.95 ± 0.03 g/cm$^3$ is reported. In an alternate view of $\rho_{meas}^{1100,acid}$, regression of the lower potential energy species $H^+$, $TMA^+$, and $\mathrm{Ru(bpy)_3^{2+}}$ yields slopes and intercepts similar to $\rho_{meas}^{1100,water}$. In Table 1, other literature density data (*4*, *14*) regressed with *E* yield slopes and intercepts comparable to the values reported in Reference (*11*).

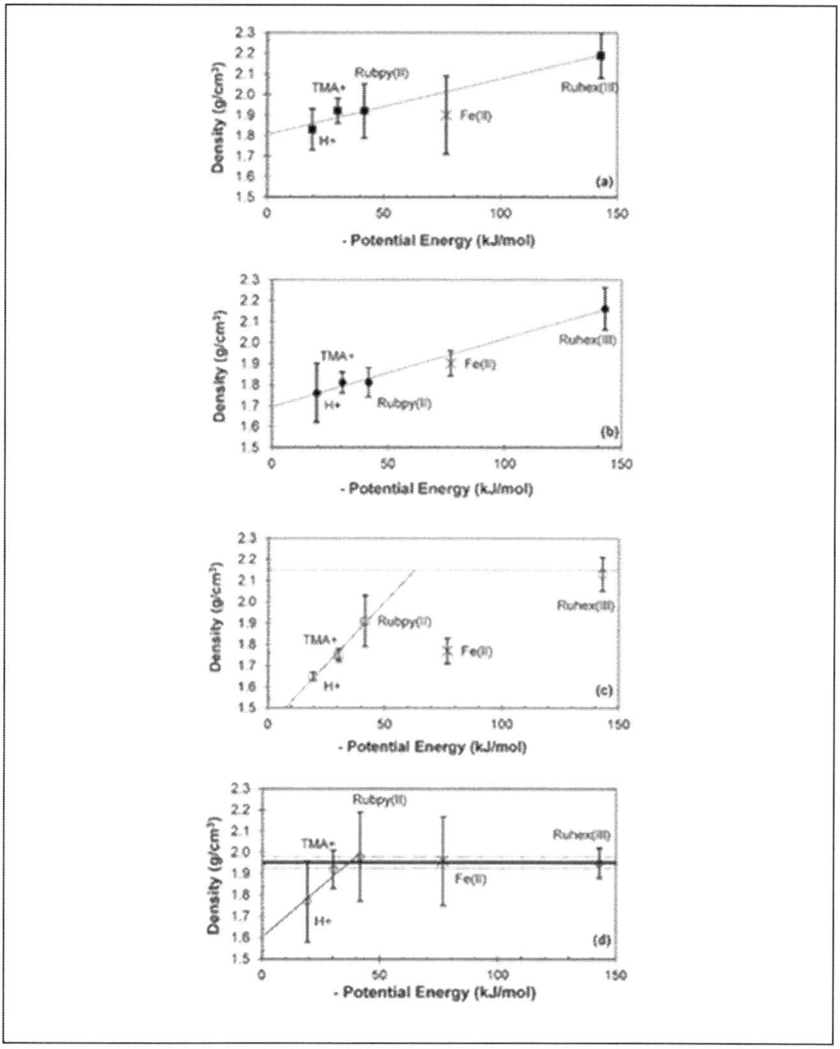


*Figure 1. Densities measured by the hydrostatic weighing method shown for Nafion 117 membrane pretreated in water (a) and strong acid (b) and for recast Nafion 1100 suspension pretreated in water (c) and strong acid (d) correlate linearly with electrostatic interactions reported as - Potential Energy (11) (used with permission). Regression statistics are reported in Table 1 but do not include $Fe^{2+}$ because ferrous ion speciation may be affected in the unbuffered electrolyte. $TMA^{+}$ is tetramethyl ammonium and Ruhex(III) is hexaamine ruthenium (III),* $Ru(NH_3)_6^{3+}$. *In panel (d), a regression line is added not reported in the original reference.*

**Table 1. Regression of Measured Density $\rho_{means}$ and Electrostatic Potential Energy $E$ for Several Nafion Preparations and Various Cations. Slopes, Intercepts, and Correlation Coefficients**

| Density<br>Nafion<br>in Solvent | Cations | Slope<br>mmol g/<br>(kJ cm$^3$) | Intercept<br><br>(g/cm$^3$) | $R^2$ |
|---|---|---|---|---|
| $\rho_{meas}^{117,water}$<br>117[a]<br>in $H_2O$ | $H^+$,$TMA^+$,<br>$Ru(bpy)_3^{2+}$,<br>$Ru(NH_3)_6^{3+}$ | – (2.7 ± 0.3) | 1.81 ± 0.02 | 0.975 |
| $\rho_{meas}^{117,acid}$<br>117[b]<br>in $H_2O$ | $H^+$,$TMA^+$,<br>$Ru(bpy)_3^{2+}$,<br>$Ru(NH_3)_6^{3+}$ | – (3.2 ± 0.2) | 1.70 ± 0.02 | 0.993 |
| $\rho_{meas}^{1100,water}$<br>1100 cast[a]<br>in $H_2O$ | $H^+$,$TMA^+$,<br>$Ru(bpy)_3^{2+}$ | – (12 ± 1) | 1.41 ± 0.05 | 0.98 |
| $\rho_{meas}^{1100,wacid}$<br>1100 cast[b]<br>in $H_2O$ | $TMA^+$,<br>$Ru(bpy)_3^{2+}$,<br>$Ru(NH_3)_6^{3+}$ | ~ 0 | 1.95 ± 0.03 | |
| | $H^+$,$TMA^+$,<br>$Ru(bpy)_3^{2+}$ | – (9 ± 2) | 1.60 ± 0.08 | 0.94 |
| $\rho_{meas}^{117,CH_3OH}$<br>117[c]<br>in $H_2O$ | $H^+$, $Li^+$,<br>$Na^+$, $K^+$ | –(14.5 ± 0.9) | 1.35 ± 0.03 | 0.996 |
| $\rho_{meas}^{117,CHCl_3}$<br>117[c]<br>in $CHCl_3$ | $H^+$, $Li^+$,<br>$Na^+$, $K^+$ | –(5.5 ± 0.5) | 1.87 ± 0.01 | 0.99 |
| $\rho_{meas}^{1155,water}$<br>1155[d]<br>in $H_2O$ | $Li^+$, $Na^+$,<br>$K^+$, $Cs^+$ | – (20 ± 3) | 1.33 ± 0.09 | 0.97 |

∝[a] pretreated in water;[11] [b] pretreated in strong acid;[11]
[c] washed with methanol;[4] [d]Ref.[14]
$r_+$ (nm, hydrated) reported in,[11,15] $H^+$ 0.45, $TMA^+$ 0.23, $Ru(bpy)_3^{2+}$ 0.70, $Ru(NH_3)_6^{3+}$ 0.34, $Li^+$ 0.30, $Na^+$ 0.23, $K^+$ 0.15, $Cs^+$ 0.13; $r_-$ ($SO_3^-$) 0.35.
All dielectric constants approximated as 20 except $CHCl_3$ where $\epsilon = 4.81$.

A notable outcome is that density varies linearly with electrostatic inter-actions between sulfonates and exchanged cations. The strong dependence on electrostatics highlights the impact of electrochemical potential on Nafion properties. Activity rather than ion concentration characterizes transport (*16*). Electrostatic interactions between sulfonates and exchanged cations establish density and so the water content of Nafion. Water content impacts Nafion conductance. Conductance is dependent on electrochemical potential.

## 1.3. Equivalent Weight: EQWT

Equivalent weight for Nafion and other cation exchange polymers is determined by first, full exchange of the weighed cation exchanger with protons; second, addition of overwhelming sodium ion concentration to exchange $Na^+$ for $H^+$ in the membrane and flush protons into the electrolyte; and third, titration of exchanged protons with base in a strong acid strong base titration. The ratio of Nafion mass to moles of protons yields the equivalent weight (grams of polymer/mole of exchange sites).This protocol only accesses exchangeable protons, which are the protons in the water domains of Nafion. The dielectric constant in fluorocarbon is too low to support ions and charge separation needed for conductivity. Only cations in the water domains support ionic conductivity.

The reported nominal equivalent weight of Nafion, 1100 g/mol of sulfonates, corresponds to $n \approx 6.5$ in the Scheme 1. Titration of Nafion films yields an equivalent weight of 996 ± 24 g/mol (*17*), which corresponds to $n \approx 5.5$. Nafion suspension has an equivalent weight 10 % higher than the nominal equivalent weight, which may reflect the process for suspension. For calculation here, the nominal equivalent weight 1100 is used.

## 1.4. Water Content and Ion Concentrations

Water content establishes ion concentrations and mechanical properties of Nafion. As water fraction decreases, ion concentrations and electrostatic interactions increase to alter activity and conductivity of Nafion. As water fraction decreases, the fluorocarbon fraction increases to improve mechanical robustness and decrease Nafion solubility.

### *1.4.1. Hydrated Volume Fraction: $f_{aq}$*

Density measures water content in Nafion. Divide Nafion into two volume fractions $f_{CF2}$ and $f_{aq}$, for the fluorocarbon domain and the hydrated domain that includes all ions, where $f_{CF2} + f_{aq} = 1$. Allow fluorocarbon density $\rho_{CF2} = 2.15$ g/cm$^3$ and hydrated domain density as $\rho_{aq} = 0.9982$ g/cm$^3$, the density of water at 25 °C. Then, measured density, $\rho_{meas} = f_{CF2}\rho_{CF2} + f_{aq}\rho_{aq}$. On substitution of $f_{CF2} = 1 - f_{aq}$, hydrated volume fraction varies linearly with $\rho_{meas}$.

$$f_{aq} = \frac{\rho_{CF_2} - \rho_{meas}}{\rho_{CF_2} - \rho_{aq}} \tag{2}$$

For inorganic cations in well-hydrated Nafion, typical $\rho_{meas}$ values of 1.6, 1.7, and 1.8 g/cm$^3$ correspond to $f_{aq}$ of about 50, 40, and 30 % water. Well-hydrated Nafion includes Nafion in contact with water, electrolyte solution, or 100 % relative humidity. If Nafion is not well-hydrated as when Nafion is used in a dry gas stream, water can be lost and density increased.

### *1.4.2. Ion Concentrations:* $c^*_{aq}$

Ion concentrations are estimated in the hydrated domains based on $f_{aq}$ and the equivalent weight. For ions to conduct, they must ionize and move. Ionization occurs in water but not in low dielectric constant fluorocarbons. The acid base titration used to find Nafion EQWT only accesses protons in the hydrated domains. Density and EQWT measure ion concentration in Nafion.

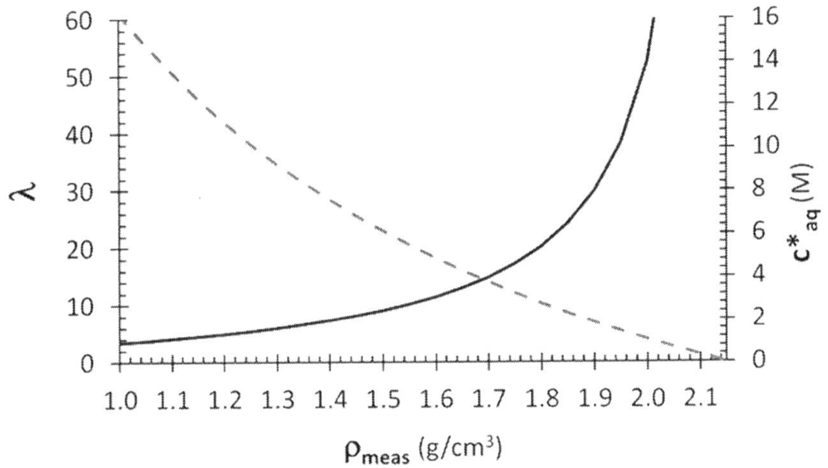


*Figure 2. Variation in the number of waters per sulfonate (and monocation) λ (- - -) and the concentration of sulfonate (and monocation)* $c^*_{aq}$*, (—) in the hydrated domain of Nafion 1100 as a function of measured membrane density* $\rho_{meas}$*. The values* $c^*_{aq}$ *and are calculated from Equations 3 and 4.*

Because of electrostatics, the concentration of positive and negative charges in Nafion must match. For the monocations, the concentration of sulfonates and monocations in the hydrated regions must be equal, $c^*_{SO3,aq} = c^*_{+,aq}$. Let $c^*_{aq} = c^*_{SO3,aq} = c^*_{+,aq}$. From the measured density $\rho_{meas}$, the concentration of ions in Nafion hydrated domains is $\rho_{meas}$/EQWT (mol/cm$^3$). To estimate $c^*_{aq}$, $\rho_{meas}$/EQWT is normalized by the hydrated volume fraction, $f_{aq}$.

$$c^*_{aq}\ (\mathrm{mol/cm^3}) = \frac{\rho_{meas}}{EQWT \times f_{aq}} = \frac{\rho_{meas}\left(\rho_{CF_2} - \rho_{aq}\right)}{EQWT\left(\rho_{CF_2} - \rho_{meas}\right)} \qquad (3)$$

As $\rho_{meas}$ increases, $f_{aq}$ decreases, and $c^*_{aq}$ increases superlinearly (Figure 2). For well-hydrated Nafion, where $\rho_{meas}$ is 1.6, 1.7, and 1.8 g/cm$^3$, ion concentration in the hydrated domain, $c^*_{aq} = c^*_{SO3,aq} = c^*_{+,aq}$ is 3.0, 4.0, and 5.4 M.

At high concentrations, charges are not well separated. In a 5 mM solution, center to center cation separation is 7 nm, but at 5 M, the separation is 0.7 nm, a few water molecule diameters. Matrix and electrostatic interactions of ions and water determine ionic conductivity. The ion environment is parameterized by activity and electrochemical potential rather than concentration.

### *1.4.3. Waters per Sulfonate: λ*

The water content of Nafion is often expressed as number of waters per sulfonate. The moles of water per volume of Nafion (mol/cm3), $c_w$ is $f_{aq}\rho_\omega$/MW ($H_2O$) where the molecular weight of water, MW ($H_2O$) is 18 g/mol. The concentration of sulfonates in Nafion is $\rho_{meas}$/EQWT. The number of waters per sulfonate, is approximated as:

$$\lambda = \frac{f_{aq}\rho_w}{MW\,(H_2O)}\frac{EQWT}{\rho_{meas}} = \frac{1}{18\ (\mathrm{cm}^3/\mathrm{mol})} \times \frac{1}{c^*_{aq}\ (\mathrm{mol/cm}^3)} \tag{4}$$

As $\rho_{meas}$ and $c^*_{aq}$ increase, λ decreases as in Figure 2. For $\rho_{meas}$ of 1.6, 1.7, and 1.8 g/cm$^3$, λ is 18, 14, and 10 waters per sulfonate (and per sulfonate-monocation pair). In Figure 2, is shown versus $\rho_{meas}$. Through Equations 2, 3, and 4, Nafion water content can be characterized by interconversion of $\rho_{meas}$, λ, $c^*_{aq}$, and $f_{aq}$

## 1.5. Dielectric Constant ε

The dielectric constant and relative permittivity estimate the polarity of a matrix, which reflects the extent to which the matrix supports ion formation. The dielectric constants for fluorocarbon (e.g., PTFE) and pure water are 2.1 and 78 at 25 °C. The low ε reflects the poor ability of fluorocarbon to support ion formation and the high ε the ready formation of ions in water. ε characterizes the extent of the solute ionization and so the conductivity that depends on ionization. The electrostatic force F between ions in a matrix of is set by separation distance r and charge $q_1$

$$F = \frac{q_1 q_2}{\epsilon r^2} \tag{5}$$

The electrostatic force between ions increases as ε and r decrease; in solution, r is approximated as $[cN_0]^{-1/3}$ where c is concentration (mol/cm$^3$). As concentration increases, electrostatic interactions increase. Dielectric constant characterizes the formation of ion pairs and other aggregates of cations and anions. Generally, matrices with $20 \leq \varepsilon \leq 40$ exhibit some degree of ion pairing. For $\varepsilon \leq 20$, ion pairing is extensive. Tetrahydrofuran and dichloromethane with ε of 7.6 and 9.1 are good electrochemical solvents despite extensive ion pairing, but solvents with < 7 generally make poor electrochemical solvents because the solvents suppress ionization. The dielectric constant is important to characterize Nafion electrochemical potential and the dependent ion conductivity.

High frequency dielectric measurements on Nafion 117 in various stages of hydration are reported as a function of λ (*13*). Membranes were pretreated with hydrogen peroxide, sulfuric acid, and water and then equilibrated to constant hydration in a fixed humidity environment. Dielectric constant decreased with λ, as shown in Table 2. λ is also represented as $\lambda_{meas}$ and $f_{aq}$. The measured values range between values for pure water and fluorocarbon, with ε decreasing with $f_{aq}$. By linear regression, = (49 ± 4) $f_{aq}$ + (2.6 ± 0.7) with $R^2$ = $0.98_4$; for f $f_{aq}$ = 0, fluorocarbon is anticipated with = 2.15, a value captured by the intercept. When $f_{aq}$ = 1, the regression yields 52 ± 8% to underestimate the anticipated ε ~ 78 of pure water.

**Table 2. Dielectric Constants Reported (*13*) for Nafion 117 at 30 °C as a Function of λ. Corresponding Values of $\rho_{means}$ and $f_{aq}$ Are Shown.**

| $\lambda$ | $\epsilon$ | $\rho_{meas}$ (g/cm$^3$) | $f_{aq}$ |
|---|---|---|---|
| 13 | 20 | 1.73 | 0.37 |
| 6 | 13 | 1.93 | 0.19 |
| 3 | 8 | 2.04 | 0.10 |
| 2 | 5 | 2.07 | 0.07 |
| 1 | 4 | 2.11 | 0.03 |

For static relative permittivity ($\varepsilon_s$ is measured at ω = 0) of aqueous 1:1 electrolytes at molar concentration c, Marcus reported (*18*) $\varepsilon_s = \varepsilon_\omega - \bar{\delta}c$ where $\bar{\delta}$ (/M) is the molar dielectric decrement. Down the alkali metals, $\bar{\delta}$ ranged from 8 to 5 and $\bar{\delta}$ = 16 for proton. Broadly, $\varepsilon_s$ decreased by 10 for each decade increase in c. In Nafion, $c_{aq}$ is several molar, even in well hydrated Nafion, which may contribute to the lower reported values. For δ ≈10 and c ≈ 4 M, an estimated ε of 40 is comparable to 52.

For λ of 13, $\rho_{meas}$ is typical of well hydrated membranes and ε is found to be 20. For electrolyte solutions, ε ~ 20 suggests some degree of ion pairing, but a matrix still well able to support ions. As λ and $f_{aq}$ decrease, decreases consistent with dried Nafion less able to support ion formation.

### 1.6. Thermal, Mechanical, Electrical, and Optical Properties

Thermal, mechanical, electrical, and optical properties of Nafion are critical in many systems. Here, optical properties do not vary with λ or exchanged cation, but thermal, mechanical, and electrical properties do vary.

### 1.6.1. Thermal Stability and Glass Transition Temperature

Yeo and Eisenberg (*19*) report glass transition temperatures $T_g$ for 1.3 mm thick Nafion 1365 membranes by dynamic studies and calorimetry, where the values are in agreement. For proton exchange membranes, values of 104 to 109 ° C were found. Dynamic studies yield $T_g$ for alkali metal exchange films, $Cs^+$ (211 ° C), $K^+$ (225 ° C), $Na^+$ (235 ° C), and $Li^+$ (217 ° C). With the exception of lithium, $T_g$ decreases with ion size.

The thermal stability of Nafion membranes of nominal 1100 equivalent weight were evaluated by thermal gravimetric analysis (*20*). Membranes were exchanged with alkali metal cations from 2 M metal chloride solutions and dried in a vacuum oven at 100 ° C for 24 hours before measurement. The thermal decomposition temperatures for the alkali metal exchanged membranes were ordered as $Li^+$ (426 °C) = $Cs^+$ (426 °C) < $Rb^+$ (430 °C) < $K^+$ (445 °C) < $Na^+$ (458 °C) where the pattern follows ion size with the exception of $Li^+$.

Thermal conductivity of Nafion 117 is reported as 6.5 J/cm ° C h (*1*), less than the thermal conductivity of water (21.6 J/cm ° C h) and PTFE (9 J/cm °C h).

### 1.6.2. Mechanical Properties

Study of the mechanical properties of Nafion 117 (183 μm) and 115 (127 μm) as well as cast Nafion membranes 211 (25.4 μm, 1100 EQWT) and 212 (50.8 μm, 1200 EQWT) are summarized in Reference (*1*). Nafion 117 and 115 and several other 1100 EQWT membranes (112, 1135, 1110) have similar mechanical properties that are reported at 50 % RH and in water at 23 and 100 ° C. There are distinctions in the properties with respect to formation of the membrane in the rolled or machine direction MD and its perpendicular or traverse direction TD. The cast membranes have distinct properties. Results are summarized in Table 3. Tensile modulus E, also called Young's modulus and the modulus of elasticity, is the ratio of tensile stress to extensional strain. For a material initially of thickness $l_0$ and area $A_0$, where force F is applied to $A_0$ and the thickness changes by $\Delta l$, $E = (F/A_0) / (\Delta l/l_0)$ where stress (pressure) is $F/A_0$ and strain (deformation) is $\Delta l/l_0$). Tensile modulus has units of pressure. For comparison, the tensile modulus for PTFE is about 400 MPa, higher than Nafion. For Nafion, higher water content is anticipated as 50 % RH (relative humidity) < water at 23 ° C < water at 100 ° C, where E decreases with water content. The maximum tensile strength is the maximum stress that can be applied to a material before it breaks. For comparison, polypropylene has values of 20 to 80 MPa and high density polyethylene is 37 MPa. In general for Nafion, as the water content increases, the mechanical properties are less robust.

Young's modulus for Nafion 117 exhibits a general increase with size of the exchanged cation. For Nafion 117, Young's modulus for $H^+$, $Na^+$, and $K^+$ are approximately 0.45, 0.72, and 1.1 MPa /%; for thinner Nafion 115, the corresponding values are 0.36, 0.67, and 0.92 MPa /% (*21*). ($\Delta l/l$ was reported as a percent.) The 117 membranes were exchanged with the corresponding salt at 80° C, conditions that increase water content. The values in Reference (*21*) are similar to the values in Table 3 for membranes in 100° C water.

**Table 3. Mechanical Properties for Nation (*1*) 1100 EQWT Membranes and Cast Membranes 211 and 212. MD is machined direction, and TD is its transverse.**

| Nafion | 117, 115, 112, 1135, 1110 | | 211 | | 212 | |
|---|---|---|---|---|---|---|
| Water Uptake at 100 °C (%) | 38 | | 50 | | 50 | |
| Linear Expansion in 100 °C Water (%) | 15 | | 15 | | 15 | |
| Tensile Modulus (MPa) | | | | | | |
| 50 % RH | 249 | | 284 | | 258 | |
| water 23 °C | 114 | | | | | |
| water 100 °C | 64 | | | | | |
| | MD | TD | MD | TD | MD | TD |
| Maximum Tensile Strength (MPa) | | | | | | |
| 50 % RH | 43 | 32 | 23 | 28 | 32 | |
| water 23 °C | 34 | 26 | | | | |
| water 100 °C | 25 | 24 | | | | |
| Elongation at Break (%) | | | | | | |
| 50 % RH | 255 | 310 | 252 | 311 | 343 | 352 |
| water 23 °C | 200 | 275 | | | | |
| water 100 °C | 180 | 240 | | | | |

### *1.6.3. Electrical and Optical Properties*

The areal resistance of Nafion 117 in 24 % NaCl is reported as 1.4 $\Omega cm^2$ in the sodium form (*1*); in 0.6 M KCl, 1.9 $\Omega cm^2$. The volume conductivity of potassium exchanged Nafion 117 is 0.01 $[\Omega cm]^{-1}$ but proton exchanged Nafion 117 is higher at 0.083 $[\Omega cm]^{-1}$.

Nafion membranes and films are water white and have no absorption in the visible spectrum. The refractive index of dry, hydrogen exchanged Nafion is 1.35 that is reduced to 1.34 on water absorption. (*1*), because the refractive index of water falls from 1.344 at 400 nm to 1.333 at 700 nm, Nafion is nearly invisible in water.

## 1.7. Interactions with Neutral Species: Swelling in Organic Solvents and Gas Solubility

Neutral species interact with Nafion. Organic solvents can swell but not dissolve Nafion. Nafion can be suspended in mixtures of alcohol and water at high temperature and pressure, but not dissolved. Solvents miscible with water sustain swelling, so swelling is likely to occur through solvent interactions with both the fluorocarbon and hydrated domains. Gases exhibit high solubility in Nafion where hydrogen and oxygen are likely concentrated in the fluorocarbon domain as gases have poor solubility in water.

### *1.7.1. Swelling in Organic Solvents*

Water in Nafion can be exchanged with other polar solvents. In Reference (*22*), Nafion 117 was exchanged from aqueous 2 M LiCl; dried overnight at 110 C; measured dimensionally to determine the initial volume V; and soaked in the swelling solvent to constant volume. Twenty nine solvents are reported. The normalized change in volume ($\Delta V/V$) expressed as a percent ranged between

From Figure 3, largest swelling occurs for $22 \lesssim$ DN $\lesssim 34$. For the sample solvents, the average and standard deviation for DN is $\overline{x} \pm s = 27._6 \pm 9._2$. In Figure 3, $\Delta V/V$ is empirically mapped as a Gaussian distribution about DN of 28. Three Gaussians are shown, where the solid line approximates all the data, the dotted line approximates solvents that exchange protons, and the dashed line characterizes all points but anticipates a greater swelling. Swelling will be limited if DN $\lesssim$ 20. Because fluorocarbon polymers are a solid, they provide greater mechanical strength than solvents. The mechanical integrity of Nafion is best maintained in water, glycerol, cyclohexanone, tetrahydrofuran, propylene carbonate, butyl acetate, dioxane, pyridine, hydrazine, and acetonitrile.

Y. Marcus reported the structureness of solvents, where solvent stiffness, openness, and ordering are considered (*23*). Solvents are deemed structured or unstructured based on the Kirkwood dipole angular correlation parameter, g, where $\mu_D$ is dipole moment, $n_D$ is solvent refractive index at sodium D-line, and volume $V$.

$$g = \frac{9k_b\epsilon_0}{N_0}\frac{VT}{\mu_D^2}\frac{\left(\epsilon - 1.1n_D^2\right)\left(2\epsilon + 1.1n_D^2\right)}{\epsilon\left(2 + 1.1n_D^2\right)^2} \tag{6}$$

For g $\lesssim$ 1.3, Marcus assigned solvents as unstructured and for g $\gtrsim$1.7, definitely structured.

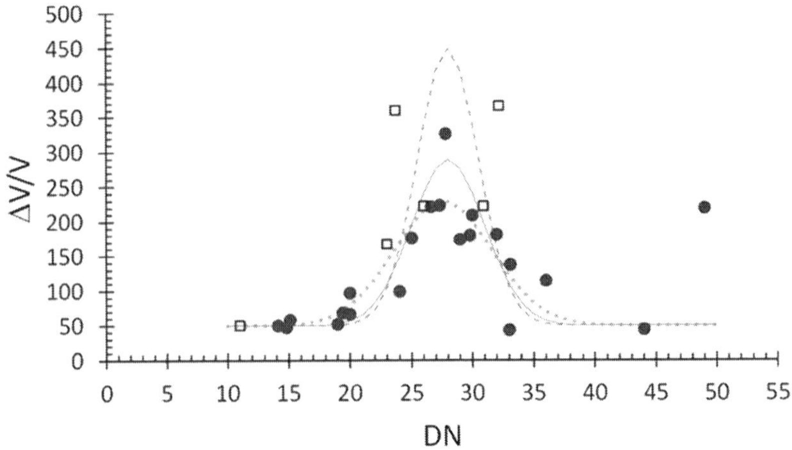


*Figure 3. For Nafion 117 swollen in solvents, the percent change in volume ($\Delta V/V$) as a function of solvent donor number (DN) is shown. The highest degree of swelling $\Delta V/V$ is about DN of 28. The behavior is similar whether the solvent can exchange protons (•) or not (□). Empirical Gaussians are set by $\Delta V/V = A_0 \exp[-(DN - 28)^2 = 2s^2] (2\pi s^2)^{-1/2}) + B_0$ where adjustable inputs are $A_0$, $B_0$, and s. For the three Gaussians, 28 or 27.6 is the mean DN and $B_0$, the offset, is 50. The Gaussians are shown for $A_0 = 1800$ and $s = 4$ (...); $A_0 = 1800$ and $s = 3$ (‒ ‒ ‒ ‒); and $A_0 = 2500$ and $s = 2{:}5$ (- - -). To avoid extensive swelling and concomitant mechanical weakening, solvents with DN ≤ 20 are best at 43 % for water and 732 % for hexamethylphosphoramide. The authors considered variation of $\Delta V/V$ with many solvent parameters, but solvent donor number DN allowed the cleanest presentation. Donor number characterizes the solvation of cations and Lewis acids by the solvent because DN measures the Lewis basicity. DN is set at zero for 1,2-dichoroethane. DN is 18 (kcal/mol) for water and 30 for ethanol and dimethyl sulfoxide (DMSO). A plot of the $\Delta V/V$ values against DN is shown in Figure 3. All reported points (22) are shown except for the largest $\Delta V/V$ = 732 % for hexaphosphoramide (DN = 38.8). Solid circles mark solvents where acid, base or autoprotolysis constants ($pK_a$, $pK_b$, or $pK_{auto}$) indicate the solvent participates in proton exchange processes. Solvents where $pK_a$, $pK_b$, and $pK_{auto}$ were not readily identified are marked by an open square. $\Delta V/V$ is not strongly correlated with whether the solvent allows protonation.*

When the percent relative volume change ($\Delta V/V$) from Reference (*22*) is plotted against g, the impact of a structured versus an unstructured solvent is apparent, as shown in Figure 4. The solvents are divided into unstructured and structured for g about 1.4 to 1.6. The unstructured solvents are shown on left by circles and the structured solvents are squares on right. For the unstructured solvents on the left marked as circles, the swelling is strongly dependent on the g for the solvent, with higher swelling as the solvent increases in stiffness, up to g of about 1.4. Once the solvent is categorized as structured, marked by squares for g $\gtrsim$ 1.6, $\Delta V/V$ is lesser. For the unstructured solvents, there are two smaller subsets. The yellow, blue ringed circles are the trialkyl phosphates, which generally coincide with the majority of the unstructured solvents, except that the change in $\Delta V/V$ for the phosphates increases more rapidly with g. The red ringed circles for g between 1 and 1.3 are unstructured solvents where $\Delta V/V$ falls below the other unstructured solvents (Figure 4). These three solvents are the only heterocyclic solvents that contain oxygen atoms, dioxane, propylene carbonate, and tetrahydrofuran. For the structured solvents, two groups are shown. The purple diamonds correspond to $\Delta V/V$ < 100 % with largely moderate values of g. The green squares denote $\Delta V/V \gtrsim$ 100 % with larger values of g. The moderately structured solvents swell Nafion less than the highly structured solvents.

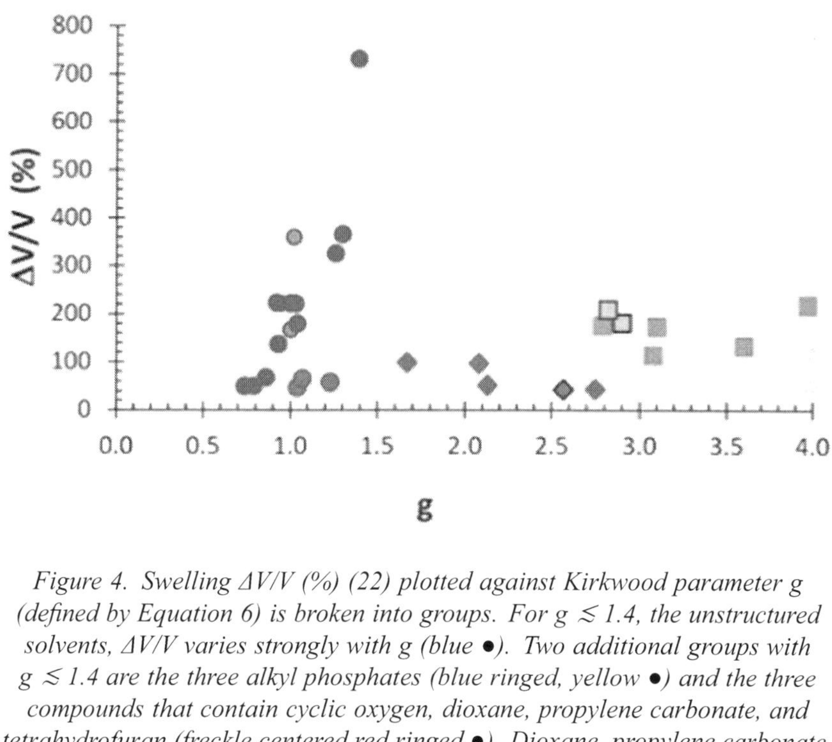


*Figure 4. Swelling $\Delta V/V$ (%) (22) plotted against Kirkwood parameter g (defined by Equation 6) is broken into groups. For g $\lesssim$ 1.4, the unstructured solvents, $\Delta V/V$ varies strongly with g (blue ●). Two additional groups with g $\lesssim$ 1.4 are the three alkyl phosphates (blue ringed, yellow ●) and the three compounds that contain cyclic oxygen, dioxane, propylene carbonate, and tetrahydrofuran (freckle centered red ringed ●). Dioxane, propylene carbonate, and tetrahydrofuran may be better classified with the structured solvents. Structured solvents are shown for limited swelling (♦ where $\Delta V/V$ < 100% and 1.6 $\lesssim$ g $\lesssim$ 2.8) and greater swelling (■ where $\Delta V/V$ < 100% and 2.8 $\lesssim$ g $\lesssim$ 4). For comparison, three solvents are highlighted: water as the black bordered purple ♦; methanol swells slightly more than ethanol, which are illustrated by the black bordered squares.*

Highly structured solvents may provide more mechanical rigidity to the structure. For the unstructured solvents, swelling may be facilitated by partial solvation of the polymer side chains in the solvated portion of the membrane. The g value may provide some intuition about solvents that can dissolve or suspend Nafion as well as identify solvents with limited swelling important to mechanical integrity. In Figure 4, several points for common Nafion solvent are black bordered; the purple diamond for water; the green pebbled squares are methanol and ethanol where methanol swells slightly more. The highest $\Delta V/V$ is for hexamethylphosphoramide and the highest g is for N-methylformamide.

### *1.7.2. Gas Solubility*

In Nafion, solubility of the technologically important gases of hydrogen and oxygen have been determined (*24*). Nafion 125 (1200 EQWT, 0.125 mm) and Nafion 117 (1100 EQWT, 0.175 mm) exchanged with $H^+$ were boiled in water for 3 h. Where exchanged with alkali metal cations, membranes were soaked in 0.5 M metal sulfate or 1 M tetramethyl ammonium chloride. Diffusion coefficients and concentrations for hydrogen and oxygen were measured by permeation and found to vary little with the water content in the membranes. Concentrations in Nafion were more similar to solubilities in polytetrafluoroethylene (PTFE) and about ten fold higher than in water. The solubilities of hydrogen and oxygen are similar. The high solubility of hydrogen and oxygen in Nafion is consistent with solubility enhanced by extraction into the fluorocarbon phase. Results are summarized in Table 4. For each, at least three replicates were run and the relative errors are about 10 %.

## **1.8. Interactions with Cations: Cation Selectivity, Ionic Conductivity, and Acidity**

Because of dielectric properties, ions exist in the hydrated domains. Ions are characterized in terms of selectivity, conductivity, and acidity.

### *1.8.1. Cation Selectivity*

Steck and Yeager (*25*, *26*) examined the selectivity of Nafion 120 membrane for alkali metal cations over proton where the protonated membrane is equilibrated with a solution of 0.01 M ionic strength at specified concentrations of metal cation $M^+$ and proton. At equilibrium,

$$H^+_{\text{Nafion}} + M^+_{\text{soln}} \rightleftharpoons M^+_{\text{Nafion}} + H^+_{\text{soln}} \tag{7}$$

where selectivity specified as

$$K^{H^+}_{M^+} = \frac{\chi_{M^+} c^*_{aq} c_{H^+}}{\chi_{H^+} c^*_{aq} c_{M^+}} = \frac{\chi_{M^+} c_{H^+}}{\chi_{H^+} c_{M^+}} \tag{8}$$

The mole fractions of proton and $M^+$ in the Nafion, $\chi_{H^+} + \chi_{M^+} = 1$ and $c_{H^+}$ and $c_{M^+}$ are concentrations in the electrolyte measured under conditions where $c_{H^+}$ $c_{M^+}$ = 0.01 M. Nafion 120 was initially washed in ethanol and then repeated exchanges with NaOH, HCl, and water. Membranes boiled in water at 96 ° C for 5 hours were identified as expanded (E). As appropriate, metal ions were exchanged from 0.1 M chloride salt except for silver nitrate. Values of $K_{M+}^{H+}$ at 25 ° C were reported without correction for activity, which is appropriate for the low solution concentrations $c_{H^+}$ and $c_{M^+}$. Also reported was λ waters per sulfonate (Table 5).

**Table 4. Hydrogen and Oxygen Solubilities (mM) in Boiled Nafion 125 and 117 and Other Matrices at 25 °C (*24*).**

| | $\lambda$ | $c^{H_2}$ (mM) | $c^{O_2}$ (mM) |
|---|---|---|---|
| **Nafion 125** | | | |
| $Li^+$ | 22.3 | 11.8 | 16.0 |
| $Na^+$ | 18.4 | 13.6 | 13.7 |
| $K^+$ | 13.3 | 9.7 | 10.9 |
| $Cs^+$ | 11.3 | 10.9 | 11.4 |
| **Nafion 117** | | | |
| $Li^+$ | 29.3 | | 13.4 |
| $Na^+$ | 24.6 | | 8.6 |
| $K^+$ | 15.9 | | 10.7 |
| $(CH_3)_4N^+$ | 15.6 | | 11.0 |
| **$H_2O$** | | 0.78 | 1.3 |
| **0.5 M $H_2SO_4$** | | 0.72 | |
| **0.5 M $K_2SO_4$** | | | 0.85 |
| **PTFE** | | 31.3 | 26.8 |

**Table 5. Selectivity of Metal Monocation over Proton $K_{M+}^{H+}$ and Waters per Sulfonate λ for Nafion 120 and Nafion 120 Expanded (E) in Boiling Water (*25, 26*). $K_{M+}^{H+}$ is not corrected for membrane activity.**

| | $K_{M+}^{H+}$ | $K_{M+}^{H+}(E)$ | $\lambda$ | $\lambda(E)$ |
|---|---|---|---|---|
| **$H^+$** | | | 16.7 | 22.3 |
| **$Li^+$** | 0.579 | 0.586 | 14.3 | 22.3 |
| **$Na^+$** | 1.22 | 1.18 | 11.9 | 18.4 |
| **$K^+$** | 3.97 | 3.48 | 8.8 | 13.3 |
| **$Rb^+$** | 6.26 | 4.71 | 7.7 | 11.8 |
| **$Cs^+$** | 9.11 | 7.06 | 6.6 | 11.3 |
| **$Tl^+$** | 6.12 | 3.83 | 8.0 | 11.7 |
| **$Ag^+$** | 1.07 | 0.90 | 12.2 | 17.6 |

The purpose of expanding Nafion 120 by boiling in water is to introduce more water into the membrane. The effect is to increase λ and to slightly decrease $K_{M+}^{H+}$. Kielland (*27*) estimates the ion size parameter $\dot{a}$ for $Rb^+$, $Cs^+$, $Tl^+$, and $Ag^+$ to be the same, 0.25 nm; $K^+$ at 0.30 nm; $Na^+$ at 0.45 nm; $Li^+$ at 0.60 nm; and $H^+$ at 0.90 nm. The general pattern is for $K_{M+}^{H+}$ to increase as decreases. For the smaller $\dot{a}$ values, expanding the membrane adds about 5 additional waters per sulfonate. For the larger hydrated ions, 6 to 8 waters per sulfonate are added.

### *1.8.2. Ionic Conductivity*

Ionic conductivity varies with membrane preparation, cation exchange process, water content λ, and cation (*28*). The values extracted depend on the measurement method and how the concentration of the exchanged cation is determined. A few reports in the literature are considered here.

Okada and coworkers (*29*) evaluated Nafion 115 and 117 (dry thicknesses of 125 and 172 µm; wet thicknesses of 160 and 220 µm) membranes that were pretreated as 2 % $H_2O_2$ at 80 ° C for 2 hours; immersed in 0.1 M HCl for 24 h; rinsed in pure water; and stored in 30 mM HCl. Alkali metal chlorides were exchanged with different molar fraction of HCl and metal chloride where the total ionic strength was 0.03 M and exchange solutions were changed at least four times. Selectivity coefficients of alkali metal and proton were determined by x-ray fluorescence and inductively coupled plasma; the selectivity coefficients were not corrected for activity in the membrane and solution. Concentrations of the cations in Nafion were based on the total wetted membrane volume. The number of waters transferred with each cation was determined by the electromotive force method and reported as λ and the water transference coefficient $t_{H2O}$. Data are summarized in Table 6. Exclusive of the proton, $t_{H2O}$ decreases monotonically with λ, Kielland's ion size parameter $\dot{a}_{H2O}$, and enthalpy of hydration $\Delta H_{hyd}$ and $t_{H2O}$ decreases monotonically with the inverse of the ionic radius absent waters of hydration and the mobility of the cation in water, $u_{H2O}$. Diffusion coefficients are calculated from the mobilities at infinite dilution as $D = uRT/|z|F$.

Xie and Okada (*30*) used streaming potential measurements to measure $t_{H2O}$ as a function of λ for Nafion 117 that was pretreated in water at 80°C for 2 hours; immersed in 0.1 M HCl for 24 h; and rinsed in pure water. Where exchanged with cations, the membranes equilibrated for a month in appropriate 0.03 M chloride electrolytes. The results are tabulated in Table 6. The patterns of behavior are similar to the first set of data in the Table. The hydrogen peroxide cleaned membranes have higher λ but similar mobilities. This suggests the amount of water transferred with the cation flux is more dependent on the interactions of the cations with water than on the amount of water in the membrane. Water bound to the cation, which varies with cation, is important in control of water flux across the membrane.

In another study (*31*), Nafion 117 was pretreated (expanded) by boiling in 3% $H_2O_2$ for 1 h; repeatedly rinsed in boiling water; boiled in 0.5 M $H_2SO_4$ for 1 h; rinsed with water; and equilibrated in 0.1 M HCl for 20 h. To exchange cations, the membranes were soaked in a 0.1 M solution of the appropriate alkali metal chloride. Diffusion coefficients were measured as a function of mole fraction of alkali metal and reported as the ratio of the diffusion coefficient for proton to alkali metal cation. Then from proton diffusion coefficient for measured by radio tracer $D_{H+}^{expand}$ of $3.5 \times 10^{-6}$ cm²/s, diffusion coefficients for the alkali metal exchanged membranes are calculated. For alkali metal cations, the diffusion coefficient decreases with $\dot{a}_{H2O}$ and $\Delta H_{hyd}$ as in Okada's work. However, the reported diffusion coefficients ranged from $2.9\times10^{-7}$ cm²/s for $Cs^+$ to $8.9\times10^{-7}$ cm²/s for $Li^+$ which are smaller than the diffusion coefficients found by Okada. Different membrane pretreatments and exchange protocols as well as measurement conditions and data analysis lead to differences in the final magnitudes reported, but the trends at the atomic level are the same.

Hongsirikarn and coworkers (*32*) reported values of cation selectivity over proton and conductivity λ for $H^+$, $Na^+$, and $NH^+_4$ and reported these with literature values of λ and $t_{H2O}$. Selectivity coefficients for ammonium and sodium over proton $K_{M+}^{H+}$ were corrected for solution activity but activity coefficients for the ions in Nafion were taken as 1. A conductivity cell was used to determine σ in deionized water at 25°C. Nafion 211 (25 _m thick, 1100 EQWT) was pretreated at 80 °C for 1 h with sequentially 3 % $H_2O_2$; 0.5 M HCl; and deionized water; rinsed and stored in deionized water. Cations were exchanged into the membranes as chloride salts where the chloride ion concentration in the electrolyte was maintained at 0.1 M. Equilibration extended over at least 10 days with shaking at room temperature and the electrolyte solution was changed periodically. For ammonium, $\dot{a}$ is 0.25 nm. The data are listed in Table 7. The conductivity of ammonium is the middle value but conductivity does not scale with λ, $t_{H2O}$, and $\dot{a}$. The paper notes that the conductivities for proton reported by Okada and coworkers are about twice as large as proton conductivities reported by several other groups.

**Table 6. Water Transference $t_{H2O}$, Mobilities $u_j$, and Diffusion Coefficients $D_j$ with λ for Nafion 115 and 117 Exchanged with H⁺ and Alkyl Metal Cations. For each ion in water, enthalpy of hydration $\Delta H_{hyd}$, mobility $u^{H_2O}$, and Kielland's ion size parameter $\mathring{a}^{H_2O}$ are reported as well as the ionic radius with no waters of hydration $r^{H_2O}$.**

| | $\lambda$ | $t_{H_2O}$ | $u$ $(10^{-4}\text{cm}^2/\text{Vs})$ | $D = \frac{RTu}{\lvert z\rvert F}$ $(10^{-5}\text{cm}^2/\text{s})$ |
|---|---|---|---|---|
| | 80 °C $H_2O_2$ pretreated[29] | | | |
| **$H^+$** | 22.0 | 2.6 | 19.4 | 4.98 |
| **$Li^+$** | 24.2 | 15.8 | 3.51 | 9.02 |
| **$Na^+$** | 20.0 | 9.2 | 4.50 | 11.6 |
| **$K^+$** | 13.1 | 4.9 | 3.45 | 8.86 |
| **$Rb^+$** | 10.9 | 5.5 | 2.59 | 6.65 |
| **$Cs^+$** | 9.7 | 5.2 | 2.20 | 5.65 |
| | 80 °C $H_2O$ pretreated[30] | | | |
| **$H^+$** | | 2.6±0.1 | | |
| **$Li^+$** | 19.5 | 15.6 | | |
| **$Na^+$** | 16.3 | 9.7 | | |
| **$K^+$** | 10.7 | 5.3 | | |
| **$Rb^+$** | 10.0 | 5.2 | | |
| **$Cs^+$** | 8.5 | 6.3 | | |

Properties of Monocations in Water

| | $\Delta H_{hyd}$ (kJ/mol) | $u^{H_2O}$ $(10^{-4}\text{cm}^2/\text{Vs})$ | $r^{H_2O}$ (nm) | $\mathring{a}^{H_2O}$ (nm)[27] |
|---|---|---|---|---|
| **$H^+$** | | 36.3 | | 0.90 |
| **$Li^+$** | 518 | 4.01 | 0.066 | 0.60 |
| **$Na^+$** | 397 | 5.19 | 0.095 | 0.45 |
| **$K^+$** | 317 | 7.62 | 0.133 | 0.30 |
| **$Rb^+$** | 288 | 7.92 | 0.148 | 0.25 |
| **$Cs^+$** | 259 | 7.96 | 0.169 | 0.25 |

**Table 7. Water Transference $t_{H_2O}$, Selectivity over Proton $K_{M+}^{H+}$, and Conductivity $\sigma$ for Nafion 211 Pretreated with $H_2O_2$ and HCl and Exchanged with $H^+$, $NH_4^+$, and $Na^+$ as a Function of λ (*32*)**

| | $\lambda$ | $t_{H_2O}$ | $K_{M^+}^{H^+}$ | $\sigma$ (/kΩcm) |
|---|---|---|---|---|
| $H^+$ | 20.1 | 2.6 | - | 105.9±1.7 |
| $NH_4^+$ | 14.0 | 7.0 | $2.5_5 \pm 0.5_6$ | 32.1±1.5 |
| $Na^+$ | 16.5 | 9.6 | 0.84±0.08 | 26.5±1.75 |

*1.8.3. Acidity*

Nafion is used as a catalyst because of its high acidity, chemical and thermal stability, and ease of separation due to Nafion insolubility. The $pK_a$ of trifluoro-methanesulfonic acid (triflic acid) of -12 (*33*) establishes $CF_3SO_3H$ as a superacid. A superacid liberates protons with activities greater than in anhydrous sulfuric acid. From the structure in Scheme 1, the sulfonate group is bound to the fluorocarbon backbone by a fluoroether linkage so that the sulfonate group is in the hydrated domain and the backbone is part of the fluorocarbon domain. Sulfonate in the water domain allows protons to be titrated and enables ionic conductivity. The sulfonate side chain and triflic acid are structurally similar and may share acidic properties. Whether sulfonic acid sites in Nafion are also superacids will depend on whether the limited water in the hydrated domains sustains the same proton activity as available for triflic acid in bulk water.

## 2. Activity, Electrochemical Potential, and Conductance: A Sketch of the Environment in Hydrated Domains of Nafion

The sketch of the environment inside the ionic domains of Nafion provides data to consider the electrochemical potential and activity in the hydrated domains. It is the activity and not the concentration of the ions that determines the properties and behaviors of ions in Nafion. This includes ionic conductivity. Gradients in the electrochemical potential describe conductance and transport.

### 2.1. A Count of Species inside (5 nm)$^3$ Hydrated Domains of Nafion

Nafion provides a unique opportunity to look at activity and electrochemical potential from "the other side". Normally, when looking at a bulk electrolyte solution, the number of solvent and solute molecules is almost beyond visualization. But within a (5 nm) $^3$ voxel of the hydrated domain of Nafion, the number of solvent and solute species is countable. In a (5 nm) $^3$ voxel of pure water, there are approximately 4200 water molecules. As other species are introduced to that volume of water, such as the sulfonates and cations, water is displaced from the fixed voxel volume. The amount of available solvent is decreased. Further, the water present may be associated with or bound to the ions. The water is then divided into free water and bound water. The free water behaves as a pure solvent but the bound water will move and derive chemical activity from its interaction with the ions. Thus, the activity of the solvent will be suppressed by interaction with the ions. Here, a crude view of what is in the (5 nm) $^3$ voxel of hydrated Nafion is estimated.

The common characteristic dimension found in experimental studies of well-hydrated Nafion is 5 nm (*8–10*). Consider a voxel of the hydrated domain embedded in fluorocarbon. Approximated the voxel as a cube 5.00 nm on a side so that the volume of the voxel $V_{vox} = (5.00 \times 10^{-7}\ cm)^3 = 1.25 \times 10^{-19}$ $cm^3$. Given $c^*_{aq}$ is the concentration of sulfonates and hydrated monocations in the voxel, the moles of each cations and sulfonates $moles_{vox} = c^*_{aq}V_{vox}$ and the number of each cations and sulfonates $N_{vox} = N_0 moles_{vox}$ are expressed. For well hydrated membranes, where $\rho_{meas}$ is 1.6, 1.7, and 1.8 g/cm3, $c^*_{aq}$ is 3.0, 4.0, and 5.4 mmol/cm$^3$ (Figure 2 and Equation 3).

Ions displace water in the voxel. Allow the volume of one water molecule $vol_{H20} = 18.0\ (g/mol)/N_0\rho_{aq} = 2{:}99\times10^{-23}\ cm^3$. The volume of the cations is set by the volume of the specific cation without waters of hydration $\upsilon_+$ and the number of waters of hydration, $h^+$. The volume of the cation without waters of hydration is cation specific and determined from the ionic radii $r_+$ as $\upsilon_+ = 4\pi r^2{}_+{+}/3$. Kielland (*27*) lists radii without hydration $r_+$ (pm) for alkyl metal cations as $Li^+$ 40, $Na^+$ 50, $K^+$ 80, $Rb^+$ 90, and $Cs^+$ 105; $H^+$ without waters of hydration is sufficiently small to be approximated as zero. The corresponding $\upsilon_+$ ($10^{-25}$ $cm^3$) are $H^+$ 0, $Li^+$ 2.7, $Na^+$ 5.2, $K^+$ 21, $Rb^+$ 31, and $Cs^+$ 49. The volume of cations occupied in the voxel is

$$Vol_{vox,+} = N_{vox}\left(h_+ vol_{H_2O} + v_+\right) \tag{9}$$

The sulfonate volume is more difficult to approximate because the extent to which the side chains extend into the hydrated volume is unclear. The sidechains are estimated as $l_{chain}$ of 1.2 nm long and the sulfonate diameter $d_{SO3-}$ as 0.3nm, which is similar to the $CF_2$ diameter 0.30 nm (*34*). From molecular dynamics simulations, the side chain extends about 0.7 nm into the hydrated domain and partially wraps a hydronium ion so that the side chain is at the fluorocarbon hydrated domain interface. At low hydration ($\lambda \leq 6$), the structure is different with side chains extending into the hydrated domain (*35*). Here, the volume of the sulfonate is taken as $4\pi(d_{SO3-}/2)^3/3 = 1.41\times10^{-23}$ $cm^3$. For the side chain approximated as a cylinder and 'chain in the hydrated domain, the volume of water that the side chain displaces is $v_{chain} = \pi(d_{SO3-}/2)^2 l_{chain} = 8.15\times10^{-23}$ $cm^3$. The volume occupied by the anions in the voxel is in two parts, the volume of a chain and the volume of the waters of hydration,

$$Vol_{vox,-} = N_{vox}\left[v_{chain} + h_- v_{H_2O}\right] = N_{vox}v_{chain} + h_- V_{vox,+}/h_+ \tag{10}$$

The volume of free (unbound) water in the voxel $V_{vox,H_2O}^{free}$ is set by the volume displaced by the chains and ions with water of hydration,

$$Vol_{vox,H_2O}^{free} = Vol_{vox} - Vol_{vox,+} - Vol_{vox,-} \tag{11}$$

$$= Vol_{vox} - N_{vox}\left(v_{chain} + (h_- + h_+)\, vol_{H_2O}\right) \tag{12}$$

The bound water in this simple sketch is $h_+ + h- = h$, so that the total volume of water in the voxel is

$$Vol_{vox,H_2O}^{total} = Vol_{vox,H_2O}^{free} + N_{vox} v_{H_2O} h \tag{13}$$

The volumes are converted to number of ions in the voxel by the same normalization: $N_{vox,+}=V_{vox,+}N0\rho_{aq}/MW(H_2O)$ and $N_{vox,+}=V_{vox,+}N0\rho_{aq}/MW(H_2O)$. Also,

$$N_{vox,H_2O}^{free} = Vol_{vox,H_2O}^{free}\frac{N_0\rho_{aq}}{MW(H_2O)} \tag{14}$$

$$N_{vox,H_2O}^{total} = Vol_{vox,H_2O}^{total}\frac{N_0\rho_{aq}}{MW(H_2O)} \tag{15}$$

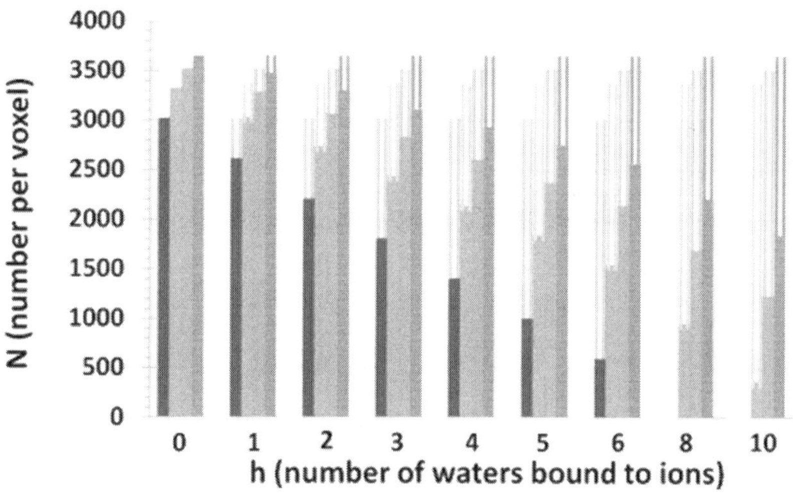


*Figure 5. For a (5 nm)³ voxel and ρ of sequentially 1.8 (blue), 1.7 (gold), 1.6 (olive), and 1.5 (blue grey) g/cm³, the calculated number waters in Nafion 1100 are shown solid bars versus h the number of waters that interact with the ions in the hydrated domain. The total number of waters in the voxel $N_{vox,H_2O}^{total}$ is fixed for a given density and represented as the high value, constant with h. The number of unbound waters $N_{vox,H_2O}^{free}$ equals $N_{vox,H_2O}^{total}$ for h = 0 and decreases as h increases. The open boxes represent the number of bound waters $N_{vox,H_2O}^{bound} = N_{vox,H_2O}^{total} - N_{vox,H_2O}^{free}$.*

The number of bound waters in the voxel $N_{vox,H_2O}^{bound} = N_{vox,H_2O}^{total} - N_{vox,H_2O}^{free}$. The number of waters varies mostly with $\rho_{meas}$ ($c_{aq}^*$) and $h$. A plot of $N_{vox,H_2O}^{free}$ and $N_{vox,H_2O}^{bound}$ and so $N_{vox,H_2O}^{total}$ is shown in Figure 5 for $\rho_{meas}$ of 1.8, 17, 1.6, and 1.5 g/cm³, densities consistent with well-hydrated Nafion. For $h = 0$, $N_{vox,H_2O}^{total} = N_{vox,H_2O}^{free}$ as shown in the Figure, whereas $\rho_{meas}$ decreases, the $N_{vox,H_2O}^{total}$ increases. The free waters $N_{vox,H_2O}^{free}$ are shown by the heavy bars and the bound waters $N_{vox,H_2O}^{bound}$ are shown by the narrow, capped bars. The total waters $N_{vox,H_2O}^{total}$ is fixed for a given $\rho_{meas}$ independent of $h$, as shown by the constant maxima at the highest value for each $\rho_{meas}$. As the number of waters bound or associated with the ions $h$ increases, $N_{vox,H_2O}^{free}$ decreases and $N_{vox,H_2O}^{bound}$ increases for each $\rho_{meas}$.

The fraction of free water is

$$f = \frac{N_{vox,H_2O}^{free}}{N_{vox,H_2O}^{total}} = \frac{Vol_{vox,H_2O}^{free}}{Vol_{vox,H_2O}^{total}} = \frac{Vol_{vox} - N_{vox}\left(v_{chain} + \left(h_{\_} + h_{+}\right) vol_{H_2O}\right)}{Vol_{vox} - N_{vox} v_{chain}} \quad (16)$$

$$= \frac{1 - N_0 c_{aq}^* \left(v_{chain} + \left(h_{\_} + h_{+}\right) vol_{H_2O}\right)}{1 - N_0 c_{aq}^* v_{chain}} \quad (17)$$

where the last term on the RHS arises through Equation 3, $N_{vox} = N_0 c_{aq}^* V_{vox}$. $N_{vox}$ depends on $c_{aq}^*$ and so $\rho_{meas}$. The number of waters per sulfonate, $\lambda$ varies with $\rho_{meas}$ and $c_{aq}^*$ as indicated in Figure 2 and calculated here as

$$\lambda = \frac{Vol_{vox,H_2O}^{total} \rho_{aq} N_0}{18.0 N_{vox}} = \frac{Vol_{vox,H_2O}^{total} \rho_{aq}}{18.0 c_{aq}^* V_{vox}} \quad (18)$$

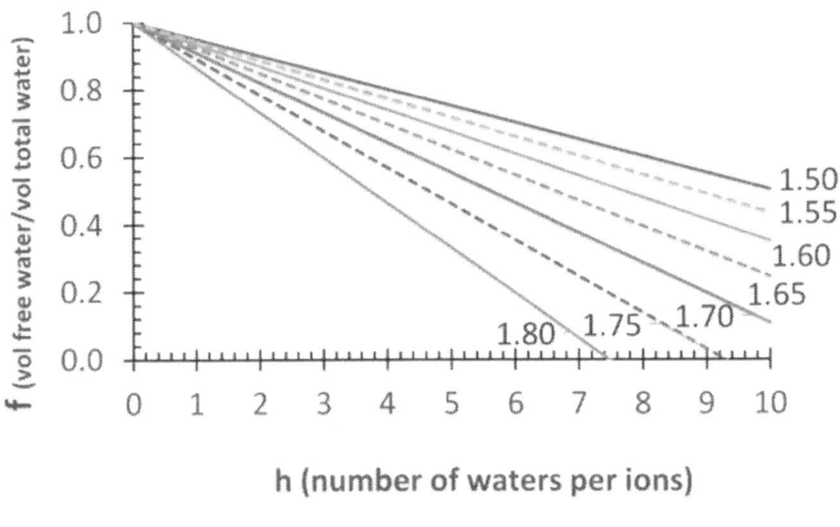


*Figure 6. The fraction of total waters in the (5 nm)³ voxel that are not associated ("bound") with the ions in the voxel, the fraction of free waters, f as a function of the total number of waters associated with the ions h. When h = 0, f= 1 as all waters are free.*

The variation in free water fraction $f$ with $\rho_{meas}$ and h is illustrated in Figure 6. For a given density $\rho_{meas}$, the fraction of free water f decreases as h increases. For $h = 0$, all densities converge to $f = 1$. For a given h, $f$ increases as $\rho_{meas}$ decreases. These results are consistent with denser, drier membranes that have fewer free waters because waters associate with the ions to help distribute charge density and shield electrostatic interactions.

The ions in Nafion associate with water to decrease the fraction of water that is free to behave as pure solvent. These effects are large because $c^*_{aq}$ is several molar, in even well hydrated Nafion. The ions affect the activity of the solvent. Further, Marcus models a decrease in dielectric constant with increasing electrolyte concentration (*18*), as consistent with measurements reported in Table 2 (*13*).

## 2.2. Impacts of Activity in Nafion

Electrochemical potential $\bar{\mu}_j^{\alpha}$ describes the equilibrium properties of species j in phase α in terms of the activity of the species $a_j^{\alpha}$, the standard electrochemical potential $\bar{\mu}_j^{\alpha,0}$, the charge on the species $z_j$ and the electrical potential φ in phase α about j. Faraday constant, F = 96485 C/mol.

$$\bar{\mu}_j^{\alpha} = \bar{\mu}_j^{\alpha,0} + RT \ln a_j^{\alpha} + z_j F \phi \tag{19}$$

The spatial gradient in the electrochemical potential describes transport of j. The activity $a^{\alpha}_j$ describes the behavior of j relative to the number density, concentration of j $c^{\alpha}_j$ in phase α through the activity coefficient, $\gamma^{\alpha}_j$.

$$a_j^{\alpha} = \gamma_j^{\alpha} c_j^{\alpha} \tag{20}$$

Activity coefficients are commonly calculated as by extended Debye Hückel equations that consider electrostatic interactions for ions of size $\dot{a}_j$ but extended Debye Hückel equations are limited to ionic strengths $I \leq 0.1$ M. From the calculations in Section 2.1, the ionic strength in Nafion for monocations is $c^*_{aq}$, which for well hydrated membrane is several molar and higher for dehydrated membranes. Extended Debye Hückel equations do not include impacts of electrostatic interactions on the activity of the solvent. The activity of a pure solvent such as pure water $a_W^0 = 1$. For low ionic strengths, this is a good approximation as the number of solvent molecules far exceeds the number of ions. As the ionic strength increases and water molecules are associated with the ions, then the fraction of solvent molecules that behave as a pure solvent are reduced and $a_w < 1$. Under conditions where the ionic strength is sufficient to decrease the activity of the solvent, the activity of the ions $a_j^{\alpha}$ tends to increase. At high ionic strength, $\gamma_j^{\alpha}$ can exceed 1.

Through activity effects, the electrochemical potential at high ionic strengths can be altered significantly. This will alter equilibrium properties as well as transport. Parameters that impact $\gamma^{\alpha}_j$ are parameters important in electrostatics charge $z_j$, size $\dot{a}_j$, and dielectric constant ε, as well as temperature T, ionic strength I, water content λ and the number of waters of hydration h. At I ≤ 1 M, activity $a_j^\alpha$ tends to increase with $z_j$, $a_j^\alpha$, I, and h. According to Y. Marcus (*18*), ε decreases as *I* increases. A decrease in ε below about 30 leads to an increase in ion pairing and other aggregation of charged species. These parameters encompass those identified by the various authors cited in Section 1. In Section 3, Physical Manipulation of Nafion Structure and Impacts on Properties, means to modify Nafion and its properties are divided into two groups: modification on length scale appropriated to establish changes in $a_j^\alpha$ and changes on a longer length scale that do not disrupt the matrix in a manner that alters $\alpha_j^\alpha$.

## 3. Physical Manipulation of Nafion Structure and Impacts on Properties

Manipulation of Nafion is scale dependent. If physical modification is on a length scale larger than the nanostructure, the mechanical properties of Nafion may be affected but electrochemical potential and associated properties such as local ionic conductivity will not be affected. If physical manipulation approaches characteristic dimensions of the nanostructure, the electrochemical potential and activity will be impacted and so the associated properties such as conductivity. The discussion is divided into macroscopic changes on longer length scales, composites that bridge the length scales between macroscopic and nanoscopic, and nanoscopic changes designed to impact activity in Nafion. The last subsection discusses fuel cell applications of Nafion.

### 3.1. Macroscopic Changes to Nafion

Macroscopic changes to Nafion include effects established on length scales too large to affect $\overline{\mu}_j^\alpha$ .and $a_j^\alpha$. The examples considered here are bilayers, multilayers, graded layers, and Nafion on a coarse support.

#### *3.1.1. Bilayer*

A bilayer might be formed by casting a polymer layer distinct from Nafion across the planar area of Nafion membrane. A layer may also be made by chemical change to the Nafion surface. Adlayers preserve Nafion bulk properties, but allow modification of interfacial properties. The water or solvent transport properties of Nafion may be limited by a polymer overlayer. High rates of proton conduction are driven by, for example, Grottius conduction (*36*). In Grottius conduction, proton transport is facilitated by propagation of protons by hydrogen bonding between adjacent solvent molecules. Protons walk from hydronium to water. On application of a macroscopic layer with properties distinct from Nafion, conduction across the Nafion surface may be increased or reduced. For example, application of a polyallyl amine layer to the outer surface of the Nafion membrane is shown to reduce solvent crossover associated with proton motion in direct methanol fuel cells (*37*). Other barriers might be envisioned to slow solvent transport, such as a monolayer of surfactant or extraction of a large zwitterion at the Nafion electrolyte interface.

#### *3.1.2. Multilayer and Graded Layers*

On extrapolation from bilayers, multilayers may be fabricated. Multilayers allow different properties to be established in layers in direct contact. For example, different equivalent weights of ion exchange polymer might control water transport. Interfaces of cation and anion exchangers would establish interesting transport regimes and perhaps allow throttling or facilitation of solvent and ion motion. Capacitive interfaces might be established.

Graded layers are an extrapolation of multilayers wherein the layers transition more smoothly so that a clear demarcation between the layers is less apparent. Such layers may control macroscopic transport properties (*38*)

#### *3.1.3. Coarse Support*

Nafion is commercially available supported on coarse Teflon mesh. While such a structure will have limited impact at a molecular level, the coarse support allows for greater mechanical stability and a membrane that can tolerate greater pressure differentials. This is a macroscopic modification that changes the bulk characteristics of the Nafion membrane without changing local transport and equilibrium properties as dependent on $\overline{\mu}_j^{\alpha}$ and $a_j^{\alpha}$

### 3.2. Composites and Disruption to the Nanostructure

Composites can be formed by casting Nafion suspension onto a structured substrate. As the characteristic dimensions of substrate approach the characteristic dimensions of Nafion, there is disruption in the nanostructure that changes $\overline{\mu}_j^{\alpha}$ and $a_j^{\alpha}$ and so the characteristics of Nafion at the nanoscopic level. Investigation of composites as a function of characteristic dimension and inert substrate geometry have been undertaken for Nafion inside cylinders and adsorbed onto microspheres. The pivotal characteristic dimension was found to be available surface area of the inert substrate normalized by the volume of Nafion in the composite, SA / Vol. As SA / Vol increased larger disruptions in the nanostructure and therefore the transport properties of Nafion were found. As interfaces in a composite increase, the properties of the composite vary (*39*, *40*).

Similar variations in the transport of transition metal complexes and organic redox probes where found for composites formed by absorption of Nafion into the cylindrical pores of neutron track etched membranes34 and for Nafion adsorbed onto polystyrene microspheres of uniform dimension (*41*, *42*). Flux of redox probes through composites of Nafion in cylindrical pores was measured by rotating disk voltammetry, where increases in flux were observed as SA / Vol increased. Transport was partitioned into an interfacial, facile transport zone and transport typical of bulk Nafion. This model of two domains applied to pores with characteristic dimensions $\geq$ 30 nm, but the pattern broke down for yet smaller pores. It was determined that in 15 nm pores, there was insufficient space for the sidechains of Nafion to pack in the same interfacial monolayer structure that was accessible in the larger pores.

For Nafion adsorbed on inert polystyrene microbeads, similar behavior was observed. For microsphere composites, flux of redox probes through the composites as measured by rotating disk voltammetry, increased with SA/Vol. Flux increased as $[\text{SA/Vol 1}]^{\mu}$, but whereas $\mu = 1$ for the cylinder composites, $\mu = 0.78$ for the microsphere composites. The non-integer exponent is characteristic of fractal behavior. A fractal dependence is observed when properties appear self-similar on different length scales, which manifests mathematically as a power law relationship where the exponent $\mu$ is non-integer. Interestingly, titration of the available sulfonic acid sites by the method outlined in Section 1.3 yields an increase in ion exchange capacity that also scales with $[\text{SA/Vol}]^{\mu}$ where $\mu = 0.78$. The increased ion exchange capacity is consistent with a monolayer of sidechains established at the surface of the microbeads such that the sulfonates are more readily assessable for proton exchange. Interactions of Nafion side chains decrease the energy needed to establish the interfacial monolayer. Thus, at short length scales (i.e., large SA/Vol), the structure of Nafion is disrupted and the properties of Nafion at the nanoscopic level are altered as measured by changes in transport.

Whether or not a composite disrupts nanostructure and $\overline{\mu}_j^{\alpha}$ and $a_j^{\alpha}$, thereby to impact Nafion properties, depends on length scales that approach Nafion characteristic length (~5 nm). A generalized, quantitative approach to interactions of ion exchange polymers with surfaces is outlined in Reference (*43*).

### *3.3. Nanoscopic and Molecular Changes to Nafion: Impacts on $\overline{\mu}_j^{\alpha}$ and $a_j^{\alpha}$*

Here, changes to Nafion that leads to changes in $\overline{\mu}_j^{\alpha}$ and $a_j^{\alpha}$ and so the nanoscopic to molecular and ionic properties of Nafion are outlined.

#### *3.3.1. Change Water Content*

The activity of the cation $a_j^{\alpha}$ in Nafion is highly dependent on water content, $\lambda$, because solvent activity is impacted by $\lambda$, in the highly concentrated electrolyte matrix of Nafion. Water content can be changed in several ways. The cation size and charge as well as the number of waters of hydration for the ion will impact water content. Based on selectivity of the cation over proton (Section 1.8.1), different ratios of electrolytes may control water content in the membrane. It is important to consider the environment where Nafion will be used. Control of water content is easier for Nafion bathed in electrolyte where equilibration maintains concentrations and for Nafion used in high humidity environments where ions cannot escape the membrane. When Nafion is subject to ion flux in a device such as a fuel cell, the steady cation flux through the cell will eventually displace any other cations extracted into Nafion by pretreatment.

Membrane pretreatment, as illustrated in various cases in Section 1, can impact the water content of the .films. For example, the membrane may be treated at elevated temperatures in various solutions that include water, hydrogen peroxide, and strong acid such as sulfuric acid.

These expanded forms membrane are not necessarily equilibrium nanostructures and can be anticipated to dissipate over time, however on the timescale of the measurements outlined in Section 1, the expanded forms are stable. For continuous operation in electrochemical devices such as fuel cells, expanded forms of the membrane are known to lose water eventually.

Water content may be impacted by exchange of cationic surfactants into Nafion. Nafion modified with alkyl ammonium cations has different activity and water content than Nafion exchanged with proton and inorganic cations (*44*).

The glass transition temperature $T_g$ for Nafion is a strong function of the exchange cation, as in Section 1.6.1. For cast films of Nafion, the structure may differ from that of membranes. Heating Nafion above $T_g$ has been proposed and investigated as a means to change the structure of proton exchanged Nafion films, as is consistent with measured flux of transition metal complexes (*42*, *45*). Conversely, freezing Nafion changes the transport properties as a function of temperature (*46*).

### *3.3.2. Change Solvent*

As in Section 1.7.1, different solvents with different properties swell Nafion to varying extents. Mixed solvents will also modify the properties of Nafion at the level of $\overline{\mu}_j^{\alpha}$ and $a_j^{\alpha}$. Solvent will also impact the mechanical and likely thermal properties of Nafion. As in Tables 6 and 7, water transference $t_{H2O}$ is altered in direct reformation fuel cells on introduction of alcoholic fuels with water. Grottius conduction and water drag increase crossover of solvent from the anode to the cathode. Autoprotolysis and acid base properties are significant factors in the increased solvent crossover.

### *3.3.3 Nanomaterial Composites and Conductivity Additives*

Composites of Nafion have been formed with various nanoparticles and materials to alter properties. Solid acid conductors and carbon nanotubes are demonstrated to alter transport properties and fuel cell performance (*47*).

For example; Nafion doped with difluoromethanediphosphonic acid exhibits enhanced conductivity, greater thermal stability, and increased water retention in low relative humidity (RH) environments (*48*).

### *3.3.4. Cocasting Polymers*

Changes in $\overline{\mu}_j^{\alpha}$ and $a_j^{\alpha}$ can be introduced by cocasting Nafion suspension with another polymer. Cocast matrixes alter water content as well as electrostatic interactions in the hydrated domain. Cocast films may change hydrophobicity. When a cationic polymer such as polyaniline or poly(allyl amine) is cocast with Nafion, electrostatic interactions between the cationic nitrogens and anionic sulfonates occur. Such electrostatic interactions may reduce the net potential energy of the matrix and thereby change water content. The use of hydrophobic perfluorinated polymers and hydrophilic polyalcohols may change the water content and mechanical properties of the cocast films.

For cocast polymers where there are no chemical bonds formed, the stability of the .lm relies on entanglement of the two different polymers. Pyrrole introduced to Nafion 117 was chemically oxidized to form polypyrrole embedded in Nafion (*49, 50*). The resulting composite of an ion and an electron conducting polymer exhibited enhanced cation transport and reduced methanol crossover.

### *3.3.5. Chemical Reaction To Modify Perfluorosulfonate*

Chemical reactions with Nafion can change the nanostructured properties and chemistries of Nafion. However, there are a few known bond-forming reactions likely to occur between either the polymer backbone or its pendant sidechains. One possibility is to exploit a photodimerization process used in crystal engineering where coordination bonds involving organosulfonants and argentophilic forces combine to build an extended supramolecular structure (*51*). In a second paper, a cationic diazo resin serves as a polycation that is electrostatically bound to the sulfonates of polystyrene sulfonate. A covalent bond is formed on irradiation of the photoreactive diazo resin. The resulting structure is more rigid and expanded than the ionic matrix (*52*). Both reactions were undertaken with hydrocarbon, not fluorocarbon substrates, although a photodimerization of a fluorocarbon with an argentophilic reagent is reported (*53*). Whether similar reactions are possible with fluorocarbons is not known, but should such reactions be possible, permanently expanded structures with tailored, Nafion-like properties may result. Expanded structures hold more solvent that can impact mechanical properties and conductivity.

### 3.4. Fuel Cell Specific Manipulation of Nafion

Several types of low-temperature fuel cells are based on Nafion separators. Nafion is a common separator in proton exchange membrane (PEM) fuel cells. In the majority of fuel cells, the oxidant is oxygen, either pure oxygen or oxygen drawn from the air. The fuels are commonly hydrogen, reformation derived from oxidation of alcohols, formic acid, and low molecular weight alcohols. The various fuels differ somewhat in the characteristics of the Nafion needed for good fuel cell performance. The discussion here is largely on hydrogen fuel cells with comments about other fuels as appropriate.

#### *3.4.1. Membrane*

1100 equivalent weight Nafion of an appropriate thickness is typically employed in PEM fuel cells. Membrane thickness is important because a balance between mechanical robustness and low resistance must be achieved. The thicker the membrane, the less likely a perforation in the membrane will lead to mixing of hydrogen and oxygen at a platinum catalyst and so to catastrophic failure. But, as the membrane becomes thicker, fuel cell performance degrades, in part because of increased resistive losses. For example, if the commonly employed separator Nafion 117 (a thickness of 0.007 in or 180 μm) has resistance of even 0.1Ω in the direction of current .flow and the current is 1 A, then the voltage drop (overpotential) under high current conditions is 0.1 V.

The theoretical voltage output for a hydrogen oxygen fuel cell at unit pressure is 1.23 V and the practical voltage maximum is ~ 0.5 V, so that a 0.1 V loss to membrane resistance is not acceptable. By appropriate control of membrane hydration, mass transport resistance drops can be minimized. As membranes dehydrate, λ decreases and resistance increases.

A challenge for effective hydrogen fuel cells is membrane stability under heavy duty cycles. Nafion membranes are sufficiently stable for steady energy output, but under heavy duty cycles where the load and power varies significantly, Nafion membranes degrade rapidly. One possible source of degradation is the generation of peroxide intermediates. At high current densities, electron transfer kinetics coupled with mass transport limitations lead to some generation of peroxides rather than water. Introduction of Nafion suspension into the membrane electrode assembly (MEA, catalyst layer) mitigates some mass transport and kinetic limitations. Effective catalysis to water for heavy duty cycles is critical.

*Pretreatment*

Various membranes pretreatment are recommended when forming a fuel cell membrane electrode assembly. As above, this includes boiling water, acid, and peroxide. Although this increases λ when the fuel cell is first formed, under continued fuel cell operation, the added water will be lost. In Figure 4, the red circles correspond to dioxane, propylene carbonate, and tetrahydrofuran. These cyclic oxygenated solvents swell Nafion to about the same extent as water and yet may differ in properties of proton transport, conductivity, and water transference. For other solvents, changes in conductivity must be balanced against possible passivation of the electrocatalyst.

*Crossover*

Crossover is a problem that arises in fuel cells once in operation. For a hydrogen fuel cell, hydrogen is oxidized at the anode to produce protons. The protons move across the Nafion separator to combine with oxygen and electrons at the cathode. Because waters are transferred with the conducted protons, water tends to be removed from the anode and collected at the cathode. Control of the water content in the fuel stream is important to prevent poor distribution of water within the membrane.

In direct reformation fuel cells (DMFCs) where an organic fuel such as methanol undergoes oxidation, crossover is a significantly larger problem. From the swelling data in Figure 4, membrane swells twice as much in alcohols. The stoichiometric fuel mixture for the anode is 50% methanol in water. Methanol and water are infinitely miscible and both undergo autoprotolysis reactions. Further, conduction of protons occurs through acid base reactions between methanol and water. In swollen membranes, conducting ions carry both water and methanol across the cell, so crossover is a larger problem in direct methanol fuel cells (DMFCs). A bilayer composite of poly(allyl amine) over Nafion limits Grottius conduction across the membrane interface and reduces solvent crossover (*37*). Control of the electrochemical potential within the hydrated domain or at the surface of Nafion can limit crossover.

### *3.4.2. Membrane Electrode Assemblies (MEAs)*

Membrane electrode assemblies (MEAs) are catalyst layers on current collectors that are formed on either side of the Nafion separator. These are complex structures designed to provide transport of gases to the catalyst site and removal of ionic products and water after electron transfer reactions occur. This necessitates a three phase contact where gas, water, and electron conductor meets at the catalyst site. To enhance gas solubility and to promote ion transport, Nafion suspension is introduced to the catalyst layer of the MEA (*54*). Addition of Nafion suspension to MEAs manifests as a better three phase contact through changes in the electrochemical potential about the catalyst site.

## 4. Summary

Properties of Nafion are known to vary with water content and the cations that charge neutralize the sulfonates. Some properties important in applications of Nafion are considered in terms of water content $\lambda$ and exchanged cation.

Physical parameters of density, equivalent weight, nanostructure, and dielectric properties evolve a sketch of the nanoenvironment in the hydrated domains of Nafion. Thermodynamic descriptors of electrochemical potential and activity are qualitatively noted. Based on length scales of micro- and nano-structure, means to modify the properties of Nafion are outlined. The focus is manip- ulation of Nafion structure by physical rather than by chemical means. At the nanoscale where the structural changes are on lengths comparable to the characteristic dimensions of the Nafion nanostructure, thermodynamic and electrostatic properties can be disrupted to change equilibrium and transport characteristic. Changes in micro- and macro-structure alter properties but on length scales well above the nanostructure. A brief discussion of Nafion in fuel cells and the role of water content and structural modification are presented.

## Acknowledgments

The financial assistance of the National Science Foundation (0809745 and 1309366) and the Obermann Fellowship at the University of Iowa are both gratefully acknowledged.

## References

1. Grot, W. *Fluorinated Ionomers*, 2nd ed.; Plastic Design Library Fluorocarbon Series; Elsevier: New York, 2011.
2. Yeager, H. Transport Properties of Perfluorosulfonated Polymer Membranes. In *Perfluorinated Ionomer Membranes*; Yeager, H., Eisenberg, A., Eds.; ACS Symposium Series 180; American Chemical Society: Washington, DC, 1982.
3. Leddy, J.; Vanderborgh, N. Modeling of the Electrode/Ionomer Interface. In *Proceedings of the Symposium on Diaphragms, Separators, and Ion Ex-change Membranes*; White, R., Kinoshita, K., Burney, H., Eds.; The Electrochemical Society: Pennington, NJ, 1986.
4. Garcia-Fresnadillo, D.; Marazuela, M. D.; Moreno-Bondi, M. C.; Orellana, G. *Langmuir* **1999**, *15*, 6451–6459.
5. Eisenberg, A.; Yeager, H. L. *Perfluorinated Ionomer Membranes*; ACS Symposium Series 180; American Chemical Society: Washington, DC, 1982.
6. Mauritz, K. A.; Moore, R. B. *Chem. Rev.* **2004**, *104*, 4535–4585.
7. Mauritz, K.; Hora, C.; Hopfinger, A. Theoretical Model for the Structures of Ionomers: Application to Nafion Materials. In *Ions in Polymers, Advances in Chemistry Series*; Eisenberg, A., Ed.; ACS Symposium Series 187; American Chemical Society: Washington, DC, 1980.
8. Gierke, T.; Hsu, W. The Cluster-Network Model of Ion Clustering in Perfluorosulfonated Membranes. In *Perfluorinated Ionomer Membranes*; Yeager, H., Eisenberg, A., Eds.; ACS Symposium Series 180; American Chemical Society: Washington, DC, 1982.

9. Gierke, T.; Munn, G.; Wilson, F. *J. Polym. Sci., Polym. Phys. Ed.* **1981**, *19*, 1687–1704.
10. Allen, F. I.; Comolli, L. R.; Kusoglu, A.; Modestino, M. A.; Minor, A. M.; Weber, A. Z. *ACS Macro Lett.* **2014**, *4*, 1–5.
11. Oberbroekling, K. J.; Dunwoody, D. C.; Minteer, S. D.; Leddy, J. *Anal. Chem.* **2002**, *74*, 4794–4799.
12. Zook, L. A.; Leddy, J. *Anal. Chem.* **1996**, *68*, 3793–3796.
13. Paddison, S. J.; Reagor, D. W.; Zawodinski, T. A. *J. Electroanal. Chem.* **1998**, *459*, 91–97.
14. Takamatsu, T.; Eisenberg, A. *J. Appl. Polym. Sci.* **1979**, *24*, 2221–2235.
15. Kielland, J. *J. Am. Chem. Soc.* **1937**, *59*, 1675–1678.
16. Bard, A.; Faulkner, L. *Electrochemical Methods*, 2nd ed.; John Wiley & Sons, Inc.: New York, 2001.
17. Chen, T.-Y.; Leddy, J. *Langmuir* **2000**, *16*, 2866–2871.
18. Marcus, Y. *J. Solution Chem.* **2013**, *42*, 2354–2363.
19. Yeo, S. C.; Eisenberg, A. *J. Appl. Polym. Sci.* **1977**, *21*, 875–898.
20. Feldheim, D. L.; Lawson, D. R.; Martin, C. R. *J. Polym. Sci., Part B: Polym. Phys.* **1993**, *31*, 953–957.
21. Kundu, S.; Simon, L. C.; Fowler, M.; Grot, S. *Polymer* **2005**, *46*, 11707–11715.
22. Gebel, G.; Aldebertt, P.; Pineri, M. *Polymer* **1993**, *34*, 333–339.
23. Marcus, Y. *J. Solution Chem.* **1992**, *21*, 1217–1230.
24. Ogumi, Z.; Kuroe, T.; Ichiro Takehara, Z. *J. Electrochem. Soc.* **1985**, *132*, 2601–2605.
25. Steck, A.; Yeager, H. L. *Anal. Chem.* **1980**, *52*, 1215–1218.
26. Yeager, H. L.; Steck, A. *Anal. Chem.* **1979**, *51*, 862–865.
27. Kielland, J. *J. Am. Chem. Soc.* **1937**, *59*, 1675–1678.
28. Evans, C. E.; Noble, R. D.; Nazeri-Thompson, S.; Nazeri, B.; Koval, C. A. *J. Membr. Sci.* **2006**, *279*, 521–528.
29. Okada, T.; Satou, H.; Okuno, M.; Yuasa, M. *J. Phys. Chem. B* **2002**, *106*, 1267–1273.
30. Xie, G.; Okada, T. *J. Electrochem Soc.* **1995**, *142*, 3057–3062.
31. Samec, Z.; Trojanek, A.; Langmaier, J.; Samcova, E. *J. Electrochem Soc.* **1997**, *144*, 4236–4242.
32. Hongsirikarna, K.; Jr, J. G. G.; Greenway, S.; Creager, S. *J. Power Sources* **2010**, *195*, 7213–7220.
33. Raamat, E.; Kaupmees, K.; Ovsjannikov, G.; Trummal, A.; Kuett, A.; Saame, J.; Koppel, I.; Kaljurand, I.; Lipping, L.; Rodima, T.; Pihl, V.; Koppel, I. A.; Leiot, I. *J. Phys. Org. Chem.* **2013**, *26*, 162–170.
34. Fang, Y.; Leddy, J. *J. Phys. Chem.* **1995**, *99*, 6064–6073.
35. Sunda, A. P.; Venkatnathan, A. *J. Mater. Chem. A* **2013**, *1*, 557–569.
36. Cukierman, S. *Biochim. Biophys. Acta* **2006**, *1757*, 876–885.
37. Dunwoody, D. C.; Leddy, J. *Poly allylamine Nafion Bilayer for Reduced Methanol Crossover*; 1997
38. Knoche, K. L. *Density Gradient Films, Lanthanide Electrochemistry, and Magnetic Field Effects on Hydrogen Evolution, Oxygen Reduction, and*

*Lanthanide Electrochemistry*, Ph.D. Thesis, University of Iowa, Iowa City, IA, 2015.

39. Leddy, J. *Langmuir* **1999**, *15*, 710–716.
40. Zook, L. A.; Amarasinghe, S.; Fang, Y.; Leddy, J. Nano- and MicroStructured Materials: The Role of the Interface in Tailoring Transport. In *Fundamentals of Materials Research Series*; Pinnavaia, T., Thorpe, M., Eds.; Plenum: New York, 1995; Vol. II.
41. Zook, L. A.; Leddy, J. *J. Phys. Chem. B* **1998**, *102*, 10013–10019.
42. Zook, L. A.; Leddy, J. Nafion - Cold Cast, Thermally Processed, and Commercial Films: Comparison of Density, Water Content, and Equivalent Weight. In *New Directions in Electroanalytical Chemistry II*; Leddy, J., Vanýsek, P., Porter, M., Eds.; The Electrochemical Society, Inc.: Pennington, NJ, 1999; Vol. 99-5.
43. Tinoco, F.; Leddy, J. *J. Phys. Chem.* **1996**, *100*, 14762–14768.
44. Klotzbach, T.; Watt, M.; Ansari, Y.; Minteer, S. D. *J. Membr. Sci.* **2006**, *282*–276.
45. Zook, L. A. *Morphological Modification of Nafion For Improved Electrochemical Flux*; Ph.D. Thesis, University of Iowa, Iowa City, IA, 1996..
46. Dunwoody, D.; Leddy, J. *ECS Transactions* **2006**, *1*, 375–381.
47. Sahu, A.; Pitchumani, S.; Sridhar, P.; Shukla, A. *Bull. Mater. Sci.* **2009**, *32*, 285–294.
48. Montoneri, E.; Boffa, V.; Bottigliengo, S.; Casciola, M.; Sganappa, M.; Marigo, A.; Speranza, G.; Minati, L.; Torrengo, S.; Alberti, G.; Bertinetti, L. *Solid State Ionics*, *181*, 578–585.
49. Easton, E. B.; Langsdorf, B. L.; Hughes, J. A.; Sultan, J.; Qi, Z.; Kaufman, A.; Pickup, P. G. *J. Electrochem Soc.* **2003**, *150*, C735–C739.
50. Xu, F.; Kameche, M.; Innocent, C. *J. Membr. Sci. Technol.* **2012**, *1*, 108–116.
51. Elacqua, E.; Groeneman, R. H.; Reinheimer, E. W.; Bucar, D.-K.; MacGillivray, L. R. *Supramol. Chem.* **2014**, *26*, 207–213.
52. Zhang, X.; Wu, T.; Sun, J.; Shen, J. *Colloids Surf., A* **2002**, *198-200*, 469–442.
53. Sinnwell, M. A.; Baltrusaitis, J.; MacGillivray, L. R. *Cryst. Growth Des.* **2015**, *15*, 538–541.
54. Raistrick, I. D. Modified gas diffusion electrode for proton exchange membrane fuel cells. In *Proceedings of the Symposium on Diaphragms, Separators, and Ion Exchange Membranes*; Zee, J. W. V., White, R. E., Kinoshita, K., Burney, H., Eds.; The Electrochemical Society, Inc.: Pennington, NJ, 1986; Vol. 86-13.

# Section III

## Chapter 7

# Catalytically Active Nanomaterials for Electrochemical Energy Generation and Storage

**P. Kolla and A. Smirnova***

**Department of Chemistry and Applied Biological Sciences, South Dakota School of Mines & Technology, Rapid City, South Dakota 57701, United States**
***E-mail: Alevtina.Smirnova@sdsmt.edu**

Electrochemical energy conversion and storage technologies such as fuel cells, electrolysers, and batteries play a significant role in sustainable energy development by decreasing consumption of fossil fuels and improving the efficiency of current energy resources. However, understanding the kinetic mechanisms on catalytic surfaces and criteria for selection of inexpensive non-precious catalysts in a specific electrochemical environment is crucial to make these technologies commercially viable. Recent advances in nanotechnology have enabled the development of catalyst nanomaterials for electrochemical energy generation and storage. A higher level of understanding in terms of the catalyst nanomaterials, their synthesis, and electrochemical kinetics at nanoscale has been accomplished. In this context, the most challenging electrochemical reactions, the reaction mechanisms on the catalyst surfaces, and the recent approaches related to different types of catalytically active nanomaterials are discussed. The emphasis is given on three electrochemical reactions namely: Oxygen Reduction Reaction (ORR), Oxygen Evolution Reaction (OER), and Methanol Oxidation Reaction (MOR). A comprehensive discussion of the recent approaches in the state-of-the art catalysis and novel catalytic nanomaterials is highlighted in regard to the corresponding electrochemical reaction mechanisms.

# Introduction

Consumption of fossil fuels is continuously increasing around the world due to the 1.2 – 2.0% population growth per year. The world population is expected to reach 12 billion by 2050 and consequently, the global energy demands are likely to grow by as much as an order of magnitude. Energy security, economic growth, and environmental protection are significantly affected by the fact that ~80% of total fossil fuel reserves are concentrated in a very few (~6-8) countries. In addition, an increase in the fossil fuel consumption further increases the effect of global warming due to excessive carbon dioxide release into the atmosphere. According to the US Environmental Protection Agency, the USA produces five times the world average emissions. Specifically, the motor vehicles in the United States account for 78% of $CO_2$ emissions (*1*). In this regard, enhanced sustainable energy technologies are required to improve the efficiency of the currently existing technologies and provide a continuous supply of energy that can replace fossil fuel based energy sources without compromising the environment and satisfy the energy demands of the current and future generations (*2*).

Renewable energy technologies, such as solar, wind, tidal, and geothermal, have potential to provide global energy demands. However, it is impossible to harvest energy continuously as they are limited by climate changes, portability, the location, and the day-and-night cycles. Therefore, efficient energy storage and conversion devices are vital to provide continuous energy supply on demand. For example, solar energy can be used to generate hydrogen in the chemical reaction of water splitting. The stored hydrogen can fuel a hydrogen fuel cell to generate continuous supply of electricity whenever it is necessary. Fuel cells, electrolysers, batteries, and supercapacitors are the examples of the energy conversion and storage technologies that are considered to be sustainable. However, higher energy density (Wh/Kg) and/or power density (W/kg) of these devices are required in order to replace the conventional fossil fuel based technologies. Therefore, hybrid technologies comprising of two or more electrochemical storage/conversion devices gained significant attention in the recent years. For instance, when fuel cell with high specific energy density is combined with supercapacitor possessing high power density, this system outperforms an internal combustion engine with much higher conversion efficiency. Nonetheless, the performance, efficiency, and commercial feasibility of these devices should be improved which is intimately related to understanding of the kinetic mechanisms, materials choices, and their design pertaining to the electrochemical reactions of interest.

Recent advancements in nanotechnology facilitate the nanoengineering approaches of different types of nanomaterials having better control of their structure, chemical composition and morphology at nanoscale. These nanotechnological approaches further enable the design of the catalysts and support nanomaterials with higher efficiency, advanced selectivity, and desired electrochemical behavior by controlling their surface energy, electrochemical surface area, and physical/chemical properties at atomic or molecular level. Therefore, broad understanding of the electrochemical mechanisms on the catalitic surfaces, the state-of-the-art catalysis related to the electrochemical processes, and the recent developments in different types of catalyst nanomaterials is required. In this regard, challenging electrochemical reactions such as methanol oxidation, oxygen reduction, and oxygen evolution reactions as well as their state-of-the-art catalysis mechanisms are presented. The emphasis is made on understanding the approaches associated with the catalyst/support nanomaterials based on noble and non-noble metals, metal alloys, complex metal oxides, carbon nanostructures, and the related electrochemical conversion reactions.

# 1. Oxygen Reduction Reactions

## 1.1. Electrochemical Mechanism of the Oxygen Reduction Reaction

The reduction reactions involving oxygen are the most common yet challenging electrochemical reactions that are important for many natural biological processes as well as for electrochemical energy conversion and storage devices such as polymer electrolyte fuel cells and metal-air batteries. A typical oxygen reduction reaction (ORR) takes place in aqueous acidic or alkaline solution involving two or four electron transfer (Figure 1) and a number of different catalytic pathways (Figure 2a). It is necessary to note, that in non-aqueous solvents, especially in relation to lithium-air battery cathodes, the reversible oxygen reduction potential strongly depends on the electrode material and the solvent. In aprotic solvents, the ORR in presence of the catalytically active metals, for example Pd, Pt, Au and Ru (*3*), highlights the metal-oxygen bond as one of the key factors that governs the catalytic activity. Weak surface bonds with adsorbed oxygen on carbon and gold surfaces are the limiting factors in electron transfer resulting in lithium superoxide formation: $Li^{+}+O_2+e^{-}\rightarrow LiO_2$. If these bonds are weak, $Li_2O_2$ is the major reaction product, while for catalysts characterized by strong interaction with adsorbed oxygen, contribution of $Li_2O$ is dominating.

The selectivity of the ORR toward two or four electron transfer is one of the most important factors that is primarily defined by the area of application. For example, the ORR in a polymer electrolyte fuel cell (PEMFCs) four-electron pathway leading to water ($H_2O$) formation is preferable in comparison to the two-electron mechanism resulting in the formation of the hydrogen peroxide ($H_2O_2$) (Figure 2a). The peroxide species formed during two-electron transfer have high reduction potential of 1.76V vs. standard hydrogen electrode (SHE, Figure 1). As a result, the high oxidative nature of $H_2O_2$ species leads to chemical deterioration of carbon supports and electrolyte membranes within a membrane electrode assembly (MEA) of the PEMFC (*4*).

The four-electron pathway is observed for highly active catalysts such as Pt, Pd, Rh, and Au, or their allys which are capable of breaking the O-O bond via dissociative oxygen reduction mechanism. On contrary to the four-electron transfer, the two-electron pathway is observed for the materials with low catalytic activity such as carbon materials and metal oxides (*5*). However, some metals oxides, for example manganese (IV) oxide is known to catalyze a chemical disproportionation reaction of peroxide to hydroxyl ions: $2HO_2^{-} \rightarrow 2OH^{-} + O_2$ with overall process similar to the four-electron ORR.

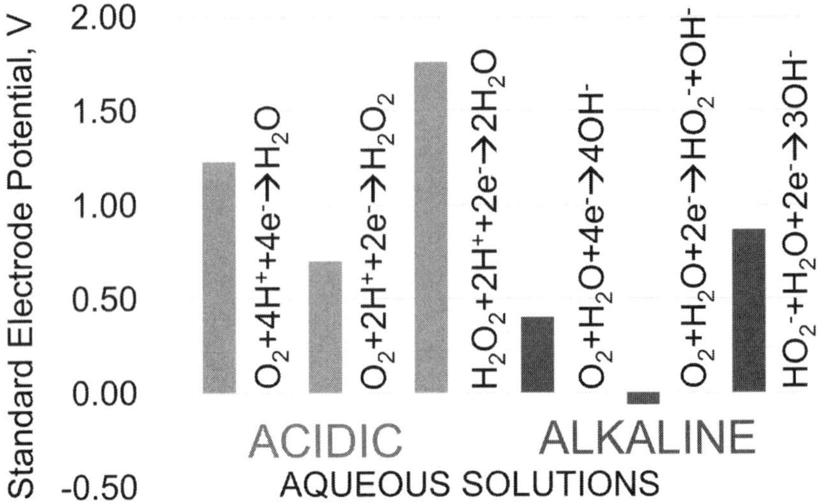


*Figure 1. Standard electrode potentials and the corresponding electrochemical reduction reactions in acidic and alkaline aqueous solutions involving oxygen.*

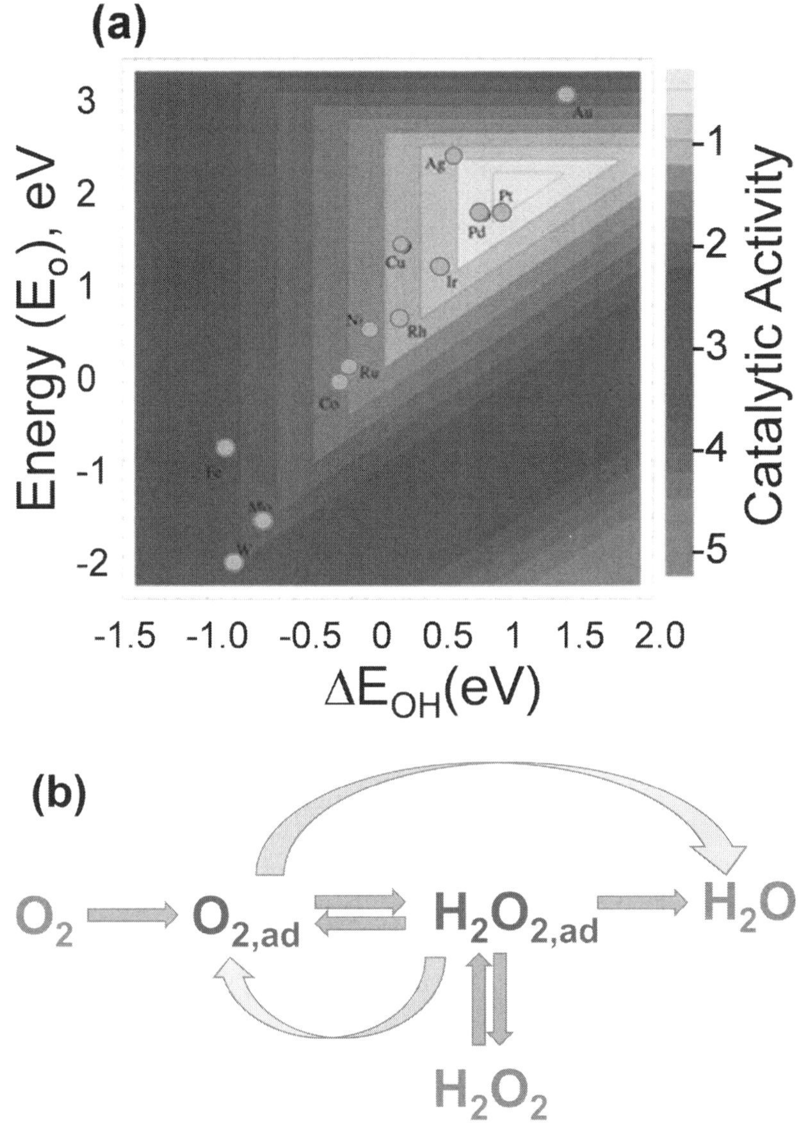


*Figure 2. Trends in ORR activity of different metals plotted against their O and OH binding strengths (a), and ORR pathway mechanisms (b). Reproduced with permission from reference (6). Copyright (2004) American Chemical Society.*

## 1.2. Nanomaterials for the Oxygen Reduction Reaction

The challenges associated with oxygen electrode or cathode are known as long-term research goals that are critical for future energy technology. The cathode reactions are intrinsically slow and significantly impede the efficiency of the rechargeable batteries that require a bi-functional cathode active for both oxygen reduction and oxygen evolution reactions.

Among many transition metal noble- and non-noble metals, Pt is known as the best ORR catalyst (Figure 2a). However, even in presence of Pt, the reversible ORR potential of 1.23 V in acidic aqueous solution (*7*) corresponding to the electrochemical reaction involving four electrons transfer cannot be achieved due to slow ORR kinetics. The ORR overpotential is related to the electron and proton transfer and explained by strongly adsorbed hydroxide or oxygen molecules on platinum surface at its equilibrium potential. Due to optimum binding strength of both hydroxyl and oxygen species, platinum has the lowest overpotential and the highest catalytic activity toward ORR compared to any other metal catalysts (Figure 2a). Therefore, platinum dispersed on carbon support (Pt/C) is the state-of-the-art catalyst for ORR. Decrease in Pt particle size increases Pt surface area, but does not increase the net ORR efficiency because of the particle size effect. This effect, causing the reaction rate to decrease at the Pt particle size of less than 2.5-3.0 nm, has been explained by $OH_{ads}$ species that block the catalyst surface and prevent the $O_2$ adsorption (*8*).

To improve the ORR kinetics and the electronic transport in the cathode layer, various commercial carbons have been employed, such as activated carbons, Super P, Vulcan XC-72, Ketjen black, graphene, and carbon nanotubes. Among all commercial carbons, KB EC600JD has both the largest surface area and the pore volume (2600 $mAhg^{-1}$) (*9*). In regard to specific capacitance important for lithium-ion batteries, the activated carbons with the highest surface area of 2100 $m^2/g$ demonstrates the lowest specific capacity (414 $mAhg^{-1}$), which is explained by the small (2 nm) pores. However, the low surface area Super P (62 $m^2/g$) reveals much higher specific capacity of 1736 $mAhg^{-1}$ (*10*) which highlights the importance of nanoengineering approaches. Yet, mesocellular carbon foams (*11*) reach even higher capacitances of 2500 $mAhg^{-1}$.

### *1.2.1. Noble Metal Based Carbon Nanostructures*

Electrospun nanofiber mats containing TiC crystallites in carbon matrix impregnated with Pt nanoparticles (Figure 3a) have been studied as a novel ORR electrocatalyst in acidic and alkaline media (*12*) using a rotating disk electrode approach. The Pt/TiC system exhibits higher electrocatalytic activities for ORR in terms of on-set potential, kinetic limiting currents (Figure 3b) and mass activity (Figure 3c) than Pt on the electrospun carbon nanofiber (Pt/ECNF) or Pt/carbon paper (Pt/CP). Furhtermore, TiC nanofiber mats without addition of noble or transition metals are also active for ORR in the alkaline solution, indicating that they can be utilized as low-cost alternative supports.

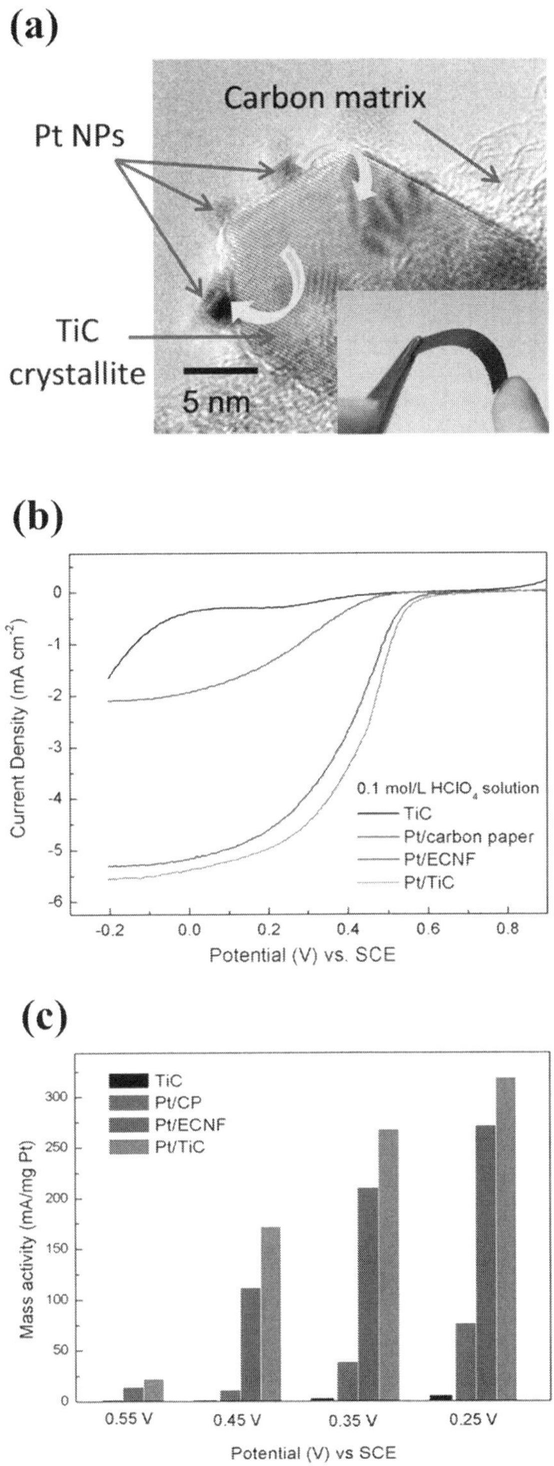

(a)
Carbon matrix
Pt NPs
TiC
crystallite
5 nm
(b)
0.1 mol/L $HClO_4$ solution
TiC
Pt/carbon paper
Pt/ECNF
Pt/TiC
Current Density (mA cm$^{-2}$)
Potential (V) vs. SCE
(c)
TiC
Pt/CP
Pt/ECNF
Pt/TiC
Mass activity (mA/mg Pt)
0.55 V
0.45 V
0.35 V
0.25 V
Potential (V) vs SCE

*Figure 3. Electrospun TiC nanofiber mats with 7 wt. % Pt nanoparticles (a) enhanced ORR activity as indicated by the onset potentials and high limiting currents of the polarization curves (b) and mass activity (c) of Pt/TiC compared to Pt/ECNF Pt/C and TiC. Reproduced with permission from reference (13). Copyright (2013) Royal Society of Chemistry.*

In comparison to pure Pt catalyst, Pt alloy nanoparticles have superior ORR kinetics. These nanoparticles can be synthesized in supercritical fluids (SCFs) that have enhanced properties (*14*, *15*) in comparison to catalysts synthesized by other techniques, such as precipitation, sol-gel approaches, chemical vapor deposition (CVD) or sputtering. By providing the mass-transport free conditions, SCFs favor the nano-particle formation within three-dimensional and electrically conductive carbon support. In the SCFs, the mass transfer rates are fast in comparison to liquids which enable fast rates of deposition, high dispersion and surface area, and, as a result, high ORR reaction rates. The degree of diffusion rates within the substrate and the partitioning of the precursor between the SCF and the substrate are controlled through solvent density adjustments (e.g. changing temperature and pressure) and the concentration of the organometallic precursor. For synthesis of metal or metal oxide nanoparticles or thin films in supercritical carbon dioxide ($scCO_2$), many organometallic complexes soluble in non-polar $scCO_2$ can be utilized. For example, the cyclooctadiene- and β-diketone complexes or fluorinated acetyl-acetonates of noble and transition metals (e.g. Pt, Ir, or Co) with metal ions shielded by the corresponding organic moieties are soluble in $scCO_2$ reaching the solubilities of $\sim 1.3\times10^{-2}$ mole/L at 50°C and 20MPa. It was demonstrated that PtIrCo/CA catalysts synthesized in $scCO_2$ (Figure 5) have enhanced ORR specific activities (~ 6.9X greater than Pt/C), and mass activities (~6.0X greater than Pt/C) at 0.6 V vs. SHE.

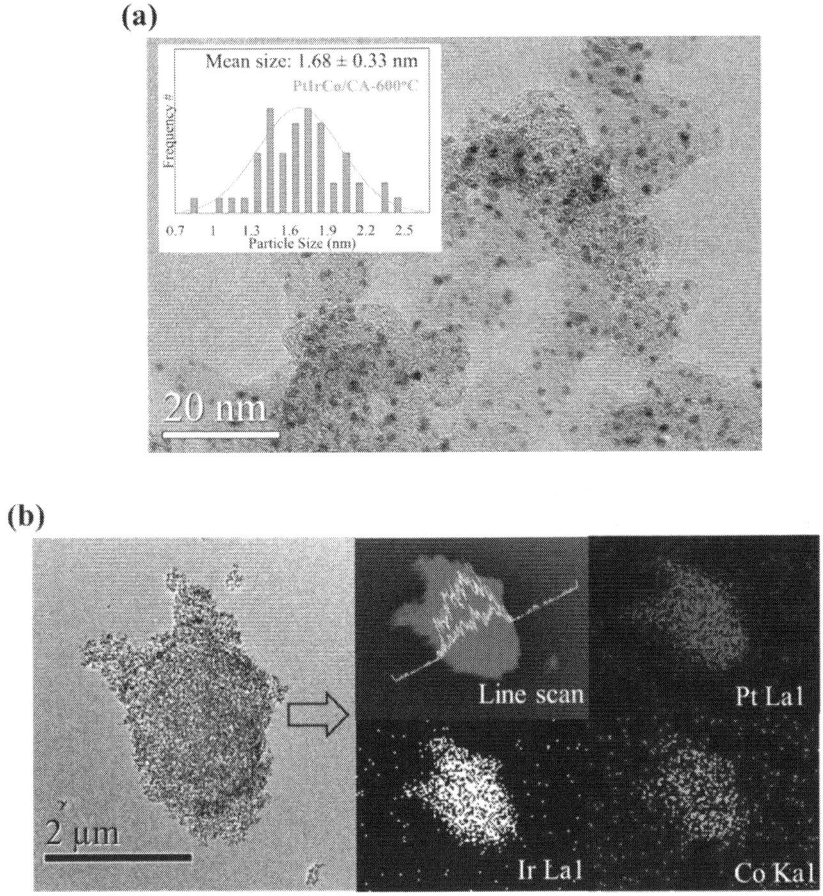


*Figure 4. (a) The TEM image of the PtIrCo alloy nanoparticles deposited within carbon aerogel support in $scCO_2$ followed by the heat-treatment at 600°C and (b) TEM/EDS elemental mapping of the Pt, Ir, and Co in PtIrCo-alloy nanoparticles. Reproduced with permission from reference (16). Copyright (2015) Elsevier.*

### *1.2.2. Metal Oxide Catalytic Materials*

Besides metals, metal alloys, metal oxides, and metal carbides, the complex metal oxides, such as perovskites and spinels, have been recently recognized as active ORR catalysts at close-to-room temperatures. The DFT calculations performed for a large group of complex metal oxides revealed the volcano plots (Figure 5) introduced earlier (*17*) for transition and noble metals. The results of the DFT calculations (Figure 5) (*18*) clearly demonstrate the changes in catalytic activity as a function of the complex metal oxide composition which are based on a *d*-band metal center model with two major insights into the ORR kinetics mechanisms on the metal surfaces. First, the electron transfer efficiency of ORR catalysts typically increases as the *d*-band center shifts positive. Second, the oxygen binding energy to these surfaces also increases with increase of the *d*-band center energy. The increase in the oxygen binding energy yields strongly adsorbed oxygen species on the catalytic surface which are difficult to reduce and thus, provides a superior ORR behavior. The catalytic activity, *d*-band center, and chemical stability are influenced by the particle shape, size and the nature of the catalyst support. As an example, a synergistic contribution of the CNT/GCA support structure and morphology toward ORR activity for iron nanoparticles (10 and 15 wt. %) incorporated into 3D CA matrix was demonstrated (Figure 6) (*48*).

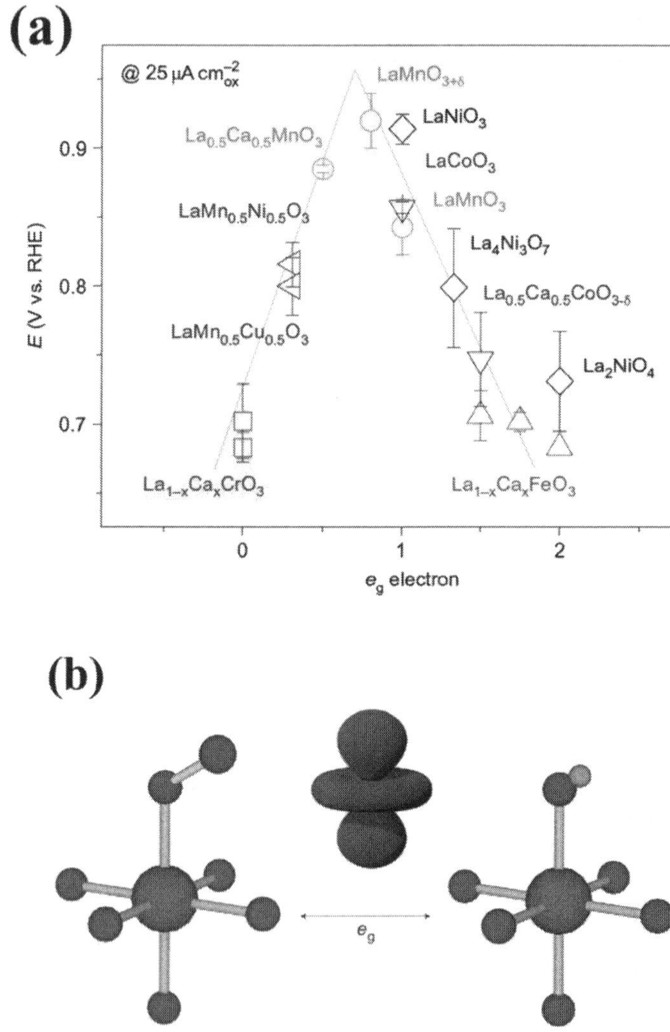


*Figure 5. (a) A volcano plot for the ORR taking place on the sufaces of the complex metal oxides and (b) Role of the $e_g$ electron in the ORR on the perovskite surface. Reproduced with permission from reference (18). Copyright (2011) Nature Publishing Group.*

In $ABO_3$ perovskites, the catalytic activity is defined by the nature of B-atom. As an example, $Ba_{0.5}Sr_{0.5}Co_{1-x}Fe_xO_{3-\delta}$ perovskites with iron-doped B-site demonstrate high catalytic behavior at elevated temberatures as SOFC cathodes (*19*). The catalytic activity of transition complex metal oxides, spinels, with chemical formula $AMn_2O_4$ (A = Co, Ni, Cu), was tested for remediation of volatile organic compounds (*20*). Another group, the layered perovskites, e.g.,$GdBaCo_2O_{5+x}$ (*21*), with oxygen vacancies located in the rare earth $[GdO]_x$ planes were shown to improve the oxygen transport at higher temperatures compared to the non-ordered perovskites. Furthermore, they have relatively high electronic conductivity above the metal-insulator transition temperature at 360K.

A comparison between the $ABO_3$ perovskites catalytic activity of and the *d*-electron number per B cation on the perovskite surface reveals an M-shaped relationship with the maximum activity near $d^4$ and $d^7$ orbitals and the catalytic activity as a function of the $e_g$-filling of B ions (*22*). Since the overlap between $e_g$ of the B-ion with O-2$p_\sigma$ is stronger than between $t_{2g}$ of the B-ion and O-2$p_\pi$, it can be stated that $e_g$ rather than $t_{2g}$ defines the rate limiting step of $O_2$ adsorption/ desorption. The extent of covalent interaction between B-site transition metal and oxygen is the second factor that affects the catalyst activity of the complex metal oxides. The volcano shape was interpreted in terms of too strong B-$O_2$ bonding for low $e_g$-filling e.g., for $La_{1-x}Ca_xCrO_3$, or too weak B-$O_2$ bonding for high $e_g$-filling. The optimal catalytic activity was predicted for $e_g$-filling close to 1, when B-$O_2$ bond is neither too weak nor too strong.

### *1.2.3. Carbon and/or Transition Metal Based Catalyst Nanomaterials*

In many examples provided above, the catalyst nanostructures (e.g. metals, metal alloys, metal oxides, or metal carbides) are discussed in terms of their activity toward oxygen reduction or oxygen evolution reactions. It was also mentioned that the second major component of the catalyst nanostructure, specifically carbon plays equally important role as the catalyst. Different types of carbon have been used as a catalyst support (*23–28*). Among them, carbon aerogels (*29, 30*) combine high surface area (up to 3000 $m^2/g$ (*31*)), interconnected pore structure, narrow pore-size distribution (±1nm), high electrical conductivity (25-100 S/cm), low cost ($10/kg), and remarkable mechanical and thermal properties (*32–35*). Carbon supports influence electrode kinetics by changing Galvani potential, raising the electronic density, and shifting Fermi energy levels (*36*). All these factors accelerate the electron transfer at the electrode-electrolyte interface and improve the electrode performance.

However, besides the beneficial properties of carbon supports, the major problem is their chemical instability resulting in corrosion and chemical degradation. The platinum group metals are known to catalyze carbon corrosion leading to significant loss of active surface area and catalytic activity (*37*). The stability of carbon support can be improved by decreasing the carbon defect sites responsible for carbon oxidation by graphitization. Furthermore, at high degree of carbon graphitization, high strength of π-sites ($sp^2$-hybridized carbon atoms) helps to anchor Pt and impede Pt sintering (*38*). Another possibility to minimize carbon degradation is to modify it with nitrogen since N-atoms increase the carbon catalytic activity (*39*, *40*) due to π-bonding, high Lewis basicity, and electron donor behavior. Furthermore, incorporated nitrogen atoms and π-bonds contribute to the basicity of carbon increasing platinum-carbon interaction (*41*, *42*). Another option for corrosion stability improvement is to deposit ultra-thin films of metal oxides, carbides, or nitrides such as $Ti_4O_7$, $V_2O_3$, TiC, ZrC, CrN/$Cr_2N$ that have electrical conductivities comparable to that of graphitized carbon ($10^3$-$10^5$S/cm) (*43–45*). They are known to interact with platinum and minimize the Pt particle growth. These materials have an intrinsic resistance to oxidation thus, preventing carbon from direct contact with metal nanoparticles. As an example, $Ti_4O_7$-supported Pt anode has the same catalytic activity as XC-72/Pt hybrid nanostructure, but much higher corrosion resistance (*46*, *47*). An important role of the carbon graphitization was demonstrated in presence of the iron nanoparticles (10 and 15 wt. %) incorporated into 3D CA matrix (Figure 6). Enhanced graphitization at 900°C, 1200°C, and 1400°C in Ar followed by catalytic CVD process in nitrogen atmosphere resulted in N-doped CNTs. Depending on CNT growth temperature (e.g. 700°C, 800°C or 900°C) in reducing environment containing nitrogen, hydrogen and acetylene, the N-doped multiwall CNTs, nanofibers, and nanoribbons (10-50 nm in diameter) were formed. Among CNT/GCA materials, CNT-800°C/GCA shows the best ORR activity at lowest onset potential of 0.5 V. The CNT-900°C/GCA nanocomposite exhibits the highest ORR mass activity with a half-wave onset potential difference of 120mV in comparison to Pt (40 wt.%)/C. It was demonstrated that the role of CA graphitization prior to CNT growth is essential for the electrochemical stability of the catalysts in acidic medium.

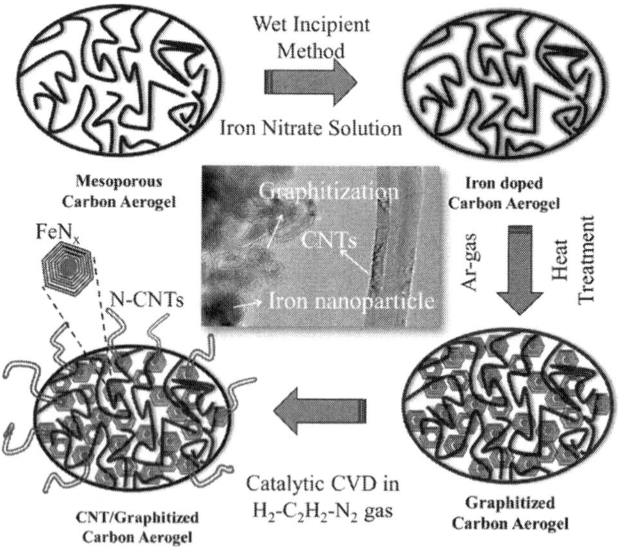


*Figure 6. Major steps in synthesis of iron modified 3D graphitized carbon aerogels (GCA) and CNT/GCA composites by catalytic CVD method. Reproduced with permission from reference (48). Copyright (2014) Elsevier.*

# 2. Oxygen Evolution Reaction

Development of electrocatalysts for splitting of water to molecular hydrogen and oxygen is essential for the large scale electrochemical energy storage systems. Platinum has been widely considered as the state-of-the-art catalyst for Hydrogen Evolution Reaction (HER). However, highly active, stable and inexpensive electrocatalysts for OER have not been recognized due to many challenges associated with OER. Theoretical DFT calculations show that the limiting step of the OER is the binding strength of the oxygen species at the catalyst surface.

## 2.1. Electrochemical Mechanism of the Oxygen Evolution Reaction

The mechanisms of catalytic activity of the mixed ionic-electronic conductors, complex ceramic oxides, and metal-ceramic composites for high temperature oxygen reduction reaction (ORR) (*49*) and internal reforming (*50*) have been discussed in numerous publications. However, in addition to the ORR, current interests are mainly focused on oxygen reduction combined with oxygen evolution, which are central for direct solar cells, electrolytic water-splitting systems, fuel cells, and metal air batteries.

Recent data focused on ORR and OER on metal oxide surfaces and interfaces strongly suggest that the catalytic activity mechanisms of ceramic oxides should be reconsidered in terms of low temperatures. Howeevr, the kinetic mechanisms and the major trends for different families of oxides at nanoscale are still to be revealed. Current technology defines kinetics of the electrochemical reactions at the nanostructured surfaces as the major driving force due to faster reaction rates, short diffusion paths compared to the bulk, and enhanced thermodynamic and electronic properties.

A large number of transition and post-transition metal oxides in comparison to the state-of-the-art irridum oxide has been evaluated (*51*) (Figure 7). Their OER in acidic (top) and alkaline (bottom) media is reperesented by a plot where the $x$-axis is the overpotential ($\eta$) required to achieve the current density of 10 $mA{\cdot}cm^{-2}$ per geometric area at time $t = 0$ and the $y$-axis is the overpotential at the same value of current density (10 $mA{\cdot}cm^{-2}$), but different time ($t = 2$ h). The maximum stability of the corresponding catalyst is achived when the values are located on the diagonal. In alkaline conditions most of the evaluated catalysts are stable (fall in the white regionnd along the diagonal) . In acidic medium, on contrary, most of the catalysts except $IrO_x$ are outside white area and cannot be considered as efficient and chemically stable OER catalysts.

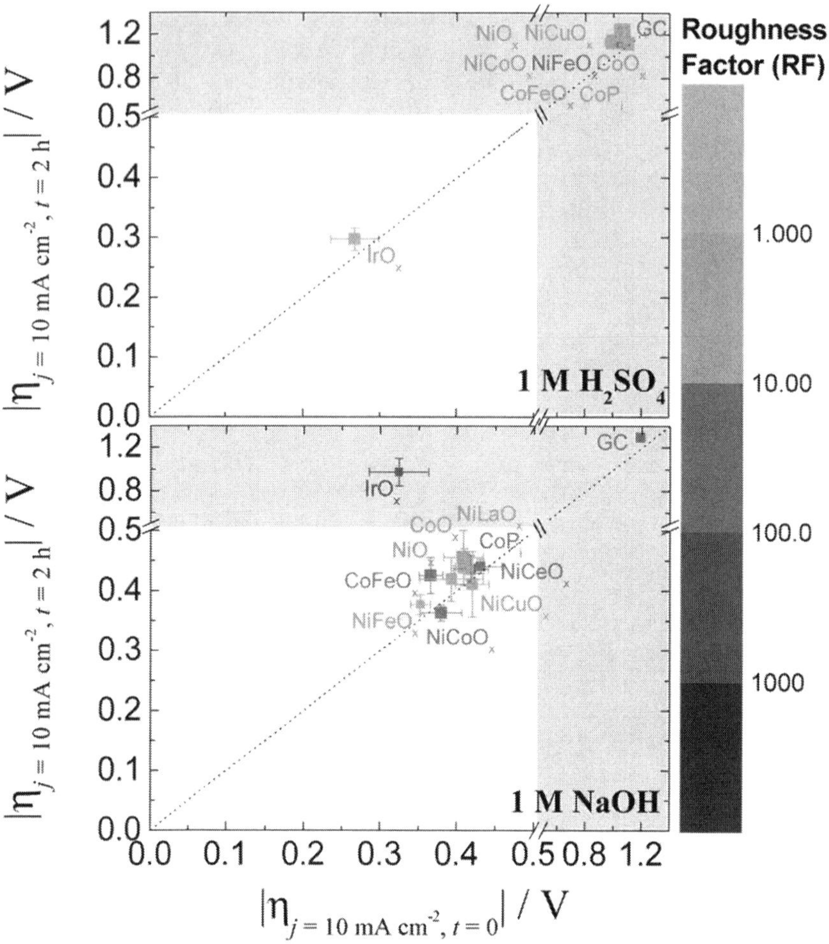


*Figure 7. Catalytic activity, stability, and the electrochemically active surface area for OER electrocatalysts in acidic (top) and alkaline (bottom) solutions. The color of each point represents the roughness factor of the catalyst. Reproduced with permission from the reference (51). Copyright (2013) American Chemical Society.*

### 2.2. Nanomaterials for the Oxygen Evolution Reaction

Similar to the ORR, the nanomaterials considered for OER span from metals and metal alloys to simple and complex metal oxides. However, due to the differences between the ORR and OER mechanists these materials demonstrate the volcano plot where Ir and Ru oxides, but not Pt, have the highest catalytic activity (*52*). A comparative study of scaling effects from bulk to nano-size for the OER activity reveales that in comparison to other nanostructures, for example Pt and Ru, iridium has the highest catalytic activity and chemical stability (*53*). High OER activity was also demonstrated for amorphous $RuO_x$, alloys of Pt, Ir and/or Ru (*54*), $RuO_2$-$IrO_2$, $IrO_2$-Pt (*55*), $Ru_xIr_{1-x}O_2$ (*56*), and Pt-Ir-Ru (*57*). However, these catalysts suffer from corrosion of carbon support and/or dissolution of catalyst nanoparticles at high electrochemical potentials and in strong acidic environment. On the other hand, irridium oxide possesing the highest OER activity in the acidic environment is not stable in the alkaline environment (Figure 7). Therefore, metal oxides such as spinels and perovskites have been studied as OER catalysts in alkaline rather than acidic medium.

The attractive feature of the transition metal oxide catalysts is their flexibility in terms of their catalytic properties. Complex metal oxides with pyrochlore or perovskite structures are bifunctional in nature and have an ability to participate in both ORR and OER. During the charging process on the cathode of the lithium-air battery the OER reaction ($4OH^- - 4e^- \rightarrow O_2 + 2H_2O$) takes place with the net reaction of the metal oxide decomposition: $2Me_2O_x \rightarrow 4Me + xO_2$.

A traditional ORR/OER bifunctional hybrid catalyst as a mixture of platinum metal nanoparticles and iridium oxide (*58*) combines the Pt activity toward ORR and the $IrO_2$ activity toward OER. However, similar to perovskite ORR catalysts (*59*), doped perovskites, e.g., $Ba_{0.5}Sr_{0.5}Co_{0.8}Fe_{0.2}O_{3-\delta}$, remarkably outperform $IrO_2$ as an OER catalyst and demonstrate that the oxygen evolution is a function of the occupancy of 3*d* electron orbital with an $e_g$ symmetry of the transition metal atom on B-site.

In comparison to metals and $ABO_3$ perovskites, $A_2B_2O_{7\pm\delta}$ pyrochlores are not well known as low temperature catalysts, though for many years they have been actively used in high temperature catalysis. For example, in methane combustion, among $Ln_2B_2O_7$ (Ln=Sm, Eu, Gd and Tb; B=Zr or Ti) the highest catalytic activity was demonstrated for $Sm_2Zr_2O_7$ (*60*). The effects of A and B substitution in $La_2Zr_2O_7$ on kinetic behavior in methane reforming were tested for $La_2Zr_{2-x}Rh_xO_{7-\delta}$, $La_2Zr_{2-x}Ni_xO_{7-\delta}$, $La_{1.95}Ca_{0.05}Zr_{2-x}Rh_xO_{7-\delta}$ and $La_2Zr_{2-x}Ni_xO_{7-\delta}$ in comparison to $La_{1.97}Sr_{0.03}Zr_2O_{7-\delta}$. It was shown that rhodium-substituted pyrochlores are more active in comparison to porous nickel. Replacement of $La^{3+}$ for $Ca^{2+}$ demonstrated improved oxygen mobility by introducing lattice oxygen defects. Substitution of metals lowers the bond energy of La–O and Zr–O resulting in release of oxygen from the lattice (*61*). In addition to catalytic activity, pyrochlores $A_2[Ru_{2-x}A_x]O_{7-y}$ ($A$ = Pb or Bi) exhibit high electrical conductivity at room temperature ranging from 10 to 1000 (ohm-cm)$^{-1}$ which is defined by the electron transfer between the large conducting segments consisting of connected networks of $RuO_6$ octahedra (*62*). $Sm_2Ti_2O_7$–$MTiO_3$ (M = Mg, Co and Ni) complex metal oxides demonstrate anionic conductivity interpreted by migration of extrinsic oxygen vacancies into the conduction pathways (*63*). Furthermore, the OER and ORR activity was demonstrated for iridium-based $Pb_2(Pb_xIr_{2-x})O_{7-y}$ and $Nd_3IrO_7$ pyrochlores (*64*). Efficient and stable bifunctional properties in alkaline medium for ORR and OER were also shown for $Pb_2Ir_2O_{7-y}$ and $PbBiRu_2O_{7-y}$ oxides (*65*). All these encouraging results indicate that pyrochlores these materials should be also considered for evaluation of the ORR and OER catalytic activities.

In regard to potential applications, the OER/ORR catalytic activity is a crucial feature of the Li-air battery cathodes. Different approaches related to Li-air battery cathodes address graphene-metal oxide composites (*66*), Pt-graphene

(4.8 Ah/g and 70 Wh/kg) (*67*), $\alpha$-$MnO_2$ nanowires and nanorods (*68*), Au-Pd nanoparticles on $MnO_2$ (*69*), Pd and $PdO_2$, nanostructured diamond-like carbon films (*70*), nitrogen-doped CNTs (*71, 72*), graphene (*73, 74*), spinels (e.g. $LiNi_{0.5-x}Co_{2x}Mn_{1.5-x}O_4$ (*75*)), and CNT buckypaper (1.1 KW/kg) (*76*). The trends in catalytic activity of perovskites based on $\sigma^*$-orbital ($e_g$) occupation and the extent of B-site transition metal-oxygen covalency serving as a secondary activity receptor have been emphasized. However, despite the well-recognized role of the oxygen reaction kinetics, the governing mechanisms remain elusive, especially on complex ceramic surfaces and interfaces. Furthermore, the trends in catalytic activity of various groups of complex oxides are not recognized.

Besides lithium-air battery application, a large number of transition metal (e.g. Cr, Mn,Fe, Co, Ni, Cu, Ru, Ir) and post-transition metal (e.g. La, Ce) electrocatalysts and their metal alloys was screened for the OER catalytic activity under conditions relevant to an integrated solar water-splitting device in aqueous acidic or alkaline solution (*77*). In acidic solution, non-noble metal catalysts did not show promising OER catalytic activity and stability, whereas in alkaline solution many OER catalysts performed with similar activity achieving 10 mA $cm^{-2}$ current densities at overpotentials of ~0.33−0.5 V. Most OER catalysts showed comparable or better specific activity per electrochemical area when compared to the state-of-the art Ir and Ru catalysts in alkaline solutions.

The organometallic bioinspired water oxidation $3d$–$4f$ catalysts $[Co^{II}_3Ln(hmp)_4(OAc)_5H_2O][Co^{II}_3Ln(OR)_4]$ (*78*), where Ln = Ho−Yb and hmp = 2-(hydroxymethyl)pyridine) were tested as homogeneous OER catalysts in photocatalytic applications. The stability and photocatalytic activity of these complexes as active catalytic promoters with flexible aqua-acetate ligands influenced by the tunable $Ln^{3+}$ centers illustrates that rapid initial kinetics in the four electron-transfer process takes place, that however, do not result in high overall $O_2$ yields.

Efficient water oxidation was achieved by using $TiO_2$/CdS/Co–Pi photo-anode with the cobalt phosphate water oxidation catalyst (Co–Pi) that accelerates the water oxidation reaction. In this system, CdS serves as light absorber for wider solar spectra harvesting, and $TiO_2$ matrix provides direct pathway for electron transport. CdS sensitized $TiO_2$ nanowire arrays for nonsacrificial solar water splitting demonstrated significantly improved photo-electrochemical stability compared to the $TiO_2$/CdS electrode, with ≈72% of the initial photocurrent retained after two hours of irradiation (*79*).

Both the oxygen evolution and oxygen reduction reactions were studied for doped praseodymium and samarium nickelate-cobaltites $PrNi_xCo_{1-x}O_{3-\delta}$ and $SmNi_xCo_{1-x}O_{3-\delta}$, respectively (*80*). It was shown that Sm-based perovskites have higher ORR activity than Pr-based perovskites in regard to their half-wave overpotentials. In the ORR electrochemical experiments involving rotating disc and rotating ring disk electrodes, Sm-based perovskite sintered at 700°C ($X_{Ni}$=0.1) demonstrated the highest catalytic activity in comparison to all other tested perovskites. However, the highest OER activity was achieved for Pr-based perovskites sintered at 900°C ($X_{Ni}$=0.1). The conclusion was made that all the Pr- and Sm-based perovskites possess higher OER activity than the state-of-the-art $IrO_2$ catalyst (Figures 8 and 9).

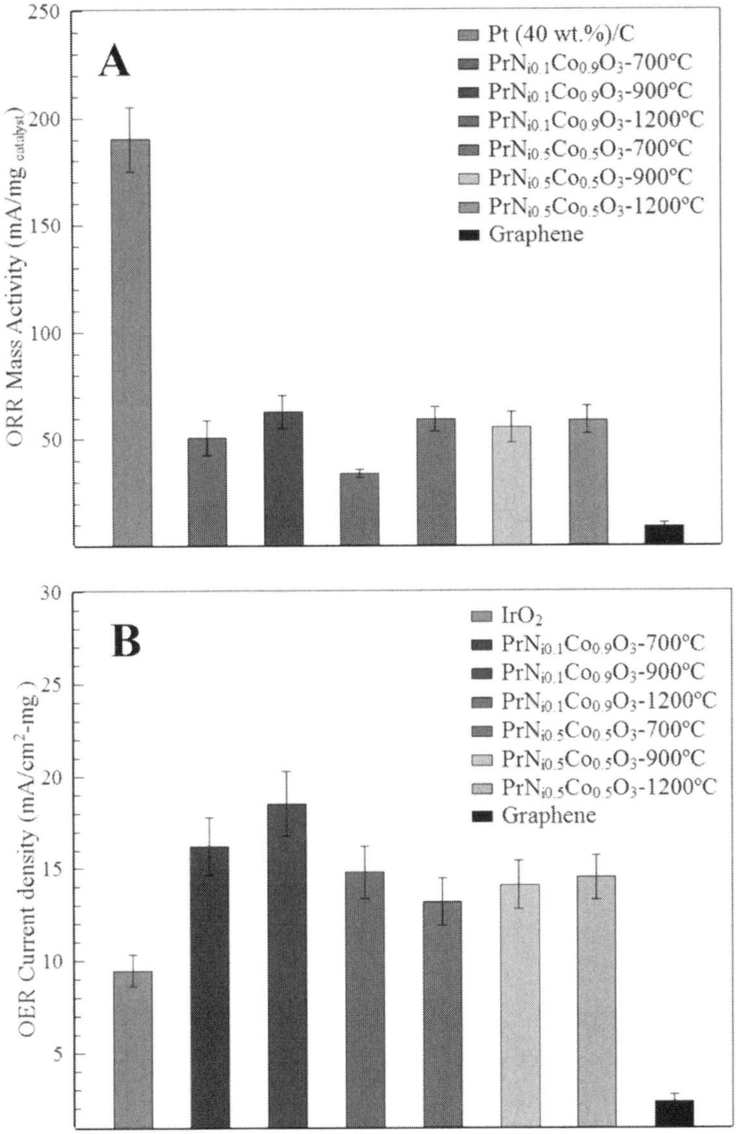


*Figure 8. Comparison of the ORR and OER catalytic activity for Pr-based Ni-Co perovskites.*

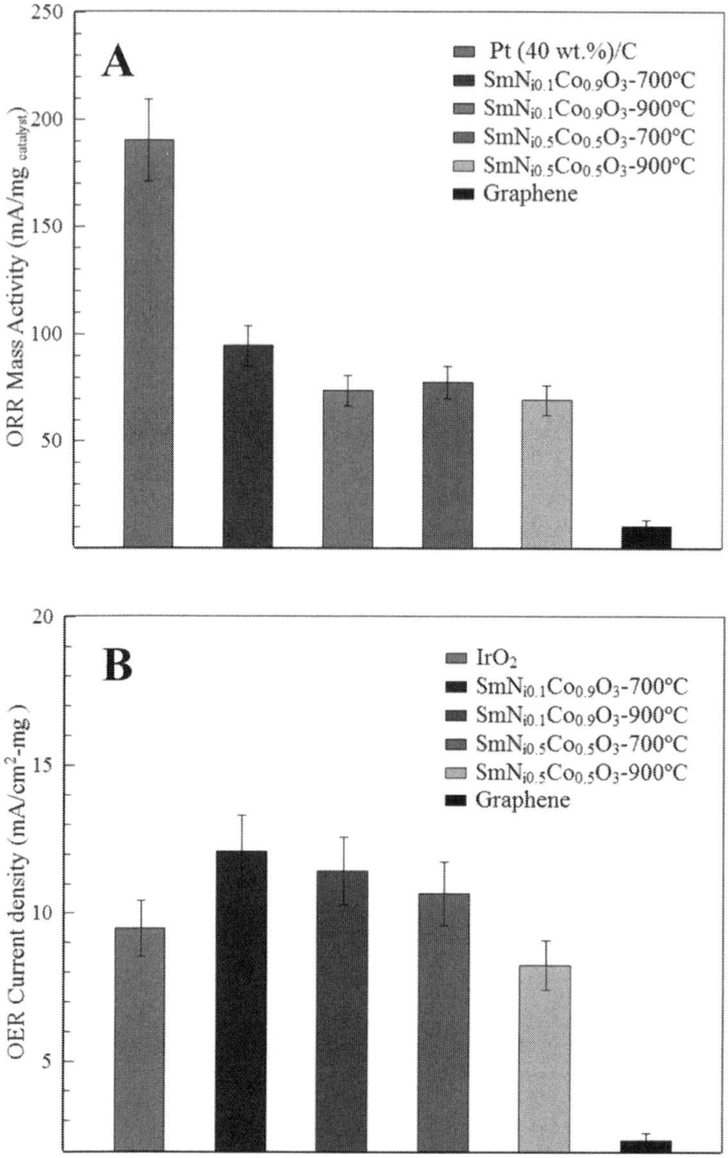


*Figure 9. Comparision of the ORR (A) and OER (B) catalytic activity for Sm-based Co-Ni perovskites.*

# 3. Methanol Oxidation Reaction

## 3.1. Methanol Oxidation Reaction Mechanism

Development of catalysts for methanol oxidation is closely related to sustainable energy economy (*81*) by generating electricity in direct methanol fuel cell (DMFC). In portable or automotive industry, the DMFC exploitation eliminates the need to store or generate hydrogen. Liquid methanol is relatively safe, less expensive, easy to transport, and can be mass-produced. In addition, galvanometric energy density of methanol (4,817 Wh/L) is an order of magnitude higher than that of compressed hydrogen (520 Wh/L at 2000 psi). Therefore, DMFC is an attractive candidate for transportation applications where space and weight savings are the priority. Moreover, renewable methanol can be produced through sunlight by photochemical reduction of $CO_2$ (*82*). On the other hand, when rechargeable batteries are replaced with DMFC in portable electronics, such as mobile phones and laptops, it gives constant energy supply by eliminating the recharging step. Recent studies have shown that methanol has greater potential for gasoline substitution and carbon dioxide mitigation than ethanol (*83, 84*). However, in a long-term perspective, one of the major goals is to enhance the kinetics of methanol oxidation and eliminate the activation DMFC losses.

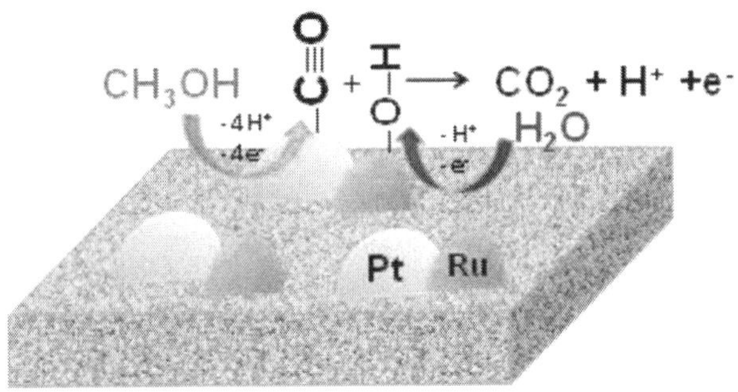


*Figure 10. Promotion mechanism of methanol electrooxidation on Pt/Ru catalyst.*

Compared to oxidation of water in water electrolysis, much lower thermodynamic potential for methanol electrooxidation (0.02 V vs. 1.23 V for water) (*85–88*) makes methanol more energy efficient (*89–91*). However, because of the large number of electrons involved ($CH_3OH + H_2O \rightarrow CO_2 + 6H^+ + 6e^-$), the equilibrium cannot be readily achieved, the reaction is slow, and the total oxidation process consists of a pattern of parallel reactions.

On pure platinum surface, complete electrooxidation to $CO_2$ takes place through two processes in separate potential regions. The first one involves adsorption of methanol molecules and requires the sites free from adsorbed hydrogen at ≥ 0.2 V vs. RHE. The second one requires dissociation of water, which is the oxygen donor for the reaction. Interaction of water with Pt surface starts at ≥ 0.45 V (*92*) and the adsorbate layer is reactive above ~ 0.7 V where the oxidation of adsorbed by-products takes place. Existence of parallel reactions depending on potential, time, and surface morphology impedes fundamental understanding of methanol electrooxidation and reaction intermediates. Furthermore, in the stepwise oxidation of $CH_3OH$ to $CO_2$ (*93*), CO adsorption, poisoning of the catalyst, and corrosion are the major problems. If these difficulties are overcome, the process of methanol oxidation offers an opportunity for lower hydrogen production costs and reduction of the infrastructure required for hydrogen distribution.

Carbon supported Pt nanoparticles have the highest activity toward MOR. However, strong adsorption of carbon monoxide (CO) as an MOR intermediate onto the platinum surface reduces the Electrochemical Surface Area (ESA) of platinum and prevents the Pt-CO sites from further methanol oxidation. The oxidation of Pt-CO occurs at higher potentials, which limits the DMFC operating voltage. Therefore, ruthenium is successfully used along with platinum as a bifunctional catalyst (Figure 10) to decrease the Pt-CO adsorption strength through ligand effect and/or to provide OH-species to platinum surface for sequential Pt-CO oxidation at lower potentials via a bifunctional mechanism (*94*, *95*). Nonetheless, high noble metal loading in the Pt/Ru alloy catalyst ($> 1$ mg/cm$^2$) is still necessary for practical DMFC operation. In this regard, development of inexpensive DMFC anode catalysts by promoting PtRu alloy nanoparticles with metal oxides and/or synthesis of CO-tolerant platinum alloys is important for DMFC commercialization.

### 3.2. Nanomaterials for Methanol Electrooxidation

#### *3.2.1. Metal Catalyst Promoters*

Pt is known as the best catalyst for methanol electrooxidation, but has low tolerance for carbon monoxide. Since CO blocks active sites on Pt surface, a second oxophilic metal (*96*), such as Se (*97*), Ni, Sn, Mo, Bi, Pb, Ni (*98*), Os, W, Ru, or Ir (*99*, *100*) is commonly added. Among these, Ru is considered as the best promoter. However, Pt/Ru alloy is known for insufficient activity and stability caused by migration of ruthenium atoms. The improvements of Pt/Ru and Pt/Sn-based anodes (*101*, *102*), are focused on addition of the third or fourth promoter forming ternary and quaternary catalysts such as $PtRuMeO_x$ (Me = W, Mo or V) (*103*) and $Pt_5Ru_4M$ (Me = Ni, Sn and Mo) (*104*).

Many transition metal oxides and carbides with and without addition of noble metals or metal alloys have been also studied as MOR catalysts. Transition metal carbides possess methanol electrooxidation activity similar to Pt group metals. As an example, Mo (*105*), V/Ti (*106*), tungsten carbides ($W_2C$ and WC) (*107*) or WC on XC-72 impregnated with Pt (*108*) showed enhanced behavior that improved with W increase (Pt/C < Pt/WC < Pt/$W_2C$) explained by a better Pt dispersion and activation of water molecules for CO removal. Other studies (*109*) indicate that only WC is stable at $\leq 0.6$ V and $W_2C$ do not have a stable region, causing immediate oxidation to $W_xO_y$ when exposed to air or during CV tests. The tolerance of Pt–WC/C to CO is explained by the DFT, confirming that WC provides additional electron density to the Pt atoms (*110*). Incorporation of WC into Pt increases the amount of $OH_{ad}$ on Pt, which accelerates $CO_{ad}$ oxidation and removal of $CO_2$. Since WC partially exists as a mixture of $W^{5+} \leftrightarrow W^{6+}$ oxides, it tends to promote the dissociation of $H_2O$ into $H^+$ and $OH^-$, and provides $OH_{ad}$

species to the catalytically active sites on the Pt surface. Similar effects were observed with PtMo which CO tolerance was higher than that for PtRu (*111*).

To explain the carbon monoxide poisoning of the catalytic surfaces during the MOR process, two mechanisms, e.g., promotion and intrinsic, and the corresponding CO tolerance effects were proposed. According to the promotion mechanism, the oxophilic metal atoms, e.g., ruthenium in the Pt/Ru alloy, break water molecules into $OH^-$ and $H^+$ at a potential lower than that of Pt. The carbon monoxide molecules chemisorbed on the surface of the platinum nanoparticle can then be oxidized by the nearby $OH^-$ species. In the second, intrinsic mechanism, Pt electronic structure is modified by forming an alloy with Ru, that results in weakening of the CO adsorption on the Pt surface (*112*, *113*). Other factors, such as the change in metal-metal interatomic distances, the number of metal nearest neighbors, platinum 5*d*-band vacancies, and the content of platinum in the Pt/Ru alloy are also important. To accelerate all the stages of methanol electrooxidation, including the rate-determining steps of water adsorption and CO oxidation (*114*), multicomponenet alloys, e.g. Pt–Ru–Os, Pt–Ru–Ir, and Pt–Ru–Os–Ir, were introduced (*115*). The results for Sn modified Pt indicate (*116*) that non-alloyed Pt-$SnO_x$ or partially alloyed Pt-Sn with 33% Sn and 66% Pt in oxide form promote methanol oxidation. Besides, Pt-Sn and Pt–$(RuO_xH_y)_m$ electrocatalysts (*m* being the atomic Ru/Pt ratio) (*117*) demonstrated improvements in MOR, however the cyclic voltammetry tests at potentials ≥0.46V vs. SCE showed the $RuO_xH_y$ dissolution. Despite numerous publicaions, the discrepancy in synthesis parameters of the MOR catalysts does not allow to perform a fair comparison of the data in terms of effective diffusion, charge transfer coefficients, reaction intermediates, and activation energies.

### *3.2.2. Metal Oxide and Hybrid Catalyst Promoters*

The surface kinetic processes for $Mo_2C$/Mo(100)] and β-$Mo_2C$ (001) evaluated by the DFT simulations (*118*), demonstrate that Mo and K provide the most favorable sites for the methanol adsorption. Experimental evidence for methanol electrooxidation improvement was achieved by adding potassium to $Mo_2C$ (*119*). Similar to carbides, transition metal oxides also possess high activity in methanol oxidation (*120*). Various combinations, such as Pt–Ru/$TiO_2$/C (*121*), $TiO_2$ nanotube/ Ni–B (*122*) were tested. Lower adsorption energy for $CO_{ad}$ and higher reactivity and mobility of reactants and intermediates was observed for PtRu and PtFe catalysts on $TiO_{2-x}$ support (*123*). The quantum mechanics calculations demonstrated strong electron delocalization for $Pt_2Fe_2$ clusters on $(TiO_2)_4$ as compared to $(TiO_{2-x})_4Pt_2Ru_2$. Mechanism of cooperative surface diffusion was proposed that, however, needs further exploration.

Nanocomposite catalysts based on metal oxides synergistically improve methanol oxidation and corrosion resistance (*124*) which is important, since the anode catalyst could be exposed to high oxidative conditions during the cell voltage reverse caused by fuel starvation. Metal oxides such as $CeO_2$ (*125*, *126*), $RuO_2$, $SnO_2$, $WO_3$, and $V_2O_5$ enhance Pt catalytic activity (*127*), however, a major problem is low surface area, resulting in a low metal dispersion and low catalytic activity.

$SnO_2$ coated CNT core-shell nanocomposite (*128*) used as catalyst supports show negligible corrosion at all potentials. Furthermore, composite materials comprising of $TiO_2$ as a core and carbon as a shell, demonstrate higher activity and stability than that of Pt/Vulcan XC-72R. The enhancement in catalytic activity of $TiO_2$–C (*129–131*), $Nb_2O_5$ (*132*) or $MnO_2$–CNT (*133*) was attributed to the interactions between metal particles or core-shell supports. The higher electrochemical stability was ascribed to the higher corrosion resistance of $TiO_2$/C compared to Vulcan XC-72R. Based on these approaches it can be concluded that the metal carbides and oxides are beneficial for enhanced methanol electrooxidation providing additional electron density to Pt atoms and initiating C-O bond breakage. Furthermore, the high surface area of the carbon support and the core-shell structure formation are essential for methanol electrooxidation.

One of the ways for the MOR catalytic activity improvements is based on the application of high surface area and catalytically active hybrid supports with supercritically deposited metal alloy nanoparticles. For enhanced chemical stability, these nanoparticles can be further encapsulated into a thin (~ 1 nm) ceramic shell. In comparison to standard procedures, this SCF approach has advantages, such as fast deposition rates of ceramic nanolayers within conformal three-dimetional (3D) structures and easy scale up. The importance in using SCFs for the synthesis of functionalized catalysts (*134*, *135*) is validated by the unique properties of SCFs, such as densities similar to liquids, diffusivities comparable to gases, and absence of mass transport limitations (*136–138*). In this approach, the "in-situ" metal or metal alloy nanoparticles within a porous support are usually formed from the organometallic precursors dissolved in supercritical $CO_2$ (*139*). To prevent the metal nanoparticle agglomeration and improve catalyst durability, ceramic shells from supercritically deposited and catalytically active complex oxides (e.g. perovskites) are formed in supercritical $H_2O$. Furthermore, a superior methanol tolerance in comparison to Pt/C leading to minimized methanol crossover effect has been detected. Interestingly, ORR specific and mass activities of the same PtIrCo/CA catalysts enhanced proportionally with decrease in *fcc*-lattice parameter. Selective adsorption of oxygen molecules in comparison to methanol molecules (associated with PtIrCo-alloy formation and electronic level interactions between Pt, Ir, and Co) are contributing to their higher methanol-tolerance towards ORR in PtIrCo/CA. The quantitative studies regarding current losses due to the mixed currents and methanol poisoning are proportional to their specific ORR activities. At the same time, the voltage losses are proportional to their corresponding Pt-ESAs. The enhanced MOR and ORR performance was explained by a weak adsorption of oxygen molecules and hydroxide species on the surface of PtIrCo-alloy compared to pure platinum nanoparticles. Much higher methanol tolerance of ORR for PtIrCo/C compared to Pt catalyst was explained by selective adsorption of $O_2$ in comparison to $CH_3OH$ on the PtIrCo-alloy surface (Figure 4b).

Hybrid catalysts (*140*) (Figure 11) comprising of both metal alloy and metal oxide nanoparticles were synthesized by the *in-situ* combustion of $TiO_2$ or $CeO_2$ within PtRu/C. The structure-dependent electrochemical behavior toward MOR, carbon dioxide (CO) tolerance, and chemical stability was studied by XRD, HRTEM, BET, EDS, cyclic voltammetry, chronoamperometry, and CO-stripping. An enhanced MOR activity of the hybrid catalysts, compared to the PtRu/C catalyst as a baseline, was explained by the changes in the electronic behavior of platinum and improved adsorption of oxophilic ions. The $CeO_2$-PtRu/C nanocomposites heat treated at 600°C, demonstrated the highest MOR mass activity of 580 mA/mg in comparison to $TiO_2$-PtRu/C (394 mA/mg) that was three times higher than that of the state-of-the-art PtRu/C.

Besides metals, the $ABO_3$ perovskites, where A=Sr, Ce, or La and B=Co, Fe, Ni, Pt, or Ru, is another promising class of MOR catalysts that has not been sufficiently studied yet. Among other perovskites, $SrRuO_3$ demonstrates the methanol oxidation activity comparable to that of Pt/Ru (*141*). Addition of 1-5wt.% Pt to the $SrRuO_3$ perovskite during the synthesis yields higher activity than if added externally at the ink preparation stage. The alloy of platinum with osmium (*142*) shows catalytic enhancement over Pt which is explained by the oxophilic behavior of osmium and the change in Os valence states ($Os^{3+} \leftrightarrow Os^{4+}$). This transformation leads to the loss of the active OH species on the catalytically active Os sites and results in the pre-peak at 0.48 V for the methanol oxidation, attributed to a potential gate effect. In other words, Os valence states and activity are determined by the potential that alter the corresponding Fermi energy levels. This and the other above presented approaches focused on the MOR catalyst development demonstrate that the presence of oxophilic metals, oxides, and perovskites is beneficial for the process of methanol electrooxidation.

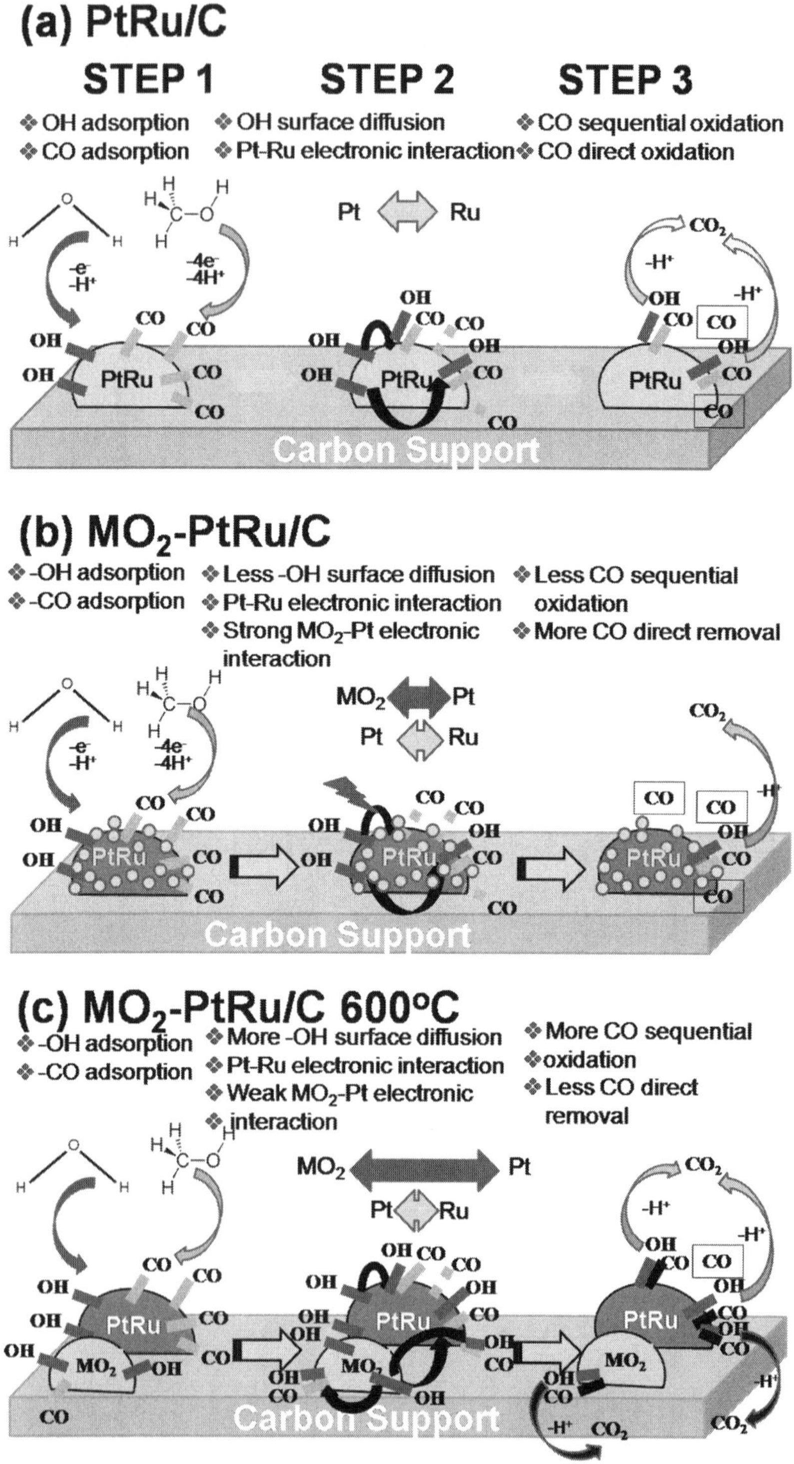
(a) PtRu/C
STEP 1
STEP 2
STEP 3
❖ OH adsorption
❖ CO adsorption
❖ OH surface diffusion
❖ Pt-Ru electronic interaction
❖ CO sequential oxidation
❖ CO direct oxidation
Pt
Ru
PtRu
Carbon Support
(b) MO2-PtRu/C
❖ -OH adsorption
❖ -CO adsorption
❖ Less -OH surface diffusion
❖ Pt-Ru electronic interaction
❖ Strong MO2-Pt electronic interaction
❖ Less CO sequential oxidation
❖ More CO direct removal
Carbon Support
(c) MO2-PtRu/C 600°C
❖ -OH adsorption
❖ -CO adsorption
❖ More -OH surface diffusion
❖ Pt-Ru electronic interaction
❖ Weak MO2-Pt electronic interaction
❖ More CO sequential
❖ oxidation
❖ Less CO direct removal
Carbon Support

*Figure 11. Mechanisms of MOR for (a) State-of-the-art PtRu/C compared to (b) PtRu/C impregnated with ceramic oxide via in-situ sol-gel combustion synthesis, and (c) PtRu/C with metal oxide heat-treated at 600°C. Reproduced with permission from reference (140). Copyright (2013) Elsevier.*

## Summary and Conculsion

Recent advances in nanotechnology paved the way to the development of the novel nanomaterials for electrochemical energy generation and storage devices with higher efficiency, enhanced selectivity, and improved endurance. Important breakthrough findings were achieved for the nanostructures catalytically active in terms of methanol oxidation, oxygen reduction, and oxygen evolution reactions. In conculsion, the highlighted recent approaches for novel catalytic nanomaterials are still not sufficient and should be improved in terms of kinetics at nanoscale, structural directionality, and fesibility of the electrochemical reactions for specific environment. These challenging goals can be achieved by choosing new nano-engineering approaches and novel groups of catalytically active materials crucial for energy security, economic growth, and environmental protection.

## List of Abbreviations and Units

| | |
|---|---|
| $ABO_3$ | General formula for perovskite crystal structure |
| $AB_2O_4$ | General formula for spinel crystal structure |
| BET | Brunauer-Emmett-Teller surface area analysis |
| CA | Carbon aerogel |
| CNT | Carbon nanotube |
| CNT/GCA | Carbon nanotube-graphitized carbon aerogel composite |
| CO | Carbon monoxide |
| $CO_2$ | Carbon dioxide |
| $CH_3OH$ | Methanol |
| DFT | Density functional theory |
| DMFC | Direct methanol fuel cell |
| ECNF | Electrospun carbon nanofibers |
| EDS | Energy dispersive spectroscopy |
| $e_g$ and $t_{2g}$ | *d*-orbital energy splitting in a transition metal atom exposed to crystal field environment (see "group theory" for details) |
| ESA | Electrochemical surface area |
| HER | Hydrogen evolution reaction |
| HRTEM | High resolution transmission electron microscopy |
| N-doped CNT | Nitrogen doped carbon nanotube |
| mAh/g | Milliamp-hour per gram (unit of specific capacitance) |
| $m^2/g$ | Square meter per gram (unit of the surface area) |
| Me | Metal |
| MEA | Membrane electrode assembly (e.g., in a fuel cell) |
| MOR | Methanol oxidation reaction |
| mA/mg | Milliamp per milligram (mass activity unit) |

| | |
|---|---|
| O | Oxygen atom or oxygen species on the surface of the catalyst particle |
| O-O | Oxygen-oxygen bond |
| OH or $OH_{ad}$ | Hydroxyl species on the surface of the catalyst particle |
| ORR | Oxygen reduction reaction |
| OER | Oxygen evolution reaction |
| PEMFC | Polymer electrolyte membrane fuel cell |
| psi | Pound per square inch (unit of pressure) |
| Pt-ESA | Electrochemical surface area of the platinum phase (platinum nanoparticles) |
| Pt/ECNF | Platinum nanoparticles deposited on electrospun carbon nanofiber |
| Pt/TiC | Platinum nanoparticles deposited on titanium carbide |
| PtIrCo | Platinum-iridium-cobalt alloy catalyst |
| PtIrCo/CA | Platinum-iridium-cobalt nanoparticles deposited on carbon aerogel |
| Pt/Ru | Platinum-ruthenium catalyst |
| PtRu/C | Platinum-ruthenium catalyst nanoparticles deposited on carbon |
| Pt-CO | Platinum-carbon monoxide (catalytically active site) |
| RF | Roughness factor of the catalyst |
| RHE | Reference hydrogen electrode |
| Ru | Ruthenium |
| S/cm | Siemens per centimeter (unit of electrical conductivity) |
| SCF | Supercritical fluid |
| $scCO_2$ | Supercritical carbon dioxide fluid |
| SHE | Standard hydrogen electrode |
| 3D | Three dimensional |
| TiC | Titanium carbide |
| V | Volt (electrical unit of voltage or potential difference) |
| Vulcan XC-72 | Type of high surface area carbon (carbon black) |
| W/kg | Watt per kilogram (power density unit) |
| Wh/kg | Watt-hour per kilogram (energy density unit) |
| Wh/L | Watt-hour per liter (energy density unit) |
| XRD | X-ray diffraction |

## References


1. Stambouli, B. A.; Traversa, E. Fuel Cells, An Alternative to Standard Sources of Energy. *Renewable Sustainable Energy Rev.* **2002**, *6* (3), 295–304.
2. Lemaire, X. *Glossary of Terms in Sustainable Energy Regulation*; August 2004.
3. Lu, Y.-C.; Gasteiger, H. A.; Shao-Horn, Y. Method Development to Evaluate the Oxygen Reduction Activity of High-Surface-Area Catalysts for Li-Air Batteries. *Electrochem. Solid-State Lett.* **2011**, *14* (5), A70–A74.

4. Inaba, M. In *Degradation Mechanism of Polymer Electrolyte Fuel Cells*, 14th International Conference on the Properties of Water and Steam, Kyoto, 2004.
5. Song, C.; Zhang, J., Electrocatalytic Oxygen Reduction Reaction. In *PEM Fuel Cell Electrocatalysts and Catalyst Layers*; Zhang, J., Ed.; Springer London: 2008; pp 89–134.
6. Nørskov, J. K.; Rossmeisl, J.; Logadottir, A.; Lindqvist, L.; Kitchin, J. R.; Bligaard, T.; Jónsson, H. Origin of the Overpotential for Oxygen Reduction at a Fuel-Cell Cathode. *J. Phys. Chem. B* **2004**, *108* (46), 17886–17892.
7. Bard, A. J.; Faulkner, L. R. *Electrochemical Methods: Fundamentals and Applications*; Wiley: 2000; p 864.
8. Lu, Y.-C.; Gasteiger, H. A.; Shao-Horn, Y. Catalytic Activity Trends of Oxygen Reduction Reaction for Nonaqueous Li-Air Batteries. *J. Am. Chem. Soc.* **2011**, *133* (47), 19048–19051.
9. Park, C. K.; Park, S. B.; Lee, S. Y.; Lee, H.; Jang, H.; Cho, W. I. Electrochemical Performances of Lithium-air Cell with Carbon Materials. *Bull. Korean Chem. Soc.* **2010**, *31* (11), 3221–3224.
10. Zhang, L.-L.; Wang, Z.-L.; Xu, D.; Zhang, X.-B.; Wang, L.-M. The Development and Challenges of Rechargeable Non-Aqueous Lithium–Air Batteries. *Int. J. Smart Nano Mater.* **2012**, *4* (1), 27–46.
11. Yang, X.-h.; He, P.; Xia, Y.-y. Preparation of Mesocellular Carbon Foam and Its Application for Lithium/Oxygen Battery. *Electrochem. Commun.* **2009**, *11* (6), 1127–1130.
12. Zhang, L.-L.; Wang, Z.-L.; Xu, D.; Zhang, X.-B.; Wang, L.-M. The Development and Challenges of Rechargeable Non-Aqueous Lithium–Air Batteries. *Int. J. Smart Nano Mater.* **2012**, *4* (1), 27–46.
13. Liu, H.-j.; Wang, F.; Zhao, Y.; Fong, H. Mechanically Resilient Electrospun TiC Nanofibrous Mats Surface-Decorated with Pt Nanoparticles for Oxygen Reduction Reaction with Enhanced Electrocatalytic Activities. *Nanoscale* **2013**, *5* (9), 3643–3647.
14. Saquing, C. D.; Kang, D.; Aindow, M.; Erkey, C. Investigation of the Supercritical Deposition of Platinum Nanoparticles into Carbon Aerogels. *Microporous Mesoporous Mater.* **2005**, *80* (1-3), 11–16.
15. Cansell; , F.; Aymonier, C. Design of Functional Nanostructured Materials Using Supercritical Fluids. *J. Supercritic. Fluids* **2009**, *47* (3), 508–516.
16. Kolla, P.; Smirnova, A. Methanol Oxidation and Oxygen Reduction Activity of PtIrCo-Alloy Nanocatalysts Supercritically Deposited within 3D Carbon Aerogel Matrix. *Electrochim. Acta* **2015**, *182*, 20–30.
17. Balandin, A. A. Modern State of the Multiplet Theory of Heterogeneous Catalysis. *Adv. Catal.* **1969**, *19*, 1–210.
18. Suntivich, J.; Gasteiger, H. A.; Yabuuchi, N.; Nakanishi, H.; Goodenough, J. B.; Shao-Horn, Y. Design Principles for Oxygen-Reduction Activity on Perovskite Oxide Catalysts for Fuel Cells and Metal–Air Batteries. *Nat. Chem.* **2011**, *3* (7), 546–550.
19. Lu, Y.-C.; Gasteiger, H. A.; Shao-Horn, Y. Method Development to Evaluate the Oxygen Reduction Activity of High-Surface-Area Catalysts for Li-Air Batteries. *Electrochem. Solid-State Lett.* **2011**, *14* (5), A70–A74.

20. Hosseini, S. A.; Niaei, A.; Salari, D.; Nabavi, S. R. Nanocrystalline $AMn_2O_4$ (A = Co, Ni, Cu) Spinels for Remediation of Volatile Organic Compounds—Synthesis, Characterization and Catalytic Performance. *Ceram. Int.* **2012**, *38* (2), 1655–1661.
21. Tarancón, A.; Morata, A.; Dezanneau, G.; Skinner, S. J.; Kilner, J. A.; Estradé, S.; Hernández-Ramírez, F.; Peiró, F.; Morante, J. R. $GdBaCo_2O_{5+x}$ Layered Perovskite as an Intermediate Temperature Solid Oxide Fuel Cell Cathode. *J. Power Sources* **2007**, *174* (1), 255–263.
22. Suntivich, J.; May, K. J.; Gasteiger, H. A.; Goodenough, J. B.; Shao-Horn, Y. A Perovskite Oxide Optimized for Oxygen Evolution Catalysis from Molecular Orbital Principles. *Science* **2011**, *334* (6061), 1383–1385.
23. Gómez de la Fuente, J. L.; Martínez-Huerta, M. V.; Rojas, S.; Terreros, P.; Fierro, J. L. G.; Peña, M. A. Enhanced Methanol Electrooxidation Activity of PtRu Nanoparticles Supported on $H_2O_2$-Functionalized Carbon Black. *Carbon* **2005**, *43*, 3002–3009.
24. Cerro-Alarcón, M.; Maroto-Valiente, A.; Rodríguez-Ramos, I.; Guerrero-Ruiz, A. Further Insights into the Ru Nanoparticles–Carbon Interactions and Their Role in the Catalytic Properties. *Carbon* **2005**, *43*, 2711–2715.
25. Chen, C.; Hsu, C.; Li, H. Growth and Characteristics of Carbon Nanotubes on Carbon Cloth as Electrodes. *Diamond Related Mater.* **2005**, *14*, 770–765.
26. Rajalakshmi, N.; Ryu, H.; Shaijumon, M. M.; Ramaprabhu, S. Performance of Polymer Electrolyte Membrane Fuel Cells with Carbon Nanotubes as Oxygen Reduction Catalyst Support Material. *J. Power Sources* **2005**, *140*, 250–255.
27. Zhang, D.; Shi, L.; Fang, J.; Li, X.; Dai, K. Preparation and Modification of Carbon Nanotubes. *Mater. Lett.* **2005**, *59*, 4044–4051.
28. Zhao, Y.; Zheng, M.; Cao, J.; Ke, X.; Liu, J.; Chen, Y.; Tao, J. Easy Synthesis of Ordered Meso/Macroporous Carbon Monolith for Use as Electrode in Electrochemical Capacitors. *Mater. Lett.* **2008**, *62*, 548–551.
29. Saquing, C. D.; Aindow, M.; Kang, D.; Erkey, C. Investigation of the Supercritical Deposition of Platinum Nanoparticles into Carbon Aerogels. *Microporous Mesoporous Mater.* **2005**, *80*, 11–16.
30. Moreno-Castilla, C.; Maldonado-Hódar, F. J. Carbon Aerogels for Catalysis Applications: An Overview. *Carbon* **2005**, *43*, 455–463.
31. Baumann, T. F.; Worsley, M. A.; Han, T. Y.-J.; Satcher, J. H., Jr. High Surface Area Carbon Aerogel Monoliths with Hierarchical Porosity. *J. Non-Cryst. Solids* **2008**, *354*, 3513–3515.
32. Liu, J.; Wang, X.; Wang, Y.; Huang, Q.; Dai, C.; Gamboa, S.; Sebastian, P. J. Structure and Electrochemical Properties of Carbon Aerogels Synthesized at Ambient Temperatures as Supercapacitors. *J. Non-Cryst. Solids* **2008**, *354*, 19–24.
33. Czakkel, O.; Marthi, K.; Geissler, E.; László, K. Influence of Drying on the Morphology of Resorcinol–Formaldehyde-Based Carbon Gels. *Microporous Mesoporous Mater.* **2005**, *86*, 124–133.
34. Wang, J.; Yang, X.; Wu, D.; Fu, R.; Dresselhaus, M. S.; Dresselhaus, G. The Porous Structures of Activated Carbon Aerogels and Their Effects on Electrochemical Performance. *J. Power Sources* **2008**, *185*, 589–594.

35. Job, N.; Marie, J. S.; Lambert, S.; Berthon-Fabry, S.; Achard, P. Carbon Xerogels as Catalyst Supports for PEM Fuel Cell Anode. *Energy Convers. Manage.* **2008**, *49*, 2461–2470.
36. Guraya, M.; Sprenger, S.; Rarog-Pilecka, W.; Szmigiel, D.; Kowalczyk, Z.; Muhler, M. The Effect of Promoters on the Electronic Structure of Ruthenium Catalysts Supported on Carbon. *Appl. Surf. Sci.* **2004**, *238*, 77–81.
37. Maass, S.; Finsterwalder, F.; Frank, G.; Hartmann, R.; Merten, C. Carbon Support Oxidation in PEM Fuel Cell Anodes. *J. Power Sources* **2008**, *176*, 444–451.
38. Coloma, F.; Sepulvedaescribano, A.; Rodriguezreinoso, F. Heat-Treated Carbon-Blacks as Supports for Platinum Catalysts. *J. Catal.* **1995**, *154*, 299–305.
39. Rajesh, B.; Karthik, V.; Karthikeyan, S.; Thampi, K. R.; Bonard, J.-M.; Viswanathan, B. Pt–$WO_3$ Supported on Carbon Nanotubes as Possible Anodes for Direct Methanol Fuel Cells. *Fuel* **2002**, *81*, 2177–2190.
40. Huang, C.; Doong, R.; Gu, D.; Zhao, D. Dual-Template Synthesis of Magnetically-Separable Hierarchically-Ordered Porous Carbons by Catalytic Graphitization. *Carbon* **2011**, *9*, 3055–3064.
41. Coloma, F.; Prado-Burguete, C.; Rodriguez-Reinoso, F. Pt-C Interaction in Catalyst Supported on a Carbon Black Subjected to Different Heat Treatments. *Stud. Surf. Sci. Catal.* **1993**, *75*, 2103–2106.
42. Shao, Y.; Yin, G.; Gao, Y. Understanding and Approaches for Durability Issues of Pt-Based Catalysts for PEM Fuel Cell. *J. Power Sources* **2007**, *171*, 558–566.
43. Kato, M.; Tani, Y.; Imamura, T.; Hirota, K.; Yoshimura, K. Metal-Insulator Transition in $Pb_{2-x}Ln_xRu_2O_{7-x}$ (Ln = Sm, Eu) Compounds. *Phys. B* **2008**, *403*, 1315–1317.
44. Lacorre, P.; Torrance, J. B.; Pannetier, J.; Nazzal, A. I.; Wang, P. W.; Huang, T. C. Synthesis, Crystal Structure, and Properties of Metallic $PrNiO_3$: Comparison with Metallic $NdNiO_3$ and Semiconducting $SmNiO_3$. *J. Solid State Chem.* **1991**, *91*, 225–237.
45. Baldomir, D.; Pardo, V.; Blanco-Canosa, S.; Rivadulla, F.; Rivas, J.; Pineiro, A.; Arias, J. E. Pressure-Induced Metal-Insulator Transition in $MgV_2O_4$. *Phys. B* **2008**, *403*, 1639–1941.
46. Ioroi, T.; Senoh, H.; Yamazaki, S-I.; Siroma, Z.; Fujiwara, N.; Yasuda, K. Stability of Corrosion-Resistant Magneli-Phase $Ti_4O_7$-Supported PEMFC Catalysts at High Potentials. *J. Electrochem. Soc.* **2008**, *155*, B321–B326.
47. Ioroi, T.; Siroma, Z.; Fujiwara, N.; Yamazaki, S-I.; Yasuda, K. Sub-Stoichiometric Titanium Oxide-Supported Platinum Electrocatalyst for Polymer Electrolyte Fuel Cells. *Electrochem. Commun.* **2005**, *7*, 183–188.
48. Kolla, P.; Lai, C.; Mishra, S.; Fong, H.; Rhine, W.; Smirnova, A. CVD Grown CNTs within Iron Modified and Graphitized Carbon Aerogel as Durable Oxygen Reduction Catalysts in Acidic Medium. *Carbon* **2014**, *79*, 518–528.
49. Jeon, M. K.; Lee, C. H.; Park, G.; Kang, K. H. Combinatorial Search for Oxygen Reduction Reaction Electrocatalysts: A Review. *J. Power Sources* **2012**, *216*, 400–408.

50. Smirnova, A. L.; Sadykov, V. A.; Mezentseva, N. V.; Bunina, R. V.; Pelipenko, V. V.; Alikina, G. M.; Krieger, T. A.; Bobrova, L. N.; Smorygo, O. L.; Van Berkel, F.; Rietveld, B. Design and Testing of Structured Catalysts for Internal Reforming of $CH_4$ in Intermediate Temperature Solid Oxide Fuel Cells. *ECS Trans.* **2011**, *35*, 2771–2780.
51. McCrory, C. C. L.; Jung, S.; Peters, J. C.; Jaramillo, T. F. Benchmarking Heterogeneous Electrocatalysts for the Oxygen Evolution Reaction. *J. Am. Chem. Soc.* **2013**, *135* (45), 16977–16987.
52. Miles, M. H.; Klaus, E. A.; Gunn, B. P.; Locker, J. R.; Serafin, W. E.; Srinivasan, S. The Oxygen Evolution Reaction on Platinum, Iridium, Ruthenium and Their Alloys at 80°C in Acid Solutions. *Electrochim. Acta* **1978**, *23*, 521–526.
53. Reier, T.; Oezaslan, M.; Strasser, P. Electrocatalytic Oxygen Evolution Reaction (OER) on Ru, Ir, and Pt Catalysts: A Comparative Study of Nanoparticles and Bulk Materials. *ACS Catal.* **2012**, *2* (8), 1765–1772.
54. Song, S.; Zhang, H.; Ma, X.; Shao, Z.; Baker, R. T.; Yi, B. Electrochemical Investigation of Electrocatalysts for the Oxygen Evolution Reaction in PEM Water Electrolyzers. *Int. J. Hydrogen Energy* **2008**, *33* (19), 4955–4961.
55. Di Blasi, A.; D'Urso, C.; Baglio, V.; Antonucci, V.; Arico', A. S.; Ornelas, R.; Matteucci, F.; Orozco, G.; Beltran, D.; Meas, Y.; Arriaga, L. G. Preparation and Evaluation of $RuO_2$–$IrO_2$, $IrO_2$–Pt and $IrO_2$–$Ta_2O_5$ Catalysts for the Oxygen Evolution Reaction in an SPE Electrolyzer. *J. Appl. Electrochem.* **2009**, *39* (2), 191–196.
56. Mamaca, N.; Mayousse, E.; Arrii-Clacens, S.; Napporn, T. W.; Servat, K.; Guillet, N.; Kokoh, K. B. Electrochemical Activity of Ruthenium and Iridium Based Catalysts for Oxygen Evolution Reaction. *Appl. Catal., B* **2012**, *111–112* (0), 376–380.
57. Rivas, S.; Arriaga, L. G.; Morales, L.; Fernández, A. M. Evaluation of Pt-Ru-Ir as Bifunctional Electrocatalysts for the Oxygen Electrode in a Unitized Regenerative Fuel Cell. *Int. J. Electrochem. Sci.* **2012**, *7*, 3601–3609.
58. Swette, L. L.; Laconti, A. B.; McCatty, S. A. Proton-Exchange Membrane Regenerative Fuel-Cells. *J. Power Sources* **1994**, *47* (3), 343–351.
59. Suntivich, J.; Gasteiger, H. A.; Yabuuchi, N.; Nakanishi, H.; Goodenough, J. B.; Shao-Horn, Y. Design Principles for Oxygen-Reduction Activity on Perovskite Oxide Catalysts for Fuel Cells and Metal-Air Batteries. *Nat. Chem.* **2011**, *3* (7), 546–550.
60. Sohn, J. M.; Kim, M. R.; Woo, S. I. The catalytic Activity and Surface Characterization of $Ln_2B_2O_7$ (Ln=Sm, Eu, Gd and Tb; B=Ti or Zr) with Pyrochlore Structure as Novel $CH_4$ Combustion Catalyst. *Catal. Today.* **2003**, *83*, 289–297.
61. Gaur, S.; Haynes, D. J.; Spivey, J. J. Rh, Ni, and Ca Substituted Pyrochlore Catalysts for Dry Reforming of Methane. *Appl. Catal., A* **2011**, *403*, 142–151.
62. Beyerlein, R. A.; Horowitz, H. S.; Longo, J. M. The Electrical Properties of $A_2[Ru_{2-x}A_x]O_{7-y}$ (A = Pb or Bi) Pyrochlores as a Function of Composition and Temperature. *J. Solid State Chem.* **1988**, *72*, 2–13.

63. Daidouh, A.; Pico, C.; Veiga, M. L.; Almontassir, A.; Abboudi, M.; Hanebali, L. Structural and Electrical Behaviour of the Pyrochlores: $Sm_{2-x}M_xTi_2O_{7-x/2}$ (M=Mg, Co, Ni). *Solid State Sci.* **2004**, *6*, 71–76.
64. Ten Kortenaar, M. V.; Vente, J. F.; Ijdo, D. J. W.; Müller, S.; Kötz, R. Oxygen Evolution and Reduction on Iridium Oxide Compounds. *J. Power Sources* **1995**, *56*, 51–60.
65. Shukla, A. K.; Kannan, A. M.; Hegde, M. S.; Gopalakrishnan, J. Effect of Counter Cations on Electrocatalytic Activity of Oxide Pyrochlores Towards Oxygen Reduction/Evolution in Alkaline Medium: An Electrochemical and Spectroscopic Study. *J. Power Sources* **1991**, *35*, 163–173.
66. Wu, Z.-S.; Zhou, G.; Yin, L.-C.; Ren, W.; Li, F.; Cheng, H.-M. Graphene/Metal Oxide Composite Electrode Materials for Energy Storage. *Nano Energy* **2012**, *1*, 107–131.
67. Yang, Y.; Shi, M.; Zhou, Q.-F.; Li, Y.-S.; Fu, Z.-W. Platinum Nanoparticle–Graphene Hybrids Synthesized by Liquid Phase Pulsed Laser Ablation as Cathode Catalysts for Li-Air batteries. *Electrochem. Commun.* **2012**, *20*, 11–14.
68. Tang, N.; Tian, X.; Yang, C.; Pi, Z.; Han, Q. Facile Synthesis of $\alpha$-$MnO_2$ Nanorods for High-Performance Alkaline Batteries. *J. Phys. Chem. Solids* **2010**, *71*, 258–262.
69. Thapa, A. K.; Shin, T. H.; Ida, S.; Sumanasekera, G. U.; Sunakara, M. K.; Ishihara, T. Gold-Palladium Nanoparticles Supported by Mesoporous $\beta$-$MnO_2$ Air Electrode for Rechargeable Lithium-Air Battery. *J. Power Sources* **2012**, *220*, 211–216.
70. Song, M.-K.; Park, S.; Alamgir, F. M.; Cho, J.; Liu, M. Nanostructured Electrodes for Lithium-Ion and Lithium-Air Batteries: The Latest Developments, Challenges, and Perspectives. *Mater. Sci. Eng., R* **2011**, *72*, 203–252.
71. Li, Y.; Wang, J.; Li, X.; Liu, J.; Geng, D.; Yang, J.; Li, R.; Sun, X. Nitrogen-Doped Carbon Nanotubes as Cathode for Lithium–Air Batteries. *Electrochem. Commun.* **2011**, *13*, 668–672.
72. Kichambare, P.; Kumar, J.; Rodrigues, S.; Kumar, B. Electrochemical Performance of Highly Mesoporous Nitrogen Doped Carbon Cathode in Lithium–Oxygen Batteries. *J. Power Sources* **2011**, *196*, 3310–3316.
73. Li, Y.; Wang, J.; Li, X.; Geng, D.; Banis, M. N.; Li, R.; Sun, X. Nitrogen-Doped Graphene Nanosheets as Cathode Materials with Excellent Electrocatalytic Activity for High Capacity Lithium-Oxygen Batteries. *Electrochem. Commun.* **2012**, *18*, 12–15.
74. Yang, Y.; Shi, M.; Zhou, Q.-F.; Li, Y.-S.; Fu, Z.-W. Platinum Nanoparticle–Graphene Hybrids Synthesized by Liquid Phase Pulsed Laser Ablation as Cathode Catalysts for Li-Air Batteries. *Electrochem. Commun.* **2012**, *20*, 11–14.
75. Jang, M.-W.; Jung, H.-G.; Scrosati, B.; Sun, Y.-K. Improved Co-Substituted $LiNi_{0.5-x}Co_{2x}Mn_{1.5-x}O_4$ Lithium Ion Battery Cathode Materials. *J. Power Sources* **2012**, 354–359.
76. Li, Y.; Huang, K.; Xing, Y. A Hybrid Li-Air Battery with Buckypaper Air Cathode and Sulfuric Acid Electrolyte. *Electrochim. Acta* **2012**, *81*, 20–24.

77. McCrory, C. C. L.; Jung, S.; Ferrer, I. M.; Chatman, S. M.; Peters, J. C.; Jaramillo, T. F. Benchmarking Hydrogen Evolving Reaction and Oxygen Evolving Reaction Electrocatalysts for Solar Water Splitting Devices. *J. Am. Chem. Soc.* **2015**, *137*, 4347–4357.
78. Evangelisti, F.; Moré, R.; Hodel, F.; Luber, S.; Patzke, G. R. 3d–4f [$Co^{II}_3Ln(OR)_4$] Cubanes as Bio-Inspired Water Oxidation. *Am. Chem. Soc.* **2015**, *137*, 11076–11084.
79. Ai, G.; Li, H.; Liu, S.; Mo, R.; Zhong, J. Solar Water Splitting by $TiO_2$/CdS/ Co–Pi Nanowire Array Photoanode Enhanced with Co–Pi as Hole Transfer Relay and CdS as Light Absorber. *Adv. Funct. Mater.* **2015**, *25* (35), 5706–5713.
80. P. Kolla, A. Smirnova *Oxygen Reduction and Oxygen Evolution Catalytic Activity of Pr and Sm Based Perovskite/Graphene Nanocomposites for Applications in Alkaline Fuel Cells and Metal-Air Batteries*, MRS 2015 Spring meeting, San Fransisco, CA (J. Power Sources – in preparation).
81. Abbasi, T.; Abbasi, S. A. Renewable Hydrogen: Prospects and Challenges. *Renewable Sustainable Energy Rev.* **2011**, *15*, 3034–3040.
82. Roy, S. C.; Varghese, O. K.; Paulose, M.; Grimes, C. A. Toward Solar Fuels: Photocatalytic Conversion of Carbon Dioxide to Hydrocarbons. *ACS Nano* **2010**, *4* (3), 1259–1278.
83. Xu, Y.; Ye, T.; Qiu, S.; Ning, S.; Gong, F.; Liu, Y.; Li, Q. High Efficient Conversion of $CO_2$-Rich Bio-Syngas to CO-Rich Bio-Syngas Using Biomass Char: A Useful Approach for Production of Bio-Methanol from Bio-Oil. *Bioresour. Technol.* **2011**, *102*, 6239–6245.
84. Hasegawa, F.; Yokoyama, S.; Imou, K. Methanol or Ethanol Produced from Woody Biomass: Which is More Advantageous? *Bioresour. Technol.* **2010**, *101* (1), S109–S111.
85. Burstein, G. T.; Barnett, C. J.; Kucernak, A. R.; Williams, K. R. Aspects of the Anodic Oxidation of Methanol. *Catal. Today.* **1997**, *38*, 425–437.
86. Cloutier, C. R.; Wilkinson, D. P. Electrolytic Production of Hydrogen from Aqueous Acidic Methanol Solutions. *Int. J. Hydrogen Energy* **2010**, *35*, 3967–3984.
87. Awasthi, A.; Scott, K.; Basu S. Dynamic Modeling and Simulation of a Proton Exchange Membrane Electrolyzer for Hydrogen Production. *Int. J. Hydrogen Energy* **2011**, *36*, 14779–14786.
88. Ito, H.; Maeda, T.; Nakano, A.; Takenaka, H. Properties of Nafion Membranes under PEM Water Electrolysis Conditions. *Int. J. Hydrogen Energy* **2011**, *36*, 10527–10540.
89. Siracusano, S.; Baglio, V.; Briguglio, N.; Brunaccini, G.; Di Blasi, A.; Stassi, A.; Ornelas, R.; Trifoni, E.; Antonucci, V.; Aricò, A. S. An Electrochemical Study of a PEM Stack for Water Electrolysis. *Int. J. Hydrogen Energy* **2012**, *37*, 1939–1946.
90. Take, T.; Tsurutani, K.; Umeda, M. Hydrogen Production by Methanol–Water Solution Electrolysis. *J. Power Sources* **2007**, *164*, 9–16.
91. Balaji, R.; Senthil, N.; Vasudevan, S.; Ravichandran, S.; Mohan, S.; Sozhan, G.; Madhu, S.; Kennedy, J.; Pushpavanam, S.; Pushpavanam, M. Development and Performance Evaluation of Proton Exchange Membrane

(PEM) Based Hydrogen Generator for Portable Application. *Int. J. Hydrogen Energy* **2011**, *36*, 1399–1403.

92. Iwasita, T. Electrocatalysis of Methanol Oxidation. *Electrochim. Acta* **2002**, *47*, 3663–3683.
93. Kobayashi, T.; Otomo, J.; Wen, C.; Takahashi, H. Direct Alcohol Fuel Cell—Relation between the Cell Performance and the Adsorption of Intermediate Originating in the Catalyst-Fuel Combinations. *J. Power Sources* **2003**, *124*, 34–39.
94. Rossmeisl, J.; Ferrin, P.; Tritsaris, G. A.; Nilekar, A. U.; Koh, S.; Bae, S. E.; Brankovic, S. R.; Strasser, P.; Mavrikakis, M. Bifunctional Anode Catalysts for Direct Methanol Fuel Cells. *Energy Environ. Sci.* **2012**, *5* (8), 8335–8342.
95. Roth, C.; Benker, N.; Theissmann, R.; Nichols, R. J.; Schiffrin, D. J. Bifunctional Electrocatalysis in Pt–Ru Nanoparticle Systems. *Langmuir* **2008**, *24* (5), 2191–2199.
96. Sun, Y.-P.; Xing, L.; Scott, K. Analysis of the Kinetics of Methanol Oxidation in a Porous Pt–Ru Anode. *J. Power Sources* **2010**, *195*, 1–10.
97. Shih, Z.-Y.; Yang, Z.; Lin, Z.-H.; Chang, H.-T. Direct Methanol Fuel Cells Using Se/Ru Core/Shell Anodes Provide High Catalytic Activity and Stability. *Int. J. Hydrogen Energy* **2011**, *36*, 7303–7309.
98. Park, K. W.; Choi, J. H.; Kwon, B. K.; Lee, S. A.; Sung, Y. E.; Ha, H. Y.; Hong, S. A.; Kim, H.; Wieckowski, A. Chemical and Electronic Effects of Ni in Pt/Ni and Pt/Ru/Ni Alloy Nanoparticles in Methanol Electrooxidation. *J. Phys. Chem. B* **2002**, *106*, 1869–1877.
99. De Oliveira, M. B.; Paula L., ; Profeti, R.; Olivi, P. Electrooxidation of Methanol on $PtM_yO_x$ (M = Sn, Mo, Os or W) Electrodes. *Electrochem. Commun.* **2005**, *7*, 703–709.
100. Bonilla, S. H.; Zinola, C. F.; Rodríguez, J.; Díaz, V.; Ohanian, M.; Martínez, S.; Giannetti, B. F. Catalityc Effects of Ruthenium and Osmium Spontaneous Deposition on Platinum Surfaces toward Methanol Oxidation. *J. Colloid Interface Sci.* **2005**, *288*, 377–386.
101. Patra, S.; Dash, S.; Anand, V.; Nimisha, C. S.; Rao, G. M.; Munichandraiah, N. Electrochemical Co-Deposition of Bimetallic Pt–Ru Nanoclusters Dispersed on Poly(3,4-ethylenedioxythiophene) and Electrocatalytic Behavior for Methanol Oxidation. *Mater. Sci. Eng. B.* **2011**, *176*, 785–791.
102. MacDonald, J. P.; Gualtieri, B.; Runga, N.; Teliz, E.; Zinola, C. F. Modification of Platinum Surfaces by Spontaneous Deposition: Methanol Oxidation Electrocatalysis. *Int. J. Hydrogen Energy* **2008**, *33*, 7048–7061.
103. Jusys, Z.; Schmidt, T. J.; Dubau, L.; Lasch, K.; Jörissen, L.; Garche, J.; Behm, R. J. Activity of $PtRuMeO_x$ (Me = W, Mo or V) Catalysts Towards Methanol Oxidation and Their Characterization. *J. Power Sources* **2002**, *105*, 297–304.
104. Kang, D. K.; Noh, C. S.; Kim, N. H.; Cho, S.-H.; Sohn, J. M.; Kim, T. J.; Park, Y.-K. Effect of Transition Metals (Ni, Sn and Mo) in $Pt_5Ru_4M$ Alloy Ternary Electrocatalysts on Methanol Electro-Oxidation. *J. Ind. Eng. Chem.* **2010**, *16*, 385–389.

105. Hwu, H. H.; Chen, J. G. Reactions of Methanol and Water over Carbide-Modified Mo(1 1 0). *Surface Sci.* **2003**, *536*, 75–87.
106. Hwu, H. H.; Chen, J. G. Comparative Studies of Methanol Decomposition on Carbide-Modified V(110) and Ti(0001). *Chem. Rev.* **2005**, *105*, 185–199.
107. Liu, N.; Kourtakis, K.; Figueroa, J. C.; Chen, J. G. Potential Application of Tungsten Carbides as Electrocatalysts: III. Reactions of Methanol, Water, and Hydrogen on Pt-Modified C/W(111) Surfaces. *J. Catal.* **2003**, *215*, 254–263.
108. Joo, J. B.; Kim, J. S.; Kim, P.; Yi, J. Simple Preparation of Tungsten Carbide Supported on Carbon for Use as a Catalyst Support in a Methanol Electro-Oxidation. *Mater. Lett.* **2008**, *62*, 3497–3499.
109. Serov, A.; Kwak, C. Review of Non-Platinum Anode Catalysts for DMFC and PEMFC Application. *Appl. Catal., B* **2009**, *90*, 313–320.
110. Cui, G.; Shen, P. K.; Meng, H.; Zhao, J.; Wu, G. Tungsten Carbide as Supports for Pt Electrocatalysts with Improved CO Tolerance in Methanol Oxidation. *J. Power Sources* **2011**, *196*, 6125–6130.
111. Li, L.; Xu, B. Q. Electrooxidation of CO Adsorbed on PtMo/C Catalyst: Effect of Catalyst Preparation. *Acta Phys.-Chim. Sin.* **2005**, *21*, 1132–1137.
112. Antolini, E. Platinum-Based Ternary Catalysts for Low Temperature Fuel Cells: Part I. Preparation Methods and Structural Characteristics. *Appl. Catal., B* **2007**, *74*, 324–336.
113. Vigier, F.; Coutanceau, C.; Hahn, F.; Belgsir, E. M.; Lamy, C. Mechanism of Alcohol Oxidation. *J. Electroanal. Chem.* **2004**, *563*, 81–88.
114. Neburchilov, V.; Wang, H.; Zhang, J. Low Pt Content Pt–Ru–Ir–Sn Quaternary Catalysts for Anodic Methanol Oxidation in DMFC. *Electrochem. Comm.* **2007**, *9*, 1788–1792.
115. Sivakumar, P.; Tricoli,, V. Pt-Ru-Ir Nanoparticles Prepared by Vapor Deposition as a Very Efficient Anode Catalyst for Methanol Fuel Cells. *Electrochem. Solid-State Lett.* **2006**, *9*, A167–A170.
116. Antolini, E.; Gonzalez, E. R. Effect of Synthesis Method and Structural Characteristics of Pt–Sn Fuel Cell Catalysts on the Electro-Oxidation of $CH_3OH$ and $CH_3CH_2OH$ in Acid Medium. *Catal. Today* **2011**, *16*, 28–38.
117. Ma, J.-H.; Feng, Y.-Y.; Yu, J.; Zhao, D.; Wang, A.-J.; Xu, B.-Q. Promotion by Hydrous Ruthenium Oxide of Platinum for Methanol Electro-Oxidation. *J. Catal.* **2010**, *275*, 34–44.
118. Pistonesi, C.; Juan, A.; Farkas, A. P.; Solymosi, F. DFT Study of Methanol Adsorption and Dissociation on β-$Mo_2C$(0 0 1). *Surf. Sci.* **2008**, *602*, 2206–2211.
119. Koós, Á.; Barthos, R.; Solymosi, F. Reforming of Methanol on a K-Promoted $Mo_2C$/Norit Catalyst. *J. Phys. Chem. C* **2008**, *112*, 2607–2612.
120. Xing, L.; Jia, J.; Wang, Y.; Zhang, B.; Dong, S. Pt modified $TiO_2$ Nanotubes Electrode: Preparation and Electrocatalytic Application for Methanol Oxidation. *Int. J. Hydrogen Energy* **2010**, *35*, 12169–12173.
121. Hirakawa, K.; Inoue, M.; Abe, T. Methanol Oxidation on Carbon-Supported Pt–Ru and $TiO_2$ (Pt-Ru/$TiO_2$/C) Electrocatalyst Prepared Using Polygonal Barrel-Sputtering Method. *Electrochim. Acta* **2010**, *55*, 5874–5880.

122. Cao, H.; Wang, Z.; Hou, G.; Zheng, G. $TiO_2$ Nanotube-Supported Amorphous Ni–B Electrode for Electrocatalytic Oxidation of Methanol. *Surf. Coat. Technol.* **2010**, *205*, 885–889.
123. Hepel, M.; Dela, I.; Hepel, T.; Luo, J.; Zhong, C. J. Novel Dynamic Effects in Electrocatalysis of Methanol Oxidation on Supported Nanoporous $TiO_2$ Bimetallic Nanocatalysts. *Electrochim. Acta.* **2007**, *52*, 5529–5547.
124. Antolini, E. Composite Materials: An Emerging Class of Fuel Cell Catalyst Supports. *Appl. Catal., B* **2010**, *100*, 413–426.
125. Tschöpe, A.; Sommer, E; Birringer, R. Grain Size-Dependent Electrical Conductivity of Polycrystalline Cerium Oxide: I. Experiments. *Solid State Ionics* **2001**, *139*, 255–265.
126. Kosacki, I.; Anderson, H. U. Grain Boundary Effects in Nanocrystalline Mixed Conducting FilmsIn *Encyclopedia of Materials: Science and Technology*; Elsevier, 2001; pp 3609−3617.
127. Augustyn, V.; Dunn, B. Vanadium Oxide Aerogels: Nanostructured Materials for Enhanced Energy Storage. *C. R. Chim.* **2010**, *13*, 130–141.
128. Du, C.; Chen, M.; Cao, X.; Yin, G.; Shi, P. A Novel CNT@$SnO_2$ Core–Sheath Nanocomposite as a Stabilizing Support for Catalysts of Proton Exchange Membrane Fuel Cells. *Electrochem. Commun.* **2009**, *11*, 496–498.
129. Song, H.; Xiao, P.; Qiu, X.; Zhu, W. Design and Preparation of Highly Active Carbon Nanotube-Supported Sulfated $TiO_2$ and Platinum Catalysts for Methanol Electrooxidation. *J. Power Sources* **2010**, *195*, 1610–1614.
130. Song, Y.-Y.; Gao, Z.-D.; Schmuki, P. Highly Uniform Pt Nanoparticle Decoration on $TiO_2$ Nanotube Arrays: A Refreshable Platform for Methanol Electrooxidation. *Electrochem. Commun.* **2011**, *13*, 290–293.
131. Park, K.-W.; Lee, Y.-W.; Oh, J.-K.; Kim, D.-Y.; Han, S.-B.; Ko, A.-R.; Kim, S. –J.; Kim, H.-S. $TiO_2$-Based Nanowire Supported Catalysts for Methanol Electrooxidation in Direct Methanol Fuel Cells. *J. Ind. Eng. Chem.* **2011**, *17*, 696–699.
132. Justin, P.; Charan, P. H. K.; Rao, G. R. High Performance Pt–$Nb_2O_5$/C Electrocatalysts for Methanol Electrooxidation in Acidic Media. *Appl. Catal., B* **2010**, *100*, 510–515.
133. Zhao, Y.; Zhan, L.; Tian, J.; Nie, S.; Ning, Z. $MnO_2$ Modified Multi-Walled Carbon Nanotubes Supported Pd Nanoparticles for Methanol Electro-Oxidation in Alkaline Media. *Int. J. Hydrogen Energy* **2010**, *35*, 10522–10526.
134. Cansell, F.; Aymonier, C. Design of Functional Nanostructured Materials Using Supercritical Fluids. *J. Supercrit. Fluids* **2009**, *47* (3), 508–516.
135. Szczurek, A.; Amaral-Labat, G.; Fierro, V.; Pizzi, A.; Masson, E.; Celzard, A. Porosity of Resorcinol-Formaldehyde Organic and Carbon Aerogels Exchanged and Dried with Supercritical Organic Solvents. *Mater. Chem. Phys.* **2011**, *129*, 1221–1232.
136. Yi, C.; Shi, J.; Xue, S. J.; Jiang, Y.; Li, D. Effects of Supercritical Fluid Extraction Parameters on Lycopene Yield and Antioxidant Activity. *Food Chem.* **2009**, *113*, 1088–1094.

137. Kawashima, A.; Watanabe, S.; Iwakiri, R.; Honda, K. Removal of Dioxins and Dioxin-Like PCBs from Fish Oil by Countercurrent Supercritical $CO_2$ Extraction and Activated Carbon Treatment. *Chemosphere* **2009**, *75*, 788–794.
138. Shin, Y. H.; Shin, N. C.; Veriansyah, B.; Kim, J.; Lee, Y.-W. Supercritical Water Oxidation of Wastewater from Acrylonitrile Manufacturing Plant. *J. Hazard. Mater.* **2009**, *163*, 1142–1147.
139. Lai, C.; Kolla, P.; Zhao, Y.; Fong, H.; Smirnova, A. Lignin-Derived Electrospun Carbon Nanofiber Mats with Supercritically Deposited Ag Nanoparticles for Oxygen Reduction Reaction in Alkaline Fuel Cells. *Electrochim. Acta* **2014**, *130*, 431–438.
140. Kolla, P.; Smirnova, A. Methanol Oxidation on Hybrid Catalysts: PtRu/C Nanostructures Promoted with Cerium and Titanium Oxides. *Int. J. Hydrogen Energy* **2013**, *38*, 15152–15159.
141. Deshpande, K.; Mukasyan, A.; Varma, A. High Throughput Evaluation of Perovskite-Based Anode Catalysts for Direct Methanol Fuel Cells. *J. Power Sources* **2006**, *158*, 60–68.
142. Zhu, Y.; Cabrera, C. Methanol Oxidation at the Electrochemical Codeposited Pt-Os Composite Electrode. *Electrochem. Solid-State Lett.* **2001**, *4*, A45–A48.

Chapter 8

# Metal–Organic Frameworks for Methane Storage

**Xuan Wang, Stephen Fordham, and Hong-Cai Zhou***

**Department of Chemistry, Texas A&M University, College Station, Texas 77843-3012, United States**
***E-mail: zhou@chem.tamu.edu**

Natural gas has demonstrated potential as an alternative fuel for vehicular application due to its abundance and low price. With the goal of practical natural gas vehicles, tremendous efforts have been dedicated to find an optimal method to store natural gas in a cheap and feasible tank with an economic driving range. The structural richness, extraordinary high porosity, and large surface area endow metal–organic frameworks as a new category of sorbents for natural gas storage. An introduction of the fundamentals of methane storage is presented in order to improve understanding of important sorption concepts. Several benchmark MOFs with high gravimetric and volumetric methane capacities are enumerated with detailed analyses on their adsorption sites and heat of adsorption. Based on insights of framework-methane interactions, a couple of optimal design characteristics are identified, which facilitate the design and synthesize of new MOFs for methane storage.

## 1. Natural Gas Storage as an Alternative Fuel

The ever-growing consumption of gasoline and diesel has raised concerns about both the sustainability of oil supplies as well as the development of environmental problems. In particular, the carbon dioxide produced by the combustion of gasoline and diesel have direct correlation to global climate change. Therefore, there is much motivation to use energy sources with lower carbon footprints, such as hydrogen and nature gas. However, hydrogen fuel is limited due to production. It is commonly known that the electrolysis of water produces hydrogen, but it is too expensive to be effectively commercialized. Moreover, most commercial production of hydrogen is through the steam reforming of methane, which make methane more direct and feasible to serve as a future energy carrier.

### 1.1. Current Status of Natural Gas Vehicles

In 2012, the consumption of gasoline in United States increased to 134 billion gallons per year compared to 117 billion gallons in 1995 (*1*). With such a huge global demand for gasoline, gasoline prices have increased dramatically year by year. Natural gas is a gaseous mixture of predominately methane, some longer chain hydrocarbons, carbon dioxide, hydrogen sulfide, nitrogen and oxygen. The abundance of natural gas in the environment yields a low price fuel with wide availability, thus natural gas displays potential to serve as an alternative fuel compared to other fuels, such as ethanol, hydrogen, electricity, all of which require energy input from external sources to generate. Even though extensive pipeline networks of natural gas have been constructed through the U.S. only ~0.1 % of natural gas is used for transportation (*1*). Much more interest for natural gas vehicles has been due to the fact that combustion of methane boasts the smallest amount of $CO_2$ per unit of heat produced among fossil fuels. In addition, methane produces a larger heat of combustion compare to that of gasoline on a gravimetric unit basis (55.7 MJ/kg for methane vs 46.4 MJ/ kg of gasoline).

### 1.2. LNG, CNG, and ANG

Currently, there are two common storage methods of natural gas—liquefied natural gas (LNG) and compressed natural gas (CNG). The LNG storage method involves the liquefaction of the natural gas below the critical temperature of methane 191K (-81°C) with a risk of boil-off losses, which requires an expensive cryogenic vessels as the carrier. On the other hand, the CNG storage method refers to the compression of natural gas in a heavy, thick-walled cylinder with pressure of 200-300 bar produced by multi-stage compressors. Current technology limits LNG and CNG to a volumetric energy density of 22.2 $MJ \cdot L^{-1}$ and 9.2 $MJ \cdot L^{-1}$, respectively. The lower energy densities of LNG and CNG (32.4 $MJ \cdot L^{-1}$ for gasoline), and requirements of specialized containers indicate a relatively short driving distances with additional limitations of cargo spaces. In order to realize the benefits of natural gas vehicles, adsorbed natural gas (ANG) has become an active research topic (*2–5*). ANG technology, using porous sorbents to store natural gas at a substantially higher concentration than that with simple compression (*6*). Possessing sufficient volumetric storage ability yields ANG sorbents as a competitive alternative to existing NG storage methods with no requirement of refrigeration methods or expensive ancillary equipment. ANG also has merits due to efficient gas filling and release for automotive applications.

### 1.3. DOE Target

Identifying the potential of ANG, the Advance Research Projects Agency-Energy (APRA-E) of the United States Department of Energy (DOE) issued a call including the ANG storage target to guide the future development of sorbents. The desired sorbent should have a volumetric capacity of 350 $cm^3$ (STP) $\cdot cm^{-3}$ (STP=standard temperature and pressure: 1 bar, 273.15 K), and a gravimetric capacity greater than 0.5 $g_{CH4}/g_{sorbent}$. Additionally, ideal sorbents are expected to demonstrate resistance to the impurities in natural gas with a cost of \$10 / $kg_{sorbents}$ (*7*).

### 1.4. Porous Materials as Methane Storage Media

Following the call for methane storage, recent advances of porous material with high surface area have shown tremendous potential with increasing density of natural gas in the form of ANG (*8–13*). At the beginning, zeolites were among the first evaluated materials for the ANG technology (*14*). Studies continues to design and understand zeolites materials as natural gas sorbents (*15–17*). However, the inherent properties of zeolites prevent their use in practical vehicle applications (*12*, *18*). First, the low microporosity of zeolites prevents the realization of surface area larger than 1000 $m^2/g$. Since methane has no permanent dipole or quadrupole moment, the ionic zeolites display less affinity to methane compared to other guest species due to the weak ion-induced dipole forces. In addition, the hydrophilic surface of zeolites enhances the preferred interactions with water rather than methane, which diminishes the overall storage capacity in the presence of wet gases.

Activated carbon represents another class of porous materials which has been widely used in industry. Difficulties in the enhancement of bulk density of methane hinders the utilization of activated carbons (*12*). At standard temperature and pressure, zeolites exhibit a methane capacity of 100 $cm^3$(STP) $\cdot cm^{-3}$ (*7*), while 100-170 $cm^3$(STP) $\cdot cm^{-3}$ are the general values for activated carbons (*19*).

Metal–organic frameworks have emerged as a class of advanced porous materials with adjustable surface area, tunable pore size, and uniform pore size distribution. MOFs are 3D networks constructed from metal-containing units (this generally refers to secondary building units (SBUs)) and organic units (multitopic carboxylates) through coordination bonds. The structural merits of MOFs provide interesting materials for applications in gas storage and separations, catalysis, magnetism and sensing (*20–32*). The first application of methane storage with MOFs was reported by Kitagawa and co-workers (*33*). From then on, MOFs has been realized as new generation sorbents for methane storage. In this book chapter, methane storage in MOFs will be introduced. Current benchmark MOFs as methane sorbents will be listed and discussed to drive the optimal design characteristics for optimal MOF sorbents.

## 2. Fundamentals Associated with ANG

As a multitude of efforts are being taken in the storage of methane, accurate measurements at high-pressure are essential to evaluate the performance of each sorbent for storage capacity. However, it is challenging to measure gas uptake at high pressure over a large pressure range, during which the gases usually demonstrate a nonlinear behavior under an increasing pressure (*19*).

### 2.1. Volumetric and Gravimetric Capacity

Measurement of the adsorption isotherm is the fundamental step. The amount of gas adsorbed at high pressure could be measured with the use of either volumetric instrument or gravimetric instrumentation. Gravimetric uptake analysis uses a microbalance to record the weight change of sample at different pressure of methane, so the gravimetric uptake is expressed as the mass of methane per unit mass of MOFs. On the other hand, the volumetric instrumentation uses the Sieverts apparatus to record the pressure change when dosing methane to a cell containing the MOF sample. Therefore, the volumetric uptake is expressed by the volume of adsorbed methane under standard temperature and pressure divided by the volume of MOF.

A gravimetric capacity is the most common standard measurements. In order to calculate the volumetric uptake, knowing density of MOFs is essential. Although differences between the practical packing density and crystallographic densities are indeed observed (*34*, *35*), the ideal crystallographic density is widely used to convert the gravimetric capacity to the volumetric capacity. The calculated volumetric uptake maximizes the possible storage capacity, which represents the maximum capacity achievable in absence of any practical packing losses. Improvement in volumetric capacity has more impact than that of gravimetric capacity for natural gas vehicles on the vehicle driving range (*19*).

### 2.2. Excess, Absolute, and Total Adsorption

In evaluating methane uptake capacity from the isotherms, terms, such as *excess*, *total*, and *absolute adsorption* are generally used. In the methane adsorption of MOFs, methane molecules may either adsorbed near the surface of MOF, or move into the bulk, where they are considered a free molecules (*19*). The excess adsorption is considered as the amount of methane interacting with the surface of MOFs, and it is usually measured experimentally. At the same time, the absolute amount is defined as the amount of methane molecules experimentally adsorbed on the surface plus the amount in the bulk region (Figure 1). Since it is difficult to experimentally determine the adsorbed regions, the total adsorption is commonly considered as the approximation of the absolute adsorption (*36*, *37*). The total adsorption is the amount of all methane in the pores, and it is the most relevant quality to compare the methane capacity between sorbents.

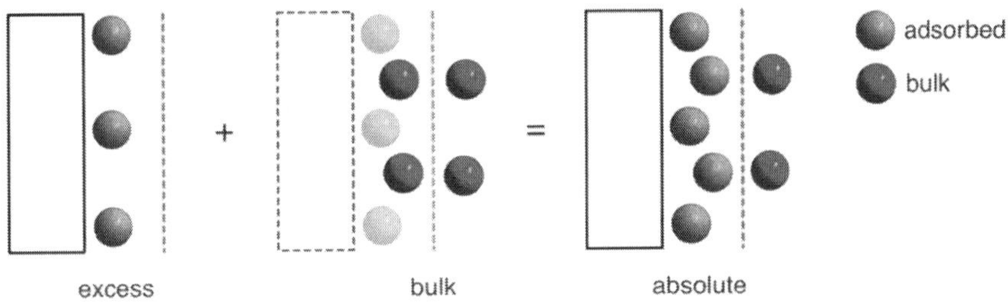


*Figure 1. The absolute adsorption is the sum of excess adsorption (green) and the bulk amount (blue). (Reproduced with permission from reference (19). Copyright 2014 The Royal Society of Chemistry). (see color insert)*

### 2.3. Deliverable Capacity

The deliverable capacity, also known as working capacity, is the key standard to determine the driving range of a methane powered vehicle. The deliverable capacity is defined as the amount of methane released when the pressure is reduced (*19*). It is generally calculated as the difference of volumetric capacity between the upper and low pressure limits at standard temperature and pressure. The lower limit of 5 bar is generally set based on the typical internal combustion engines powered by natural gas, while the upper limit is set at 35 or 65 bar, which could be achieved by a single-stage or two-stage compressor (*38*, *39*). The deliverable capacity can further improved by reaching higher storage pressure (65 bar and up) and lower engine intake pressure (1-2 bar).It is apparent that maximized capacity at upper pressure limits and minimized capacity at lower pressure limits will produce the best deliverable capacity. In this case, the maximum deliverable methane capacity between 5 bar to 35 or 65 bar will enable natural gas vehicles with on-board storage and at-home refueling in light-duty tanks with current compressor technology.

### 2.4. Isosteric Heat of Adsorption

As the adsorption is a consequence of surface energy, isosteric heat of adsorption is another key thermodynamic variables to gain insight into the adsorption process (*40*). It represents the binding enthalpy of adsorbed methane on the surface of the MOF through mainly weak van der Waals interactions. A larger isosteric heat of adsorption means a greater affinity to binding the methane molecules; this is not always true in terms of the deliverable capacity. The large heat of adsorption will increase the methane capacity at high pressure, while it also retains the amount of methane released at lower pressure. As a result, an optimal isosteric heat of adsorption is desired to achieve the maximum deliverable capacity.

## 3. Benchmark MOFs as Methane Storage Sorbents

The applications of MOFs in methane storage have not received as much attention as those of hydrogen storage or carbon capture. Due to increasing awareness of the great potential of natural gas as an alternative energy fuel, the importance of methane storage in MOFs has been gradually realized, and some MOFs exhibit comparable or even better methane capacity compared to the best activated carbons. Several benchmark MOFs will be enumerated in the following section.

### 3.1. IRMOFs

The isorecticular MOFs (IRMOFs) represents a novel MOF family with facile functionalization and tunable pore spaces. The choice of linear ditopic carboxylates and $Zn_4O(COO)_6$ SBUs afford this family with potential in various applications (*41*). Isorecticular chemistry has been widely applied in the design and synthesis of new MOF materials (*42–47*). With the perspective of methane storage, the binding sites of IRMOF-1 (MOF-5, Figure 2a), composed of 1,4-benendicarboxylate and $Zn_4O(COO)_6$ SBUs, are evaluated in detail (*48*). The primary binding sites were determined near the $Zn_4O$ cluster (Figure 2b). The well-defined orientation of methane indicates a strong interaction with methane molecules. At a high loading of methane, the extra methane molecules start to either gradually occupy the hex sites and $ZnO_2$ sites or become trapped inside the cavity (Figure 2b-d).

Another evaluation of three IRMOFs shows that the high surface area and hydrophobic nature of $C_2H_4$ endow IRMOF-6 with high excess methane uptake of 155 v(STP)/v at 36.5 bar (*41*). Extended Raman spectroscopic studies on IRMOFs reveals that the organic linker plays a critical role in the methane storage at high pressure (*49*). In the Raman study, the interactions between methane and the framework downshift the symmetric stretching band of the methane molecule. Interestingly, the frequency shits are highly linker-dependent. Therefore, the organic linker is assigned as the predominant factor controlling the methane uptake in IRMOFs. The realization of linker effects also spurred a comprehensive computational simulation of many hypothetical IRMOFs. Based on the simulation, IRMOF-993, consisting of 9, 10-anthracene dicarboxylate linkers with an expected pore size of 6.3 x 6.3Å, is predicted to be a better sorbent than IRMOF-6. In fact, experimentally this gives rise to a distortion of $Zn_4O(COO)_6$ SBUs in the crystal structure. As a consequence, a microporous MOF synthesized as the experimental realization of the simulated IRMOF-993, namely PCN-13, displays a limited methane capacity within the confined pore spaces (~3.5 Å) (*50*).

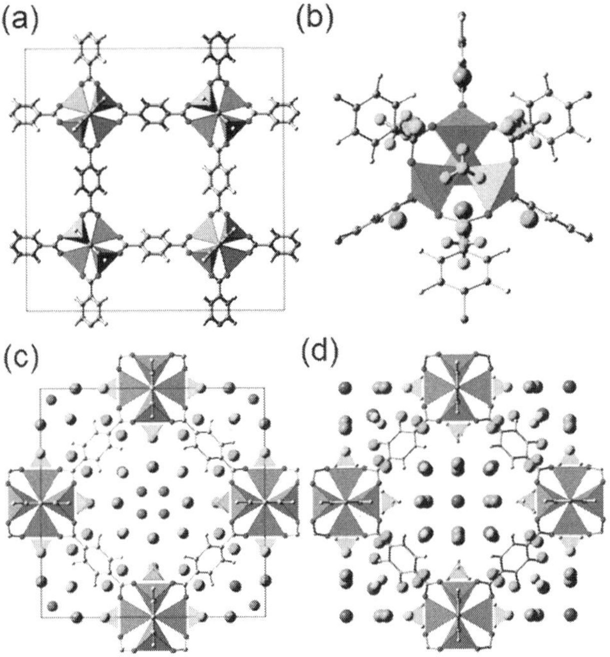


*Figure 2. (a) Crystal structure of MOF-5. C in gray and D in white. (b) The methane adsorption sites in MOF-5: cup sites (cyan) and secondary hex (yellow) and $ZnO_2$ sites (green) (c) [001] view of the I4/mmm structure of MOF-5 with more CD4 loading at 80 K, extra methane (pink) were observed to populate near the center of the pores. (d) [001] view of the P4mm structure of MOF-5 with higher CD4 loading below 60 K, confined methane sites (orange) were observed to align themselves along the c axis and further lower the symmetry to P4mm. Orientationally disordered methane molecules are shown as spheres for clarity. (Reproduced with permission from reference (48). Copyright 2009 American Chemical Society). (see color insert)*

### 3.2. M-MOF-74/CPO-27-M

The M-MOF-74 series (M=Mg, Mn, Co, Ni, Zn), also known as CPO-27-M are a family of MOFs with high density of coordinative unsaturated metal centers (UMCs) (*51–56*). This isostructural series is composed of the infinite helical metal chain SBUs and 2, 5-dioxido-1,4,-benzenedicarboxylate (dhtp) with a 1D channel of 13.6 Å in diameter. The coordinated solvent molecules on the metal SBUs could be easily removed to expose the UMCs, which serve as the primary adsorption sites to the methane molecules through Coulomb interactions (*51*). Among the excess adsorption measurements of this series of materials, Ni-MOF-74 exhibits the high value of ~190 $cm^3$ (STP) $\cdot cm^{-3}$ (*57*) (Figure 3a). All five MOFs have similar heat of adsorption in the range of 18 to 20 kJ $mol^{-1}$ (Figure 3b), which indicates no metal preference on the interactions between the UMCs and methane gas. This is ascribed to the fact that the large pore size increases the distance between the surface and methane but decreases the interaction potential between the two. Even though the large heat of adsorption is likely preferred for a stronger interaction, this trend does not apply to the overall deliverable capacity. A large amount of methane retention is witnessed at lower pressure, which is unfavorable to methane delivery. As a result, the deliverable capacity can only be achieved that is less than half of the storage amount (*51*). In addition to the primary binding site by UMCs (Figure 4a), a secondary binding site is determined, where the methane molecules interact with both the framework and adjacent methane molecules (Figure 4b).

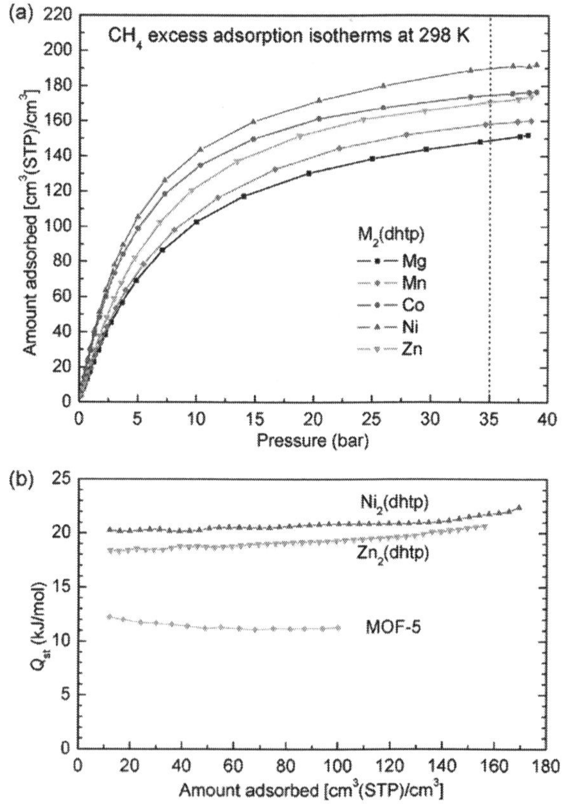


*Figure 3. (a) Excess $CH_4$ adsorption isotherms of $M_2$(dhtp) at 298 K. (b) The experimental $Q_{st}$ of $Ni_2$(dhtp) and $Zn_2$(dhtp). The $Q_{st}$'s of $Mg_2$(dhtp), $Mn_2$(dhtp), and $Co_2$(dhtp) fall between the two curves and, thus, are not shown for clarity. The $Q_{st}$'s of MOF-5 (from ref (58)) are also plotted for comparison. (Reproduced with permission from reference (51). Copyright 2009 American Chemical Society). (see color insert)*

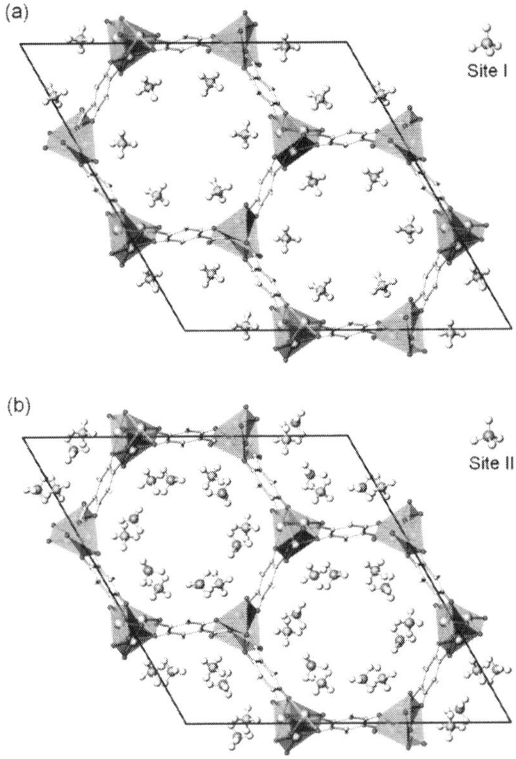


*Figure 4. Crystal unit cell of Mg-MOF-74 with methane adsorbed on site I (a) and site II (b). (Reproduced with permission from reference (51). Copyright 2009 American Chemical Society). (see color insert)*

### 3.3. PCN-14

PCN-14 was widely cited as the record-holding MOF material for methane storage until 2013 with the total capacity of 230 $cm^3$ (STP) $\cdot cm^{-3}$ at 298K and 35 bar (*59*). PCN-14 is constructed through self-assembling of dicopper paddlewheel SBUs and 5,5′-(9, 10-anthracenediyl)diisophthalate. It is composed of squashed cuboctahedral cages and other small octahedral cages. The removal of terminal coordinated solvent of the paddlewheel yields nanoscopic cages with a significant amount of UMCs, which is a source of this material's high methane capacity (*60*).

Computational studies based on Density Function Theory (DFT) have been applied to determine the potential methane adsorption sites of PCN-14 (*61*). Four major adsorption sites are identified: open copper sites, small cage window sites, small cage bottom sites, and small cage side window sites (Figure 5). In addition, the large cage corner sites also make a contribution to the total methane capacity. Notably, another simulation study demonstrates that the methane position at the open metal sites is temperature-dependent (*62*). The open metal site is densely populated at lower pressure but less populated at room temperature, which indicates that no energy barrier exists between the strong and weak adsorption sites.

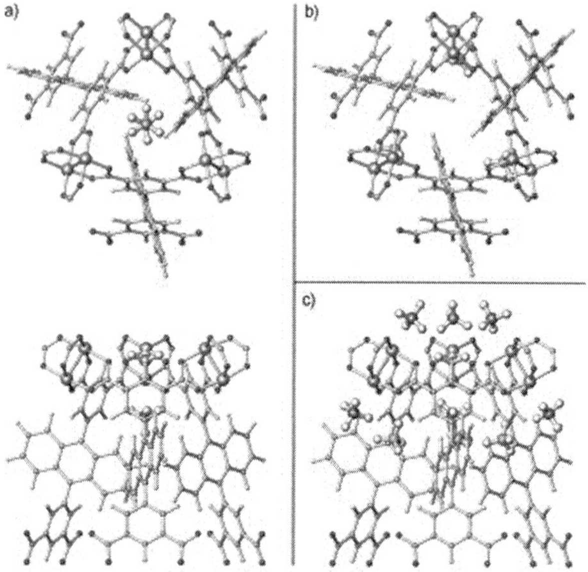

*Figure 5. Partial structure of the DFT-optimized PCN-14 crystal with $CH_4$ molecules adsorbed at a) the small cage window site and the small cage bottom site (top and side views), b) the small cage side window site, and c) all four major adsorption sites, including the open Cu sites. (Reproduced with permission from reference (61). Copyright 2010 John Wiley and Sons).*

### 3.4. UTSA-20

Immobilizing high-density of open metal sites and constructing suitable pore spaces theoretically enhance the performance of methane storage. Realization of these two design principles to maximize the methane storage in new MOFs is an essential goal of this field. Based on these two principles, one experimental attempt produced UTSA-20 with high methane storage density ($0.222 g\cdot cm^{-1}$) (*63*). UTSA-20 is constructed from dicopper paddlewheel SBUs and 3,3′,3″,5,5′,5″-benzene-1,3,5-triyl-hexabenzates ($bhb^{6-}$). UTSA-20 contains one rectangular 1D channels of about 3.4 x 4.8 Å and a 1D cylinder of 8.5 Å in diameter along the c axis with exposed UMCs pointing into the pore. Because of the interplay of high density of UMCs and optimal pore spaces, the total volumetric methane storage capacity can reach 195 $cm^3\cdot cm^{-3}$ at 300K and 35 bar. Further computational simulation reveals the channel site is an additional primary adsorption site (Figure 6a). It is found that the size of the linker channel pore is suitable enough to sandwich the methane molecules between the surfaces of the two linkers (Figure 6b).

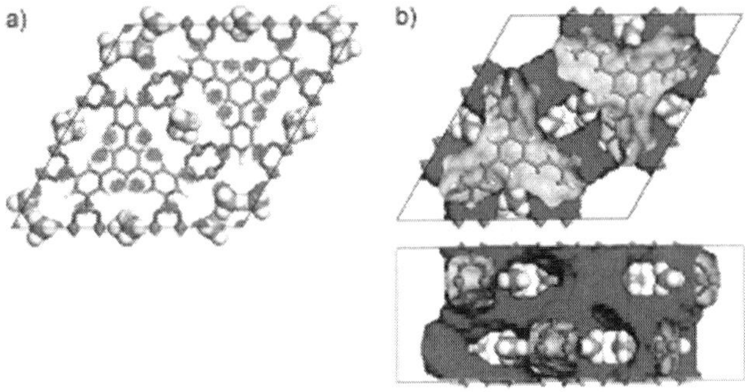

*Figure 6. (a) Probability distribution of the $CH_4$ center of mass in UTSA-20 ([0 0 1] view), obtained from Grand Canonical Monte Carlo simulation at 298 K and 10 bar. The red regions represent the places where methane molecules are heavily populated in the MOF structure. Note that the open Cu site is pre-occupied with $CH_4$ molecules to focus our effort on the search for other strong methane adsorption sites. b) The pore surface of the interconnected channel pores in UTSA-20 with adsorbed methane at the linker channel site. The channel width along the c axis matches well with the size of the adsorbed methane molecules, leading to enhanced van der Waals interaction. (Reproduced with permission from reference (63). Copyright 2011 John Wiley and Sons). (see color insert)*

### 3.5. PCN-6X and NU-111

At the stage of exploring new material for gas storage, it is generally believed that increasing the surface area of MOF materials will correspondingly offer more interaction potential surfaces and enhance the gas storage capacity. Extensive efforts have been dedicated to the synthesis of new materials through linker extension. However, linker extension can also produce several effects that can lower the suitability of the resulting MOF for methane storage, such as the undermining of stability and possible interpenetration. It has been suggested that use of metal–organic polyhedra as the building block of MOFs will not only increase the surface area and pore volume but also stabilize the large pore voids (*64*). The *rht*-type frameworks (Figure 7a) are one classical MOF family following the cage-based synthetic strategy. Cuboctahedral cages, assembled from 12 dinuclear paddlewheel SBUs and 24 isophthalate moieties, are commonly found in this *rht* family. Using this type of topology as a template, examples likes PCN-6X (*43*), NU-11X (*65–67*), and NOTT-10X (*68*, *69*) demonstrate high gas uptake with hierarchical pore cages containing UMCs.

The PCN-6X (X=1, 6, 8) series is realized by the combination of extended hexatopic carboxylates and dicopper paddlewheel SBUs. The size of T-$O_h$ and T-$T_d$ (Figure 7b) is enlarged through linker extension, but this approach was not successful for cub-$O_h$ (Figure 7b). As expected, the Langmuir surface areas of the MOFs impressively increased to 6033 $m^2$ $g^{-1}$(PCN-68) (*43*). At low pressure (<20bar), PCN-61 demonstrates the highest methane uptake due to the fact that small pore size provides more methane-framework overlapping. The volumetric excess methane uptakes of PCN-6X series are calculated to be 145, 110, and 99 cm$^3$ (STP) ·cm$^{-3}$ at 298K and 35 bar, respectively. As the pressure is increased, the effects of surface area and pore volume begin to dominate, with PCN-68 possessing the highest methane uptake.

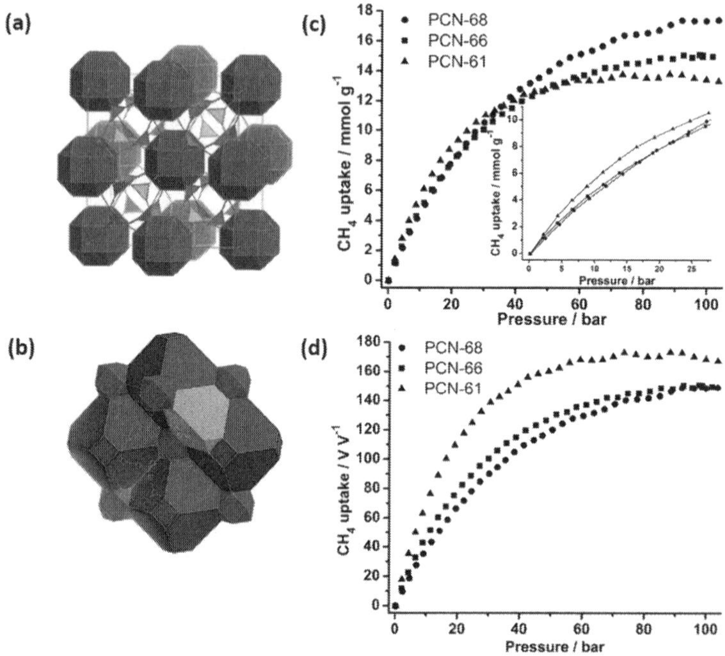


*Figure 7. (a) rht-type framework and (b) 3D polyhedra of PCN-68; (c) gravimetric and (d) volumetric methane capacities in PCN-6X at 298K. The inset (c) shows the medium-pressure region enlarged. (Reproduced with permission from reference (43). Copyright 2010 John Wiley and Sons). (see color insert)*

Continuous work on linker extension has been done to reach the record-high surface area with the design and synthesis of the NU-11X series (*65–67*). Utilizing triple bonds as a spacer in the linker design, NU-111 was synthesized with a high BET surface area of 5000 m$^2$ g$^{-1}$. Strikingly, NU-111 displays equally high total gravimetric and volumetric methane capacities of 0.36 g g$^{-1}$ and 206 cc/cc at 298K and 65 bar (*60*).

### 3.6. HKUST-1

HKUST-1, one of the earliest MOFs, is reported as a 3D network consisting of dicopper paddlewheel SBUs and 1, 3, 5-benzenetricarboxylate linkers (*70*). In the structure of HKUST-1, there is a ~1nm-size channel and three different types of cages (Figure 8a). The small octahedral cages with a diameter of ~5Å, a middle cuboctahedron cage with diameter ~10Å, and a large cuboctahedral cage with diameter of ~11Å are interconnected. Inconsistent high-pressure methane adsorption has been reported due to the variation of sample quality and the activation methods (*19*, *34*, *60*, *61*). Recently, HKUST-1 was discovered to demonstrate a room-temperature total volumetric methane uptake of 230 cc(STP)/cc at 35 bar and 270 cc(STP)/cc at 65 bar (*60*). The calculated deliverable capacity of HKUST-1 is 190 cc (STP)/cc (Figure 8b).

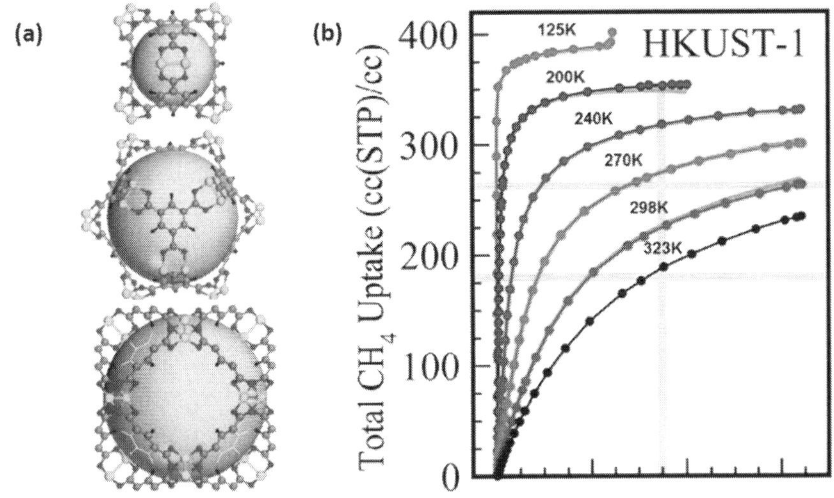


*Figure 8. (a) the nanocages and crystal structures HKUST-1. The gray, black, red, cyan, and blue spheres represent carbon, hydrogen, oxygen, copper, and nitrogen atoms, respectively. (b) Total volumetric uptake isotherms at various temperatures for HKUST-1. The gray horizontal lines indicate the DOE's old and new volumetric targets, respectively. (Reproduced with permission from reference (60). Copyright 2013 American Chemical Society). (see color insert)*

The analysis from high-resolution neutron powder diffraction (NPD) reveals that the methane adsorption sites are sequentially filled at distinct preferred adsorption sites, and the open copper sites are still the primary adsorption position at low temperature (*71*). The comprehensive study based on the NPD measurements and computation simulations discovers a second primary adsorption site at the window site of the small octahedral cage (*61*). The close distance at 2.7~3.2 Å between the frameworks provides strong interactions to methane molecules.

### 3.7. USTA-7X(X=5, 6, 7, 8, 9)

Shortly after the realization of the high methane storage capacity in HKUST-1, more efforts on functionalization of isostructural MOFs have been made to improve the methane storage capacities. Significant improvement of methane capacity has been demonstrated on the incorporation of Lewis basic sites on the linkers of UTSA-7X (X=5,6, 7, 8, 9) (*72*, *73*). The UTSA-7X materials are all isostructural to NOTT-101 (*68*, *69*), which were constructed from copper paddlewheel and nitrogen-functionalized tetratopic carboxylate linkers (Figure 9a, b). Variation of the number of nitrogen sites, from pyridine, pyridazine, to pyrimidine groups on the designed linkers did not make any significant differences on the pore volume, BET surface area or crystal densities. Moreover, the open copper sites still serve as the primary adsorption sites at 5 bar. In this case, the effect on the methane storage capacities at 5 bar can be neglected. Strikingly, the incorporation of Lewis basic nitrogen sites on the linkers significantly increase the methane uptake at 65 bar (Figure 9c), which consequently enclose UTSA-76 with the record methane storage deliverable capacities of 197 $cm^3 \cdot cm^{-3}$. DFT calculations reveals that the Lewis basic nitrogen sites (pyrimidine) of UTSA-76 provides a shallower energy barrier for hydrogen atoms to rotate in the center ring compare to that of benzene ring in NOTT-101 (*69*). These findings are all in good agreement to the results from neutron scattering studies, in which the rotation motion of the USTA-76 central ring requires much lower temperature compare to that of NOTT-101 (*69*). As a result, the Lewis basic nitrogen sties in UTSA-76 provides an enhanced van der Waals interaction at higher pressure and displays a higher $CH_4$ affinity.

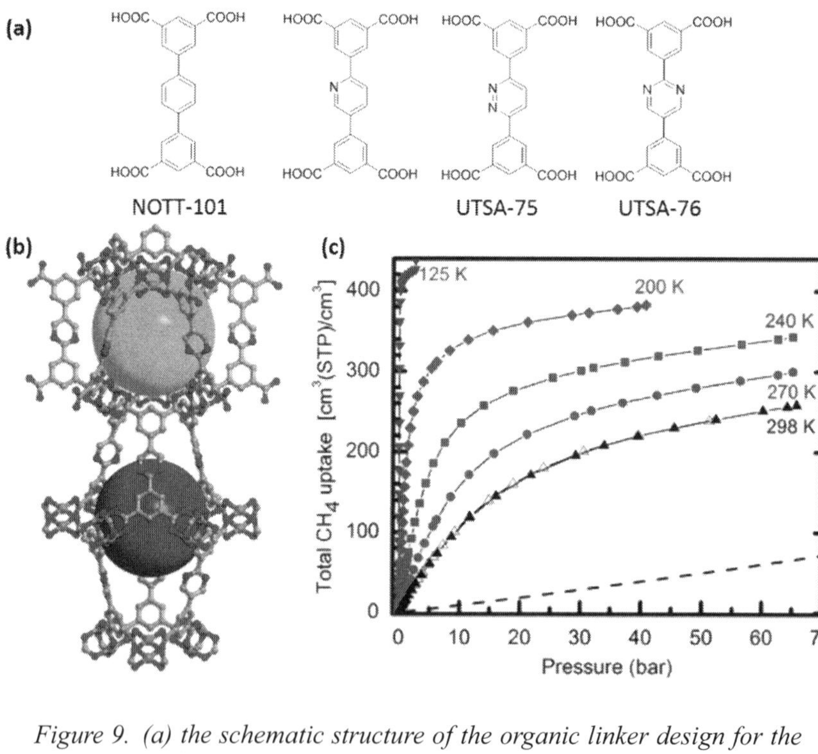


*Figure 9. (a) the schematic structure of the organic linker design for the construction of NOTT-101 and UTSA-7X; (b) Two cages of about 10.2 and 9.6×22.3Å, respectively in UTSA-76; (c) temperature-dependent high-pressure methane sorption isotherms of UTSA-76 (black dash curve represents the stored methane in a high pressure gas tank). (Reproduced with permission from reference (72). Copyright 2014 American Chemical Society). (see color insert)*

### 3.8. Zr-MOFs

When subjecting the porous materials to numerous adsorption/desorption cycles, material stability and resistance to impurities of natural gas (water, hydrogen sulfide, and other hydrocarbons), must be considered. The discovery of the zirconium-based MOF UiO-66 has inspired the investigation of stable MOFs due to the high inertness of Zr-oxoclusters (*74*). Their excellent thermal, chemical, and mechanical stability afford Zr-based MOFs great potential for methane storage. Computational simulation reveals UiO-67 shows the highest storage capacity of 146 $cm^3$ (STP) $\cdot cm^{-3}$ in a series of isoreticular Zr-based MOFs (*75*). With the goal to identify and synthesize a stable zirconium-based MOFs, another screening test is done on the MOFs featuring $(Zr_6O_4)(OH)_4(CO_2)_n$ SBUs (*76*). Among those hypothetical MOFs, the ones with alkyne groups adjacent to the zirconium corner allow efficient packing of methane and ultimately provide high methane capacities. Typically, NU-800, using a 1,4-benzenedipropiolate linker, exhibits volumetric deliverable capacities of 167 cc(STP)/cc, which is higher than those of PCN-14 (*38*) and UTSA-20 (*38*). In addition, no sample degradation is observed over a large number of gas adsorption cycles. The superior stability definitively shows the promise of Zr-based MOFs as methane sorbents.

## 4. Optimal Design Characteristic for Methane Sorbents

As more porous MOFs are being realized with improved gravimetric and volumetric methane capacity, there is a strong need to consider the factors that can dramatically affect the ultimate performance of the material for methane storage. The deliverable capacity is the dominant factor for the ultimate adoption and use of MOFs for on-board vehicle natural gas storage.

Computational simulations based on DFT can generally provide some insights directing the design of new MOF sorbents. Pore sizes centered around ~4Å and ~8 Å in MOFs are optimal for methane storage at 298K and 35 bar based on the high-throughput computational simulations (*77*). The pore size of MOFs has a strong influence in the determination of the deliverable capacity through its correlation with the heat of adsorption and pore volume available for methane occupancy (*78*). In general, the pore size ~4Å is just right to provide strong enough affinity to uptake methane during the adsorption process yet weak enough to release it during desorption. As the pore size of MOFs go beyond ~4Å, the wasted volume increase until the pore size of ~8Å, which is suitable to accommodate two methane molecules. In addition, increasing the volumetric surface area exhibits a positive correlation to the raise of volumetric methane capacity (*39*, *47*, *60*). Additionally, the optimal void fraction for MOFs with high methane uptake is suggested about 0.8 (*77*). Larger void fraction provide less overlap between the internal surface and the methane molecules.

From the perspective view of experimental studies, cage-based MOFs commonly demonstrate strong framework-methane interactions. If taking advantages of the fact that a structural priori could be conceived through proper modulation of SBUs and linkers, topology-guided synthesis offers opportunities to achieve a number of cage-based MOFs. However, proper design of linker and SBUs are also required to prevent possible interpenetration, which not only reduce the pore volume and surface area but also increase the crystalline density (*78*). Besides, the functionalization of the linker with lipophilic or small and polar groups are desired to enhance the methane performance (*77*, *79*, *80*). In the adsorption process, uncoordinated metal centers display the major role in the contribution of the total methane capacity. Therefore, design MOFs with high density of UMCs showcase the success to achieve high methane capacity. However, a portion of methane will be retained at low pressure and cannot be removed during the desorption process. As a consequence, the deliverable capacity will certainly decrease. Thus, a clear tradeoff between UMCs and deliverable capacity must be taken into consideration.

## 5. Conclusion and Summary

This chapter has provided a brief overview of fundamental properties of materials for methane storage and enumerated current benchmark MOFs sorbents. After detailed analysis of the adsorption sites in each MOFs, insights from the studies of those MOFs can be gained and further translated to a couple of optimal design characteristics for future development. Based on the performance demonstrated above, MOFs definitely show promise in methane storage. However, the research area of MOFs in the application of methane storage is still widely underexplored. Considering the complete ANG storage system, factors such as mechanical stability, heat capacity, and thermal conductivity and stability are critical to the practical application of MOFs as methane sorbents (*19*). Additonally, the dynamic adsorption process based on experimental measurements over a large range of pressure has not been extensively studied. It is apparent that collaboration between chemists, engineers and even industrial experts is necessary to drive this technology into further stages of development with the goal of widespread use for natural gas vehicles.

## Acknowledgments

X.W. and S.F. acknowledge the financial supports of Welch Foundation under Award Number A-1725. H.-C. Zhou was funded by Texas A&M University. We would also like to acknowledge Mat Bosch, Zachery Perry, and Yu Fang for assistance in the preparation of this chapter.

## References

1. Beckner, M.; Dailly, A. *Appl. Energy* **2015**, *149*, 69.
2. Collins, D. J.; Ma, S.; Zhou, H.-C. In *Metal−Organic Frameworks*; John Wiley & Sons, Inc., 2010; p 249.
3. Düren, T.; Sarkisov, L.; Yaghi, O. M.; Snurr, R. Q. *Langmuir* **2004**, *20*, 2683.
4. Lin, X.; Champness, N.; Schröder, M. *Top. Curr. Chem.* **2010**, *293*, 35.
5. Zhou, W. *Chem. Rec.* **2010**, *10*, 200.

6. Menon, V. C.; Komarneni, S. *J. Porous Mater.* **1998**, *5*, 43.
7. *Advanced Research Project Agency-Energy*, 2012.
8. *Carbon Capture and Sequestration: Integrating Technology, Monitoring and Regulation*; Blackwell Publishing: Ames, IA, 2007.
9. Yang, W.; Davies, A. J.; Lin, X.; Suyetin, M.; Matsuda, R.; Blakc, A. J.; Wilson, C.; Lewis, W.; Parker, J. E.; Tang, C. C.; George, M. W.; Hubberstey, P.; Kitagawa, S.; Sakamoto, H.; Bichoutskaia, E.; Champness, N. R.; Yang, S.; Schroder, M. *Chem. Sci.* **2012**, *3*, 2993.
10. Graetz, J. *Chem. Soc. Rev.* **2009**, *38*, 73.
11. Murray, L. J.; Dincă, M.; Long, J. R. *Chem. Soc. Rev.* **2009**, *38*, 1294.
12. Makal, T. A.; Li, J.-R.; Lu, W.; Zhou, H.-C. *Chem. Soc. Rev.* **2012**, *41*, 7761.
13. Schlapbach, L.; Züttel, A. *Nature* **2001**, *414*, 353.
14. Munson, R. A.; Clifton, R. A. *Natural Gas Storage with Zeolites*; U.S. Department of the Interior: Washington, DC, 1971; Vol. 38.
15. Babarao, R.; Jiang, J. *J. Am. Chem. Soc.* **2009**, *131*, 11417.
16. Berlier, K.; Olivier, M.-G.; Jadot, R. *J. Chem. Eng. Data* **1995**, *40*, 1206.
17. Kishima, M.; Mizuhata, H.; Okubo, T. *J. Phys. Chem. B* **2006**, *110*, 13889.
18. He, Y.; Zhou, W.; Qian, G.; Chen, B. *Chem. Soc. Rev.* **2014**, *43*, 5657.
19. Mason, J. A.; Veenstra, M.; Long, J. R. *Chem. Sci.* **2014**, *5*, 32.
20. Corma, A.; García, H.; Llabrés i Xamena, F. X. *Chem. Rev.* **2010**, *110*, 4606.
21. Cui, Y.; Yue, Y.; Qian, G.; Chen, B. *Chem. Rev.* **2012**, *112*, 1126.
22. Horcajada, P.; Gref, R.; Baati, T.; Allan, P. K.; Maurin, G.; Couvreur, P.; Ferey, G.; Morris, R. E.; Serre, C. *Chem. Rev.* **2012**, *112*, 1232.
23. Kreno, L. E.; Leong, K.; Farha, O. K.; Allendorf, M.; Van Duyne, R. P.; Hupp, J. T. *Chem. Rev.* **2012**, *112*, 1105.
24. Li, J.-R.; Sculley, J.; Zhou, H.-C. *Chem. Rev.* **2012**, *112*, 869.
25. Suh, M. P.; Park, H. J.; Prasad, T. K.; Lim, D.-W. *Chem. Rev.* **2012**, *112*, 782.
26. Sumida, K.; Rogow, D. L.; Mason, J. A.; McDonald, T. M.; Bloch, E. D.; Herm, Z. R.; Bae, T.-H.; Long, J. R. *Chem. Rev.* **2012**, *112*, 724.
27. Wu, H.; Gong, Q.; Olson, D. H.; Li, J. *Chem. Rev.* **2012**, *112*, 836.
28. Yoon, M.; Srirambalaji, R.; Kim, K. *Chem. Rev.* **2012**, *112*, 1196.
29. Zhang, W.; Xiong, R.-G. *Chem. Rev.* **2012**, *112*, 1163.
30. Yuan, S.; Lu, W.; Chen, Y.-P.; Zhang, Q.; Liu, T.-F.; Feng, D.; Wang, X.; Qin, J.; Zhou, H.-C. *J. Am. Chem. Soc.* **2015**, *137*, 3177.
31. Fordham, S.; Wang, X.; Bosch, M.; Zhou, H.-C. In *Lanthanide Metal−Organic Frameworks*; Cheng, P., Ed.; Springer: Berlin, Heidelberg, 2015; Vol. 163, p 1.
32. Wang, X.; Makal, T. A.; Zhou, H.-C. *Aust. J. Chem.* **2014**, *67*, 1629.
33. Kondo, M.; Yoshitomi, T.; Matsuzaka, H.; Kitagawa, S.; Seki, K. *Ange. Chem., Int. Ed.* **1997**, *36*, 1725.
34. Senkovska, I.; Kaskel, S. *Microporous Mesoporous Mater.* **2008**, *112*, 108.
35. Seki, K. *Chem. Commun.* **2001**, 1496.
36. Dincă, M.; Dailly, A.; Liu, Y.; Brown, C. M.; Neumann, D. A.; Long, J. R. *J. Am. Chem. Soc.* **2006**, *128*, 16876.
37. Poirier, E.; Dailly, A. *J. Phys. Chem. C* **2008**, *112*, 13047.
38. Peng, Y.; Srinivas, G.; Wilmer, C. E.; Eryazici, I.; Snurr, R. Q.; Hupp, J. T.; Yildirim, T.; Farha, O. K. *Chem. Commun.* **2013**, *49*, 2992.

39. He, Y.; Zhou, W.; Yildirim, T.; Chen, B. *Energy Environ. Sci.* **2013**, *6*, 2735.
40. Mohr, R.; Rao, M. B. *J. Phys. Chem. B* **1999**, *103*, 6539.
41. Eddaoudi, M.; Kim, J.; Rosi, N.; Vodak, D.; Wachter, J.; O'Keeffe, M.; Yaghi, O. M. *Science* **2002**, *295*, 469.
42. Ma, S.; Sun, D.; Simmons, J. M.; Collier, C. D.; Yuan, D.; Zhou, H.-C. *J. Am. Chem. Soc.* **2008**, *130*, 1012.
43. Yuan, D.; Zhao, D.; Sun, D.; Zhou, H.-C. *Angew. Chem., Int. Ed.* **2010**, *49*, 5357.
44. Banerjee, R.; Furukawa, H.; Britt, D.; Knobler, C.; O'Keeffe, M.; Yaghi, O. M. *J. Am. Chem. Soc.* **2009**, *131*, 3875.
45. Xue, M.; Liu, Y.; Schaffino, R. M.; Xiang, S.; Zhao, X.; Zhu, G.-S.; Qiu, S.-L.; Chen, B. *Inorg. Chem.* **2009**, *48*, 4649.
46. Bellarosa, L.; Castillo, J. M.; Vlugt, T.; Calero, S.; López, N. *Chem. – Eur. J.* **2012**, *18*, 12260.
47. Kong, G.-Q.; Han, Z.-D.; He, Y.; Ou, S.; Zhou, W.; Yildirim, T.; Krishna, R.; Zou, C.; Chen, B.; Wu, C.-D. *Chem. – Eur. J.* **2013**, *19*, 14886.
48. Wu, H.; Zhou, W.; Yildirim, T. *J. Phys. Chem. C* **2009**, *113*, 3029.
49. Siberio-Pérez, D. Y.; Wong-Foy, A. G.; Yaghi, O. M.; Matzger, A. J. *Chem. Mater.* **2007**, *19*, 3681.
50. Ma, S.; Wang, X.-S.; Collier, C. D.; Manis, E. S.; Zhou, H.-C. *Inorg. Chem.* **2007**, *46*, 8499.
51. Wu, H.; Zhou, W.; Yildirim, T. *J. Am. Chem. Soc.* **2009**, *131*, 4995.
52. Rosi, N. L.; Kim, J.; Eddaoudi, M.; Chen, B.; O'Keeffe, M.; Yaghi, O. M. *J. Am. Chem. Soc.* **2005**, *127*, 1504.
53. Dietzel, P. D. C.; Morita, Y.; Blom, R.; Fjellvåg, H. *Angew. Chem., Int. Ed.* **2005**, *44*, 6354.
54. Dietzel, P. D. C.; Panella, B.; Hirscher, M.; Blom, R.; Fjellvag, H. *Chem. Commun.* **2006**, 959.
55. Dietzel, P. D. C.; Besikiotis, V.; Blom, R. *J. Mater. Chem.* **2009**, *19*, 7362.
56. Caskey, S. R.; Wong-Foy, A. G.; Matzger, A. J. *J. Am. Chem. Soc.* **2008**, *130*, 10870.
57. Zhou, W.; Wu, H.; Yildirim, T. *J. Am. Chem. Soc.* **2008**, *130*, 15268.
58. Zhou, W.; Wu, H.; Hartman, M. R.; Yildirim, T. *J. Phys. Chem. C* **2007**, *111*, 16131.
59. Ma, S.; Sun, D.; Simmons, J. M.; Collier, C. D.; Yuan, D.; Zhou, H.-C. *J. Am. Chem. Soc.* **2007**, *130*, 1012.
60. Peng, Y.; Krungleviciute, V.; Eryazici, I.; Hupp, J. T.; Farha, O. K.; Yildirim, T. *J. Am. Chem. Soc.* **2013**, *135*, 11887.
61. Wu, H.; Simmons, J. M.; Liu, Y.; Brown, C. M.; Wang, X.-S.; Ma, S.; Peterson, V. K.; Southon, P. D.; Kepert, C. J.; Zhou, H.-C.; Yildirim, T.; Zhou, W. *Chem. – Eur. J.* **2010**, *16*, 5205.
62. Lucena, S. M. P.; Mileo, P. G. M.; Silvino, P. F. G.; Cavalcante, C. L. *J. Am. Chem. Soc.* **2011**, *133*, 19282.
63. Guo, Z.; Wu, H.; Srinivas, G.; Zhou, Y.; Xiang, S.; Chen, Z.; Yang, Y.; Zhou, W.; O'Keeffe, M.; Chen, B. *Angew. Chem., Int. Ed.* **2011**, *50*, 3178.
64. Zhao, D.; Yuan, D.; Sun, D.; Zhou, H.-C. *J. Am. Chem. Soc.* **2009**, *131*, 9186.

65. Zheng, B.; Yang, Z.; Bai, J.; Li, Y.; Li, S. *Chem. Commun.* **2012**, *48*, 7025.
66. Zheng, B.; Bai, J.; Duan, J.; Wojtas, L.; Zaworotko, M. J. *J. Am. Chem. Soc.* **2011**, *133*, 748.
67. Farha, O. K.; Özgür Yazaydın, A.; Eryazici, I.; Malliakas, C. D.; Hauser, B. G.; Kanatzidis, M. G.; Nguyen, S. T.; Snurr, R. Q.; Hupp, J. T. *Nat. Chem.* **2010**, *2*, 944.
68. Yan, Y.; Yang, S.; Blake, A. J.; Lewis, W.; Poirier, E.; Barnett, S. A.; Champness, N. R.; Schroder, M. *Chem. Commun.* **2011**, *47*, 9995.
69. Yan, Y.; Telepeni, I.; Yang, S.; Lin, X.; Kockelmann, W.; Dailly, A.; Blake, A. J.; Lewis, W.; Walker, G. S.; Allan, D. R.; Barnett, S. A.; Champness, N. R.; Schröder, M. *J. Am. Chem. Soc.* **2010**, *132*, 4092.
70. Chui, S. S.-Y.; Lo, S. M.-F.; Charmant, J. P. H.; Orpen, A. G.; Williams, I. D. *Science* **1999**, *283*, 1148.
71. Getzschmann, J.; Senkovska, I.; Wallacher, D.; Tovar, M.; Fairen-Jimenez, D.; Düren, T.; van Baten, J. M.; Krishna, R.; Kaskel, S. *Microporous Mesoporous Mater.* **2010**, *136*, 50.
72. Li, B.; Wen, H.-M.; Wang, H.; Wu, H.; Tyagi, M.; Yildirim, T.; Zhou, W.; Chen, B. *J. Am. Chem. Soc.* **2014**, *136*, 6207.
73. Li, B.; Wen, H.-M.; Wang, H.; Wu, H.; Yildirim, T.; Zhou, W.; Chen, B. *Energy Environ. Sci.* **2015**.
74. Cavka, J. H.; Jakobsen, S.; Olsbye, U.; Guillou, N.; Lamberti, C.; Bordiga, S.; Lillerud, K. P. *J. Am. Chem. Soc.* **2008**, *130*, 13850.
75. Yang, Q.; Guillerm, V.; Ragon, F.; Wiersum, A. D.; Llewellyn, P. L.; Zhong, C.; Devic, T.; Serre, C.; Maurin, G. *Chem. Commun.* **2012**, *48*, 9831.
76. Gomez-Gualdron, D. A.; Gutov, O. V.; Krungleviciute, V.; Borah, B.; Mondloch, J. E.; Hupp, J. T.; Yildirim, T.; Farha, O. K.; Snurr, R. Q. *Chem. Mater.* **2014**, *26*, 5632.
77. Wilmer, C. E.; Leaf, M.; Lee, C. Y.; Farha, O. K.; Hauser, B. G.; Hupp, J. T.; Snurr, R. Q. *Nat. Chem.* **2012**, *4*, 83.
78. Martin, R. L.; Simon, C. M.; Smit, B.; Haranczyk, M. *J. Am. Chem. Soc.* **2014**, *136*, 5006.
79. Zhao, X.; Sun, D.; Yuan, S.; Feng, S.; Cao, R.; Yuan, D.; Wang, S.; Dou, J.; Sun, D. *Inorg. Chem.* **2012**, *51*, 10350.
80. Wang, Y.; Tan, C.; Sun, Z.; Xue, Z.; Zhu, Q.; Shen, C.; Wen, Y.; Hu, S.; Wang, Y.; Sheng, T.; Wu, X. *Chem. – Eur. J.* **2014**, *20*, 1341.

Chapter 9

# Synthesis and Performance of New Organic Dyes and Functional Fullerenes for Organic Solar Cells

**T. Jin,[1] Md. Akhtaruzzaman,[2] and Y. Yamamoto*,[1,3]**

[1]WPI-Advanced Institute for Materials Research, Tohoku University, Katahira 2-1-1, Aobaku, Sendai 980-8577, Japan
[2]Solar Energy Research Institute (SERI), University Kabangsaan Malaysia, Bangi-43600, Selangor Darul Ehsan, Malaysia
[3]State Key Laboratory of Fine Chemicals, Dalian University of Technology, 2 Linggong Road, High Tech Zone, Dalian 116023, China
*E-mail: yoshi@m.tohoku.ac.jp, yoshi@dlut.edu.cn

Photophysical and photovoltaic properties of dye-sensitized solar cells using AK-01 with indoline as donor, Ba01~03 with fused dithienothiophene as π-spacer, YS-01~04 with dibenzosilole as π-spacer, and K-dyes or JH-dyes with thieno[2,3-*a*]carbazole or thieno[3,2-*a*]carbazole as donor have been investigated. Practical and large scale synthetic methods for functional fullerenes, which are potentially useful acceptors in bulk heterojunction solar cells, have been developed, and devices employing some of them exhibit significantly high photovoltaic performances.

## Introduction

Solar cells have already proven to be the decisive source of clean energy. Since the introduction of solar cells over five decades ago, many methodologies have been under investigation to achieve the best conditions for harnessing energy efficiently from the sun. Based on the abundance and technology availability, silicon based single layer p-n junction diode solar cells were developed. These silicon wafer based solar cells are the first generation solar cells and are still dominant in market. Rising cost for silicon ingot processing, together with expensive and quite complicated cell fabrication process, has lured scientists to find cheaper alternatives. For effective fabrication in minimalized cost, thin film technology has been adopted in solar energy research. Solar cells prepared by thin film technology have been standardized as the second generation solar cells which contain amorphous silicon, poly-crystalline silicon, micro-crystalline silicon, cadmium telluride and copper indium selenide/sulphide (*1*, *2*). In comparison to the first generation solar cells, thin film based solar cells have shown reduced cost with few limitations. They require vacuum conditions and temperature control which makes them unattractive for large scale production. Moreover, the materials being used are finite in this world. Some of such drawbacks are overcome by the use of organic materials in dye-sensitized solar cells (DSSCs) (*3*) and bulk heterojunction solar cells (BHJs) (*4*). These types of solar cells are solution processable, lighter in weight, flexible in shape and capable of being processed in non-vacuum conditions. Furthermore, the synthetic methods of dyes for DSSCs, as well as those of $C_{60}$-derivtives as acceptors for BHJs are versatile and flexible. Therefore, a number of promising new materials can be obtained and tested for performance evaluation rather easily. At present, the highest conversion efficiencies of 12 % for DSSCs (*5*, *6*) and 10 % for BHJs (*7–9*), based on an organic dye and functional $C_{60}$ acceptor, have been reported. The fundamental working principles of efficient DSSCs and BHJs have been well evaluated and verified by their corresponding electrical, optical properties and advanced characterization techniques (*10–15*). In the following sections, we mention our design, synthesis and the conceptual advancement of novel organic dyes in DSSCs and functional fullerenes in BHJs.

## 1.1. Molecular Architecture of Organic Dye for DSSCs

A dye-sensitized solar cell is composed of nanocrystalline semiconductor oxide film electrode, dye sensitizers, electrolytes, counter electrode and a transparent conducting substrate (Figure 1). Typically, a DSSC is comprised with a dye-coated nanocrystalline titanium dioxide ($TiO_2$) film deposited on a transparent conducting oxide (TCO) glass, a platinum (Pt) counter-photocathode, and an electrolyte solution with a dissolved triiodiate/iodiate ($I^{3-}/I^-$) ion redox couple. The mechanism of a DSSC at ground state is shown in Figure 1. Under the irradiation of sunlight, the dye molecule becomes photo-excited and radically injects an electron into the conduction band of the semiconductor electrode. Consequently, the original state of the dye is restored by electron donation from the electrolyte; usually the solution of an organic solvent or ionic liquid solvent contains the $I^{3-}/I^-$ redox system. The regeneration of the sensitizer by iodide intercepts the recapture of the conduction band electron by the oxidized dye. The iodide is regenerated successively by reduction of triiodide at the counter electrode. The circuit is completed by an external load. The voltage generated under illumination corresponds to the difference between the Fermi level of the electron in the semiconductor electrode and the redox potential of the electrolyte.

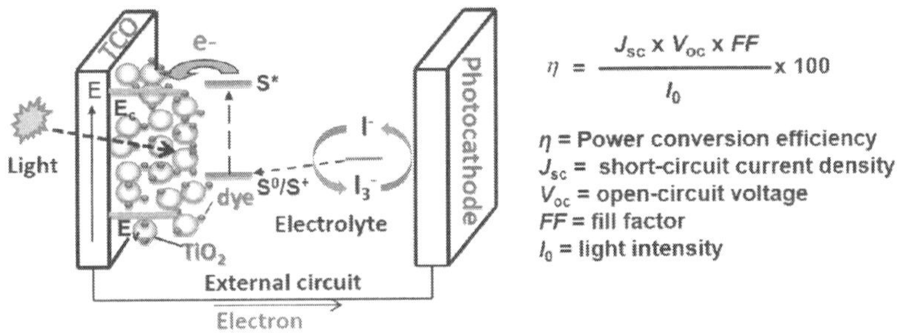


*Figure 1. Schematic diagram of a dye-sensitized solar cell (DSSC) with operational principles.*

Until today, the most challenging task for materials chemists and engineers is to design and to synthesize organic dyes for efficient DSSCs to capture the sunlight from visible to NIR (near to infrared) region. Two major types of organic dyes have been used in DSSCs; 1) functional ruthenium(II)–polypyridyl complexes such as black dye (Figure 2a) (*11*, *13*, *16*, *17*), and 2) metal-free organic dyes (Figure 2b) (*12*, *13*, *18*). Ruthenium compounds are expensive and require careful synthesis process with purification techniques. Whilst metal free organic dyes are much inexpensive and can be prepared easily by synthetic technologies. They also provide a huge advantage of having tunable electrochemical properties and absorption through molecular engineering.

Therefore, metal free organic dyes have attracted considerable attention in recent years due to their high molar absorption coefficients, tunable optical properties through by changing their functional groups, low production costs and eco-friendly characteristics. Till today, many metal-free organic sensitizers have been developed and some of their DSSCs performances are comparable with those of Ru-polypyridyl based dyes (*18*). The most common strategy that has been followed to design the dye is based on donor–(π-spacer)–acceptor (D–π–A) conjugated system as shown in Figure 2b (*12*, *19–23*).

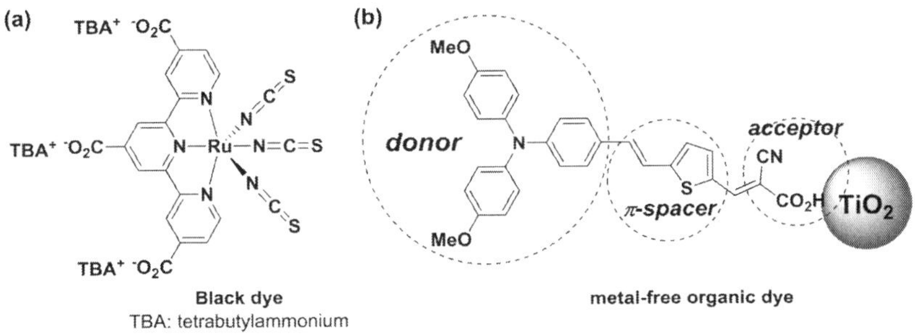


*Figure 2. Representative molecular structures of (a) Ru-based dye of black dye, and (b) metal-free organic dyes, for DSSCs applications.*

In this conventional structure, the π-spacer unit acts as a linker between the donor and acceptor moieties to transfer photo-excited electron from donor to acceptor effectively. It also plays an important role to tune the photophysical properties through the adjustment of highest occupied molecular orbital (HOMO) and lowest unoccupied molecular orbital (LUMO) energy levels. The most common π-linkers that have been used in the design of organic dyes in DSSCs are polyene, oligothiophene, oligophenylenevinylene, benzene, furan, etc. Another crucial part of the dye is the donor moieties (*12*, *13*). During last several years many efforts have been focused on design and synthesis of various organic donors such as squaraine, carbozole, triphenylamine, fluorene, indoline, phenothiazine, oligothiophene, etc. to improve the DSSCs's efficiency but none of them have had yet shown superior performances to be comparable with Ru-based dyes. Besides donor and π-spacer group, cyanoacrylic acid/rhodamine-3-acetic acid unit has been used commonly as accepting part (*23*). Finally, the dye is adsorbed on semiconducting photoanode surface (e.g. titania, $TiO_2$), through an anchoring group such as carboxylic acid (-COOH), via covalent bond or any other strong intermolecular interactions. Other anchoring groups such as sulfonate ($-SO_3$), silane, pyridyl, etc. have also been investigated on $TiO_2$ surface but the –COOH containing dye works efficiently compared to the other anchoring dyes (*25*). In addition, the synthesis and purification method of carboxylic acid containing dyes is much easier than those of dyes with other anchoring groups.

Usually a double layer $TiO_2$ photoelectrode is prepared by screen printing method on a fluorine-doped tin oxide (FTO) coated glass. For our fabricated devices, we fabricated 15 μm of thick $TiO_2$ photoelectrodes. It was composed of a 10 μm thick nanoporous layer and a 5 μm thick scattering layer. The dye solution ($3\times0^{-4}$ M) was prepared in a mixture solvent of acetonitrile and *tert*-butyl alcohol (1:1 v/v). Then the $TiO_2$ films were immersed and soaked in the dye solution for 30 h at 25 °C. The dye deposited $TiO_2$ films were enclosed by platinum coated glass and separated by a surlyn spacer. A solution of electrolyte containing 0.6 M of 1,2-dimethyl-3-propyl-imidazolium (DMPII), 0.05 M of iodine ($I_2$, 0.1 M of lithium iodide, LiI, 0.5 M of *tert*-butylpyridine (TBP) and 20 mM of deoxycholic acid (DCA)) in anhydrous acetonitrile was injected into the cell by syringe. Finally, the DSSC was sealed by heating the polymer frame. Upon successful fabrication, the device was characterized under AM 1.5G simulated sun light at 100 mW/cm light intensity. The photovoltaics properties, such as open-circuit voltage ($V_{oc}$), short-circuit current ($J_{sc}$), fill factor ($FF$), and power conversion efficiency ($\eta$), were calculated from $I$-$V$ graph under illumination.

## 1.2. Synthesis, Photophysical, and Photovoltaic Properties of New Organic Dyes

In general, for improving DSSCs performances, the organic dyes should have high $J_{sc}$, high $V_{oc}$, and high $FF$ (*12*). During the design and evaluation of the DSSCs performances, it was observed that the dyes containing minimal driving force of 0.2 electron Volt (eV) with $E_{0-0}$ energy gap of 1.4 eV (minimum ground-state oxidation potential ($S^{+/0}$) of -5.4 eV, maximum excited-state oxidation potential ($S^{+/*}$) of -4.0 eV) must be desirable for efficient electron injection and regeneration process of the DSSCs devices. The $E_{0-0}$ energy gap of the black dye is 1.6 eV which shows the highest conversion efficiency of 11.1 % (*26*). However, most of organic dyes have an energy gap of $E_{0-0}$ >2.0 eV. Another shortcoming of organic dyes is the formation of π- π aggregation on $TiO_2$ surface which is responsible for reducing the electron injection rate into the $TiO_2$ electrode, thus the $V_{oc}$ has also been decreased. Although several attempts have been taken to improve the $V_{oc}$ with adding some additives, such as DCA, DMPII and TBP, in the electrolyte but the $J_{cs}$ decreased simultaneously. Therefore, proper molecular design for finding the structure-property relationship is required to improve the DSSCs efficiency and stability.

During our systematic studies on organic dyes, we found that the reported dye of **TA-St-CA** showed excellent DSSCs performances as single dye and the authors claimed to achieve the efficiency of 9.1 % (Figure 3a) (*27*). However, our own examination showed that the incident photon-to-current conversion efficiency (IPCE) spectrum of **TA-St-CA** was below 700 nm with $E_{0-0}$ energy gap of 2.25 eV. Consequently, we replaced the triphenylamine donating part of **TA-St-CA** to a strong electron donor of indoline group (**AK01**) as shown in Figure 3a (*28*). It is well known that the indoline containing compounds exhibit excellent electron donating properties and also have the ability to reduce the dye aggregation on $TiO_2$ surface as well as minimize the charge recombination process. The $E_{0-0}$ energy gap of **AK01** was lower and showed red-shifted absorption maxima with higher molar absorption coefficient, compared with **TA-St-CA** (Table 1 and Figure 3b). The similar red-shifted absorption maxima was also observed on $TiO_2$ surface of dye **AK01** though the absorption peak slightly blue-shifted (Figure 3b, inset) compared to the solution state. The blue shift absorption maxima of **AK01** on $TiO_2$ was due to the strong binding on $TiO_2$ nanoparticles through the –COOH group. In general, wider absorption maxima, higher molar extinction coefficient, and appropriate matching of the HOMO-LUMO energy level with triiodide and $TiO_2$ are favorable for improving the DSSCs performances. Hence, the efficiency of **AK01** (6.2 %) was higher than that of **TA-ST-CA** (5.4 %) (Table 2). This work also shows that the IPCE spectrum of **AK01** has red-shifted around 100 nm compared to **TA-ST-CA** as shown in Figure 3c. Although we succeeded in increasing the $J_{sc}$, the $V_{oc}$ was quite lower than that of **TA-St-CA** as shown in Table 2 and Figure 3d (*I-V* spectrum).

To further improve the DSSC performances, we replaced the π-conjugated spacer unit of phenylenevinylene to fused dithienothiophene group (**Ba**-dyes) as shown in Figure 4a to enhance the electron injection capability into the $TiO_2$ film. All compounds were synthesized via conventional synthetic methods (*29*). To compare photovoltaic performances, we introduced a carbazole donor group (**Ba-01**). Table 3 shows the photophysical properties of Ba-dyes. The absorption maximum of carbazole donor (**Ba-01**) blue shifted compared to those of indoline based compounds (**Ba-02** and **Ba-03**) due to the lower electron-donating ability of carbazole donor in **Ba-01** (Figure 4b). Upon adsorbed dyes on $TiO_2$ nanoparticles, the dyes **Ba-02** and **Ba-03** showed broad absorption compared to **Ba-01** as shown in Figure 4c, which indicated that the **Ba-02** and **Ba-03** were more feasible to harvest sunlight to convert photocurrent. It was also observed that the DSSCs efficiency of **Ba-02** and **Ba-03** reached to up to 6.38 % which was slightly higher than that of **AK01** (Table 4). Although the $J_{sc}$ of dyes **Ba-02** and **Ba-03** were similar to that of **AK01** but the $V_{oc}$ further decreased compared to **AK01** (Tables 2 and 4). However, the IPCEs of **Ba-02** and **Ba-03** covered wider wavelength region (400-600 nm) with the quantum efficiency of around 80 %. In this design, the overall photovoltaic performance of carbazole based dye (**Ba-01**) showed lower efficiency ($\eta$) of 5.64 % compared to the efficiency of indoline based dye **Ba-03** of 6.38 % (Table 4).

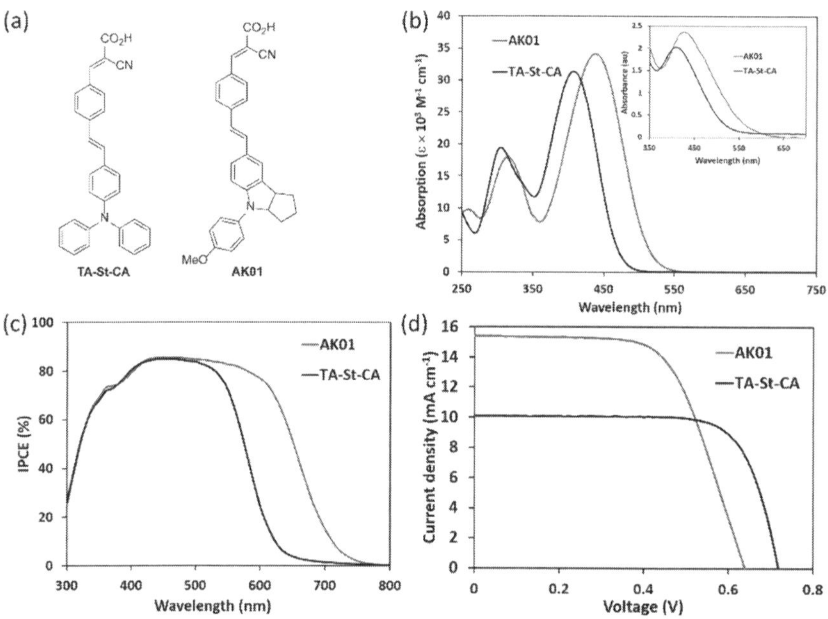


*Figure 3. (a) Molecular structure of dyes **TA-St-CA** and **AK01**. (b) UV-vis absorption spectra of compounds **TA-St-CA** and **AK01** in ethanol and on the sensitized nanocrystalline TiO₂ surface (inset). (c) IPCEs and (d) I-V (right) characteristics of nanocrystalline TiO₂ film sensitized by AK01 and TA-St-CA. Adapted from reference (28) with permission from The Royal Society of Chemistry.*

**Table 1. Photophysical properties of AK-01 and TA-St-CA. Adapted from reference (*28*) with permission from The Royal Society of Chemistry.**

| *Dye* | $\lambda_{max}$ *(nm)* $(\varepsilon \times 10^{-4}\ M^{-1}\ cm^{-1})^a$ | $\lambda_{max}$ *(nm)*[b] *on* $TiO_2$ *film* | *IP (eV)*[c] | $E_{0\text{-}0}$ *(eV)*[d] | *LUMO (eV)*[e] |
|---|---|---|---|---|---|
| **AK01** | 440 (3.4) | 426 | -5.56 | 2.03 | -3.53 |
| **TA-St-CA** | 412 (3.1) | 407 | -5.76 | 2.25 | -3.50 |

[a] Absorption maxima, measured in chloroform at room temperature. [b] Absorption measured on a transparent 4 μm $TiO_2$ film. [c] Ionization potential (IP) of absorbed dyes on the nanocrystalline $TiO_2$ film was determined by using the photoemission yield spectrometer (Riken Keiki, AC-3E). [d] Optical bandgap ($E_{0\text{-}0}$) was determined from the onset of the absorption in chloroform. [e] The LUMO energy levels were calculated from the expression of LUMO = IP + $E_{0\text{-}0}$.

**Table 2. DSSC performances of dye AK-01 and TA-St-CA[a]. Adapted from reference (*28*) with permission from The Royal Society of Chemistry.**

| *Dye* | $J_{sc}$ *[mA cm*$^{-1}$*]* | $V_{oc}$ *[V]* | *FF* | $\eta$ *[ %]* |
|---|---|---|---|---|
| **AK01** | 15.4 | 0.639 | 0.631 | 6.20 |
| **TA-St-CA** | 10.1 | 0.718 | 0.746 | 5.41 |

[a] Measurements were performed under AM 1.5 irradiation on the DSCs devices with 0.25 $cm^2$ active surface area defined by a metal mask.

**Table 3. Photophysical properties of Ba-01-03. Reproduced with permission from reference (*29*). Copyright 2013 Elsevier.**

| *Dye* | *$\lambda_{max}$ (nm) ($\varepsilon \times 10^{-4}$ $M^{-1}cm^{-1}$)[a]* | *$\lambda_{onset}$ (nm)[b] on $TiO_2$ film* | *IP (eV)[c]* | *$E_{0-0}$ (eV)[d]* | *LUMO (eV)[e]* |
|---|---|---|---|---|---|
| **Ba-01** | 428 (3.5) | 610 | -5.39 | 1.94 | -3.45 |
| **Ba-02** | 481 (4.0) | 660 | -5.31 | 1.88 | -3.43 |
| **Ba-03** | 485 (4.1) | 665 | -5.26 | 1.86 | -3.40 |

[a] Absorption maxima, measured in ethanol at room temperature. [b] Absorption measured on a transparent 4 μm $TiO_2$ film. [c] Ionization potential (IP) of absorbed dyes on the nanocrystalline $TiO_2$ film was determined by using the photoemission yield spectrometer (Riken Keiki, AC-3E). [d] Optical bandgap ($E_{0-0}$) was determined from the on-set of the absorption in chloroform; [e] The LUMO energy levels were calculated from the expression of LUMO = IP + $E_{0-0}$.

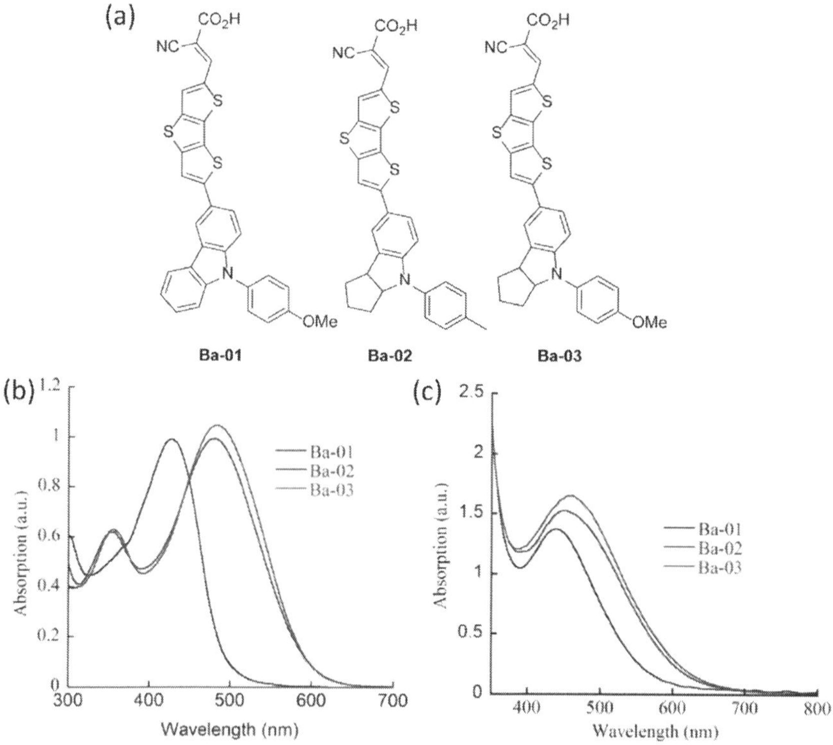


*Figure 4. (a) Chemical structure of new Ba-dyes. UV-vis absorption spectra of Ba-dyes (b) in ethanol, and (c) on the sensitized nanocrystalline $TiO_2$ surface. Reproduced with permission from reference (29). Copyright 2013 Elsevier.*

[a] Absorption maxima, measured in ethanol at room temperature.

**Table 4. DSSC performances of dye AK-01 and TA-St-CA[a]. Reproduced with permission from reference (*29*). Copyright 2013 Elsevier.**

| *Dye* | $J_{sc}$ *[mA cm⁻¹]* | $V_{oc}$ *[V]* | *FF* | *η [ %]* |
|---|---|---|---|---|
| **Ba-01** | 13.45 | 0.618 | 0.679 | 5.64 |
| **Ba-02** | 15.85 | 0.589 | 0.671 | 6.11 |
| **Ba-03** | 16.17 | 0.595 | 0.663 | 6.38 |

[a] Measurements were performed under AM 1.5 irradiation on the DSCs devices with 0.25 $cm^2$ active surface area defined by a metal mask.

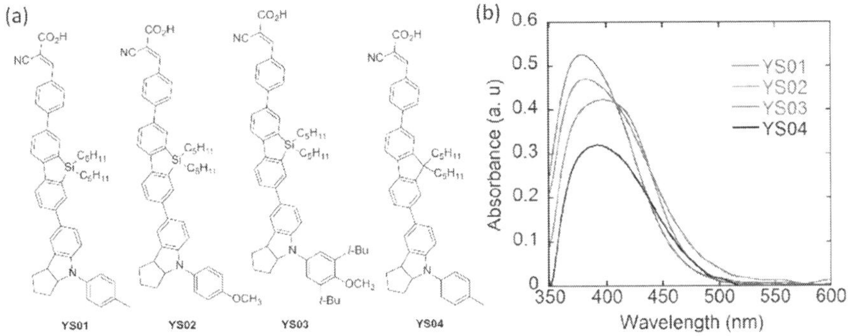


*Figure 5. (a) Molecular structure YS-dyes. (b) UV-vis absorption spectra of YS-dyes on the sensitized nanocrystalline $TiO_2$ surface. Adapted from reference (30) with permission from The Royal Society of Chemistry.*

We later investigated a few novel D–π–A sensitizers using dibenzosilole unit as a π-conjugated linker between indoline donor and cyanoacrylic acid acceptor moiety (YS-dyes, Figure 5a) (*30*). This study seemed to be promising, since the study of silole linker have been very limited. In addition, the optical properties of silole based compounds can be easily tuned due to their low LUMO energy level which depends on both silicon-carbon (Si-C, σ*-π*) and carbon-carbon (C-C, π*) anti-boding orbitals (*31, 32*). All compounds have been synthesized through the standard procedures (*33*). Table 5 showed optical characteristics of YS-dyes. The absorption maxima of YS-dyes were in 385-391 nm wavelength region with high molar extinction coefficiency ($4.5\text{-}5.4 \times 10^4$ $M^{-1}$ $cm^{-1}$). In addition, upon adsorption on $TiO_2$ surface, the YS-dyes showed broad absorption with extending up to 550 nm (Figure 5b), which was highly expected to harvest more photons in DSSCs. The absorption spectrum in UV-visible wavelength region is desirable for indoor light harvesting application. The ionization potentials (IP) of silole-derivatives (**YS01~YS03**) moved to positive direction compared to the fluorene containing compound **YS04** (Table 5) which implied that silole containing dyes were more feasible than fluorene based dyes to furnish more photocurrent. These dyes were thermodynamically favorable in the dye-regeneration process in DSSCs as their ground state oxidation potential

($S^{+/-}$) (-5.59 to -5.68 eV) was sufficiently lower to the redox potential of triiodide solution (-5.20 eV). As shown in Table 6, the silole containing dyes (**YS01~YS03**) were much better than non-silicon containing dyes (**YS04**) in DSSC devices. In addition, the IPCEs of **YS01~YS03** dyes reached >85 % in the 350-650 wavelength region. Although the $V_{oc}$ (maximum 0.8 V) was high, the solar to power conversion efficiency was quite low (maximum 5 %) due to the low $J_{sc}$ value (maximum 9.1 $mAcm^{-2}$). The lower $J_{sc}$ attributed to the high $E_{0\text{-}0}$ energy gap (2.43-2.53 eV) of YS-dyes. Therefore, this works would be potentially significant to future design and synthesis of new efficient silole based sensitizers.

**Table 5. Photophysical properties of YS01-04[a]. Reproduced from reference (*30*) with permission from The Royal Society of Chemistry.**

| *Dye* | *$\lambda_{max}$ (nm) ($\varepsilon \times 10^{-4}$ $M^{-1}$ $cm^{-1}$)[a]* | *$\lambda_{max}$ (nm)[b] on $TiO_2$ film* | *IP (eV)[c]* | *$E_{0\text{-}0}$ (eV)[d]* | *LUMO (eV)[e]* |
|---|---|---|---|---|---|
| **YS01** | 386 (4.5) | 382 | -5.66 | 2.48 | -3.18 |
| **YS02** | 390 (4.5) | 379 | -5.60 | 2.46 | -3.15 |
| **YS03** | 391 (4.8) | 392 | -5.59 | 2.43 | -3.16 |
| **YS04** | 385 (5.4) | 396 | -5.68 | 2.53 | -3.15 |

[a] Absorption maxima, measured in chloroform at room temperature. [b] Absorption measured on a transparent 4 μm $TiO_2$ film. [c] Ionization potential (IP) of absorbed dyes on the nanocrystalline $TiO_2$ film was determined by using the photoemission yield spectrometer (Riken Keiki, AC-3E). [d] Optical bandgap ($E_{0\text{-}0}$) was determined from the on-set of the absorption in chloroform; [e] The LUMO energy levels were calculated from the expression of LUMO = IP + $E_{0\text{-}0}$.

**Table 6. DSSC performances of dyes YS01-04[a]**

| *Dye* | *$J_{sc}$ [mA $cm^{-1}$]* | *$V_{oc}$ [V]* | *FF* | *$\eta$ [ %]* |
|---|---|---|---|---|
| **YS01** | 8.23 | 0.767 | 0.735 | 4.64 |
| **YS02** | 7.69 | 0.798 | 0.740 | 4.55 |
| **YS03** | 9.14 | 0.778 | 0.712 | 5.06 |
| **YS04** | 5.10 | 0.772 | 0.732 | 2.88 |

[a] Measurements were performed under AM 1.5 irradiation on the DSCs devices with 0.25 $cm^2$ active surface area defined by a metal mask.

For understanding the structure-property relationship of naphthalene based donor-π-acceptor dyes, **YF01-04** were synthesized (Figure 6) (*34*). These sensitizers were designed based on naphthalene-linked pyrrolidine or *N,N*-diphenylamine as an electron donor, oligo-phenylenevinylene as a π-linkage, and cyanoacrylic acid as an electron acceptor/anchor group. Figure 7a showed absorption maxima of compounds **YF01-04** in ethanol at 404-423 nm wavelength region which was responsible for the $\pi$-$\pi^*$ electron transition. The red-shifted absorption maxima of **YF04** was due to the stronger pyrrolidine electron donor and higher molecular planarity compared to dyes **YF01-03** (Figure 7a and Table 7). The YF-dyes actively participated in efficient dye regeneration and electron injection process in DSSCs operation, since their ground states oxidation potentials ($S^{+/-}$) (-5.71 to -5.54 eV) and excited states oxidation potentials ($S^{+/-}$) (-3.45 to -3.63 eV) matched well with energy level of $I^-/I^{3-}$ (-5.20 eV) and $TiO_2$ conduction level (-4.2 eV), respectively (Table 7). Figures 7c and 7d showed the IPCE and *I-V* spectra of YF-dyes which clearly indicated the structural effects on DSSCs performances due to the attached donor group on the naphthalene ring. It was also observed that **YF02** and **YF04** having a donor moiety of diphenyamine or pyrrolidine at the 6-posiiton of naphthalene moiety showed higher conversion efficiencies compared to **YF01** and **YF03** bearing those donors at the 2-position of the naphthalene moiety, as shown Table 8. We ascribed the high efficiencies of **YF02** and **YF04** to their better charge separation and photocurrent. However, the *I-V* graphs (Figure 7d) showed that the diphenylamine containing dyes **YF02** and **YF03** achieved higher $V_{oc}$ (0.8 V) compared to the pyrrolidine containing donor, which would be due to their more twisted geometry. The pyrrolidine containing dyes **YF01** and **YF04** have higher planar structures hence stronger $\pi$-$\pi$ interactions compared to diphenylamine containing dyes **YF02** and **YF03**. The strong $\pi$-$\pi$ interactions of organic sensitizers on $TiO_2$ surface may cause self-quenching to decrease the DSSC performances. We also found that the diphenylamine containing dye **YF02** reached higher $J_{sc}$ (9.19 mA) compared to **YF03** (6.49 mA) due to the higher planarity of the former. Although low planar molecule increases the $V_{oc}$ in DSSCs through the blocking of $I^{3-}$ ions towards the $TiO_2$ surface but at the same time the efficient intramolecular charge transfer decreases, resulting in low $J_{sc}$.

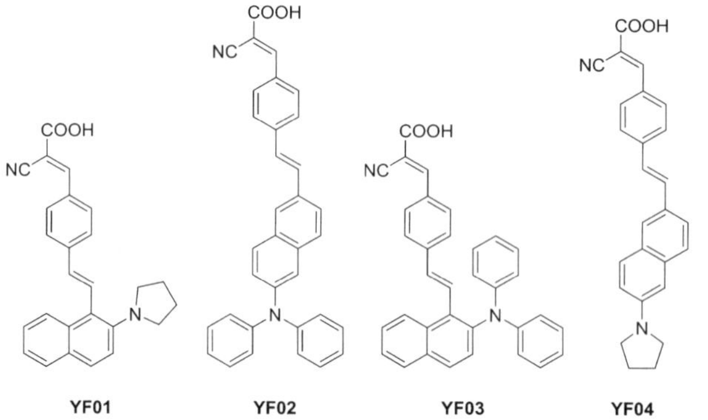


*Figure 6. Molecular Structure of dyes YF-dyes. Reproduced from reference (34) with permission from The Royal Society of Chemistry.*

**Table 7. Photophysical properties of YF01-04. Reproduced from reference (*34*) with permission from The Royal Society of Chemistry.**

| *Dye* | *$\lambda_{max}$ (nm) ($\varepsilon$ ×$10^{-4}$ $M^{-1}cm^{-1}$)[a]* | *$\lambda_{onset}$ (nm)[b] on $TiO_2$ film* | *IP (eV)[c]* | *$E_{0-0}$ (eV)[d]* | *LUMO (eV)[e]* |
|---|---|---|---|---|---|
| **YF01** | 407 (1.1) | 575 | -5.71 | 2.16 | -3.55 |
| **YF02** | 406 (3.2) | 550 | -5.70 | 2.25 | -3.45 |
| **YF03** | 404 (2.7) | 520 | -5.91 | 2.39 | -3.52 |
| **YF04** | 422 (2.1) | 650 | -5.54 | 1.91 | -3.63 |

[a] Absorption maxima, measured in ethanol at room temperature. [b] Absorption measured on a transparent 4 μm $TiO_2$ film. [c] Ionization potential (IP) of absorbed dyes on the nanocrystalline $TiO_2$ film was determined by using the photoemission yield spectrometer (Riken Keiki, AC-3E). [d] Optical bandgap ($E_{0-0}$) was determined from the onset of the absorption in chloroform. [e] The LUMO energy levels were calculated from the expression of LUMO = IP + $E_{0-0}$.

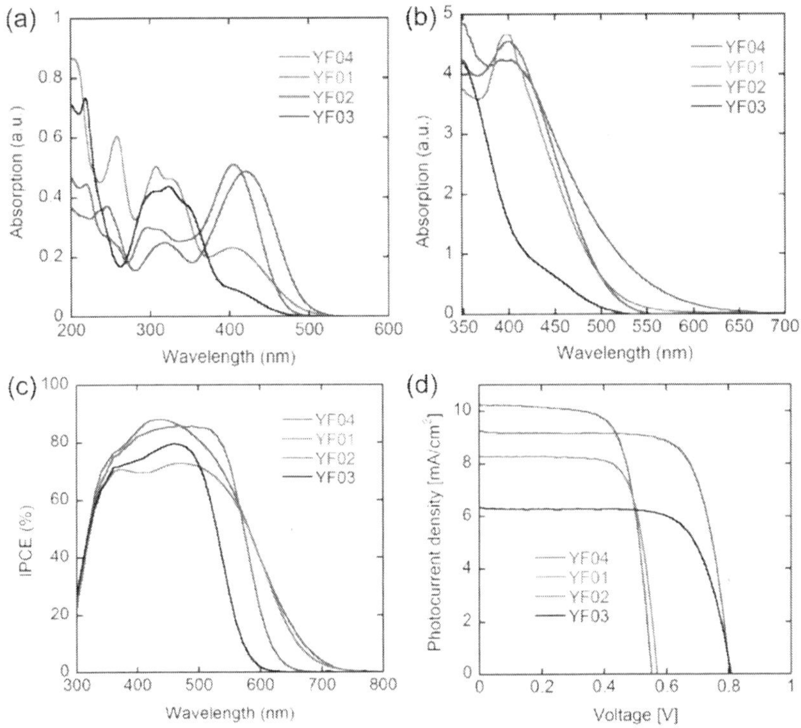


*Figure 7. Absorption spectra of YF-dyes (a) in ethanol and (b) on a transparent $TiO_2$ film. (c) IPCEs of nanocrystalline $TiO_2$ film sensitized by YF-dyes. (d) I–V characteristics of YF-devices. Reproduced from reference (34) with permission from The Royal Society of Chemistry.*

**Table 8. DSSC performances of dyes YF01-04[a]. Reproduced from reference (*34*) with permission from The Royal Society of Chemistry.**

| *Dye* | $J_{sc}$ *[mA cm⁻¹]* | $V_{oc}$ *[V]* | *FF* | *η [ %]* |
|---|---|---|---|---|
| **YF01** | 8.33 | 0.574 | 0.747 | 3.57 |
| **YF02** | 9.19 | 0.799 | 0.721 | 5.29 |
| **YF03** | 6.494 | 0.807 | 0.700 | 3.67 |
| **YF04** | 10.24 | 0.552 | 0.712 | 4.03 |

[a] Measurements were performed under AM 1.5 irradiation on the DSCs devices with 0.25 $cm^2$ active surface area defined by a metal mask.

New dyes, thieno[2,3-*a*]carbazole derivatives, were tested in K-dyes for DSSCs (Figure 8) (*35, 36*). Thienocarbazole unit may increase the molecular planarity between thieno[2,3-*a*]carbazole and oligothiophene linkage. The synthetic procedures of new K-dyes are shown in Scheme 1. The intermediate compounds **2a** was prepared with gold-catalyzed cascade cyclization method (Scheme 1a) (*37*). After lithiation and iodination of the 2-position of thienyl group, **3** was obtained by the Suzuki-Miyaura coupling reaction. Further lithiation-iodination of **3** followed by Suzuki-Miyaura coupling afforded **4** (or **5**). Knoevenagel condensation reactions of **4** (or **5**) produced **K-1** (or **K-2**) in high yields (85-95 %) (*35*). The similar procedures were used for the synthesis of **K-3**, **K-4**, and **K-5** dyes as shown in Schemes 1b and 1c (*36*), where after successful boronation of the α-position of thienyl moiety, the desired aldehydes **7a-c** were obtained by the Suzuki-Miyaura coupling reaction with the corresponding reactants **6a** and **6b**. The final compounds **K-3** to **K-5** dyes were obtained by Knoevenagel condensation in high yields.

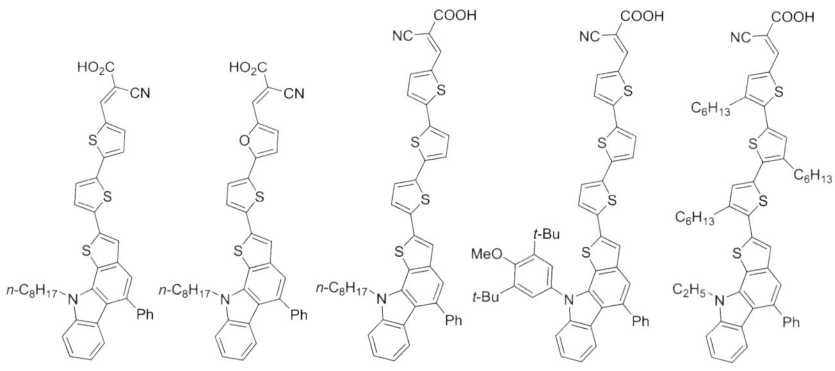


*Figure 8. Molecular structure of K-dyes.*

The optical properties of K-dyes are summarized in Table 9. All dyes exhibited broader absorption wavelength at 350-600 nm region with high molar absorption coefficient in solution. There was no significance difference in absorption and IP between **K-1** and **K-2** dyes. The intramolecular charge transfer absorption of **K-3** to **K-5** dyes red-shifted compared to the **K-1** and **K-2** dyes both in solution and on $TiO_2$ nanocrystallines due to the longer oligothiophene chain in **K-3**~**K-5** dyes (Figures 9a and 9b). The photovoltaics properties of **K-1** and **K-2** were almost similar as of Table 10 and good power conversion efficiencies of 6.62 % and 6.72 % were achieved. However, the **K-3** and **K-4** dyes having a terthiophene π-linker exhibited higher $J_{sc}$ values compared to the **K-1** cell, the $V_{oc}$ values were lower than that of the **K-1** cell, resulting in lower conversion efficiencies of 5.9 % and 6.5 %, respectively (Figure 9d). We ascribed the lower $V_{oc}$ to the increased intermolecular π-π interaction of the **K-3** and **K-4** dyes due to their extended π-linker. The **K-5** dye having three hexyl groups on the terthiophene π-spacer showed the highest $J_{sc}$ (13.90 mA) and $V_{oc}$ (0.74 V), resulting in the highest power conversion efficiency of 7.0 % among the K-dyes. The highest $J_{sc}$ of **K-5** was ascribed to its widest IPCE response (Figure 9c) and the increased $V_{oc}$ was due to the introduction of hexyl groups into the π-spacer which efficiently reduced the charge recombination between the $TiO_2$ conduction band and the electrolyte, leading to the increased electron life time. The similar observation has also been reported by Yamashita et.al., in 2010 which suggested that the molecular planarity would be a another effective way for blocking the interactions between dyes and acceptor molecules in DSSCs (*38*). The DFT calculation revealed that this thienocarbazole donor unit controlled the molecular planarity of D-π-A type dyes, thus enhancing the electron injection into the conduction band of $TiO_2$ nanoparticles.

Next, a new series of JH-dyes based on thieno[3,2-*a*]carbazole were synthesized according to the Scheme 2 (*39*). The new thieno[3,2-*a*]carbazoles **8a-c** were prepared through gold-catalyzed cascade cyclization method (*37*), in which the bromo-substituted **8b** was converted to the *N,N*-diphenylamine-substituted **8b′**. After borylation of **8**, the aldehyde derivatives **10a-c** were obtained in high yields by the Suzuki-Miyaura coupling reaction with the π-spacer units **6b** and **6c**. Finally, the desired JH-dyes were obtained in good to high yields under the Knoevenagel condensation conditions.

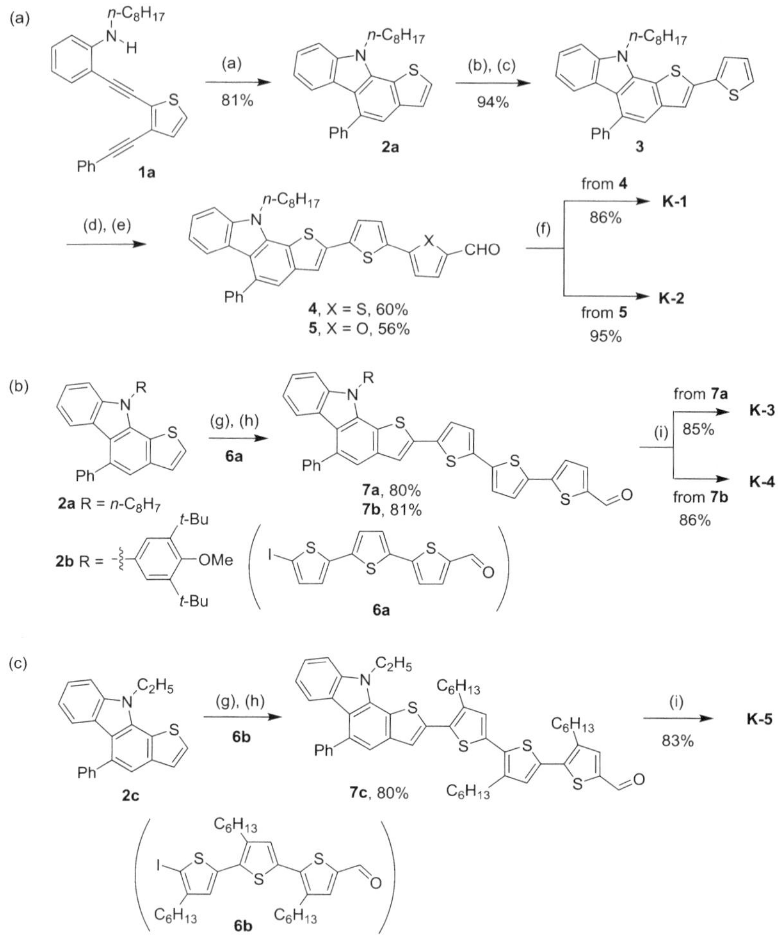


*Scheme 1. Synthesis of K-dyes. (a) $NaAuCl_4$, Ethanol, 100 °C. (b) n-BuLi, $I_2$, THF, -78°C. (c) thiophen-2-ylboronic acid, $Pd(PPh_3)_2Cl_2$, $K_2CO_3$, $DME/H_2O/EtOH$, reflux. (d) n-BuLi, $I_2$, THF, -78 °C. (e) 5-formylthiophen-2-ylboronic acid or 5-formylfuran-2-ylboronic acid, $Pd(PPh_3)_2Cl_2$, $K_2CO_3$, $DME/H_2O/EtOH$, reflux. (f) $NCCH_2CO_2H$, $NH_4OAc$, $HOAc/CH_3CN$ 100 °C. (g) n-BuLi, 2-isopropoxy-4,4,5,5-tetramethyl-1,3,2-dioxaborolane, THF, -78 °C, (h) $Pd(PPh_3)_2Cl_2$, $K_2CO_3$, $DME/H_2O$, **6a** for synthesis of **7a** and **7b**, and **6b** for synthesis of **7c**, reflux. (i) $NCCH_2CO_2H$, piperidine, $CHCl_3/CH_3CN$, reflux. Adapted with permission from reference (35). Copyright 2014 Elsevier. Adapted from reference (36) with permission from The Royal Society of Chemistry.*

**Table 9. Absorption and electrochemical properties of K-dyes. Adapted with permission from reference (*35*). Copyright 2014 Elsevier. Adapted from reference (*36*) with permission from The Royal Society of Chemistry.**

| *Dye* | $\lambda_{max}$ *(nm)* ($\varepsilon \times 10^4$ $M^{-1}cm^{-1}$)[a] | $\lambda_{onset}$ *(nm)* on $TiO_2$[b] | *IP* *(eV)*[c] | $E_{0-0}$ *(eV)*[d] | *LUMO* *(eV)*[e] |
|---|---|---|---|---|---|
| **K-1** | 476 (4.2) | 650 | -5.48 | 2.04 | -3.44 |
| **K-2** | 472 (4.5) | 645 | -5.48 | 2.08 | -3.40 |
| **K-3** | 486 (4.2) | 670 | -5.70 | 1.93 | -3.77 |
| **K-4** | 488 (4.3) | 700 | -5.66 | 1.87 | -3.79 |
| **K-5** | 464 (3.4) | 700 | -5.45 | 1.87 | -3.58 |

[a] Absorption maxima, measured in chloroform at room temperature. [b] Absorption measured on a transparent 4 μm $TiO_2$ film. [c] Ionization potential (IP) of absorbed dyes on the nanocrystalline $TiO_2$ film was determined by using the photoemission yield spectrometer (Riken Keiki, AC-3E). [d] Optical bandgap ($E_{0-0}$) was determined from the on-set of the absorption in chloroform. [e] The LUMO energy levels were calculated from the expression of LUMO = IP + $E_{0-0}$.

**Table 10. DSSC performances of K-dyes[a]**

| *Dye* | $J_{sc}$ *[mA cm$^{-1}$]* | $V_{oc}$ *[V]* | *FF* | $\eta$ *[ %]* |
|---|---|---|---|---|
| **K-1** | 12.40 | 0.70 | 0.76 | 6.62 |
| **K-2** | 12.49 | 0.71 | 0.75 | 6.73 |
| **K-3** | 13.20 | 0.63 | 0.70 | 5.90 |
| **K-4** | 13.10 | 0.68 | 0.73 | 6.50 |
| **K-5** | 13.90 | 0.74 | 0.68 | 7.00 |

[a] Measurements were performed under AM 1.5 irradiation on the DSCs devices with 0.25 $cm^2$ active surface area defined by a metal mask.

The **JH01** and **JH02** dyes combining the thieno[3,2-*a*]carbazole with the bithiophene π-linker had more planar configuration compared to the **JH03** dye having a hexyl-substituted terthiophene π-linker. The JH-dyes showed almost similar intramolecular charge transfer absorption maxima in the range of 400-600 nm in solution, though the **JH03** dye was slightly blue-shifted due to its twisted molecular geometry as shown in Figure 10a and Table 11. The absorption spectra of all JH-dyes on $TiO_2$ films appeared in wider wavelength region up to 700 nm, in which the **JH03** dye showed the most red-shifted absorption (Figure 10b), which are favorable for sunlight harvesting and higher photocurrent generation. The $V_{oc}$ of **JH01** (0.691 V) was lower than that of **JH02** (0.724 V) and **JH03** (0.777 V) (Table 12 and Figure 10d), indicating the presence of high charge recombination process for the **JH01** molecule. It was clear that **JH02** having a bulky *t*-butyl-substituted triphenylamine moiety and **JH03** having hexyl groups on the terthiophene moiety sufficiently reduced the charge recombination process on the $TiO_2$, and hence increased the $V_{OC}$. The IPCE spectra of the JH-dyes covered around 400-600 nm wavelength region with the maximum quantum efficiency of 86 % for **JH03** dye which was higher than **K-5** dye (Figures 10c and 9c). As a result, the highest $J_{SC}$ of 14.08 mA for **JH03** was achieved among the JH-dyes and K-dyes. It was concluded that the $J_{SC}$ and $V_{OC}$ of **JH02** and **JH03** were higher than the **JH01** dye (Figure 10d and Table 12), indicating the structural effect of diphenylamine into the thienocarbazole donor and hexyl group into the terthiophene π-linker, respectively. Therefore, the highest efficiency of 8.04 % was achieved using **JH03** dye. Using thieno[3,2-*a*]carbazole as a new donor, red shifted absorption on $TiO_2$ and a highest IPCE plateau of 86 % over the most visible region were successfully demonstrated.

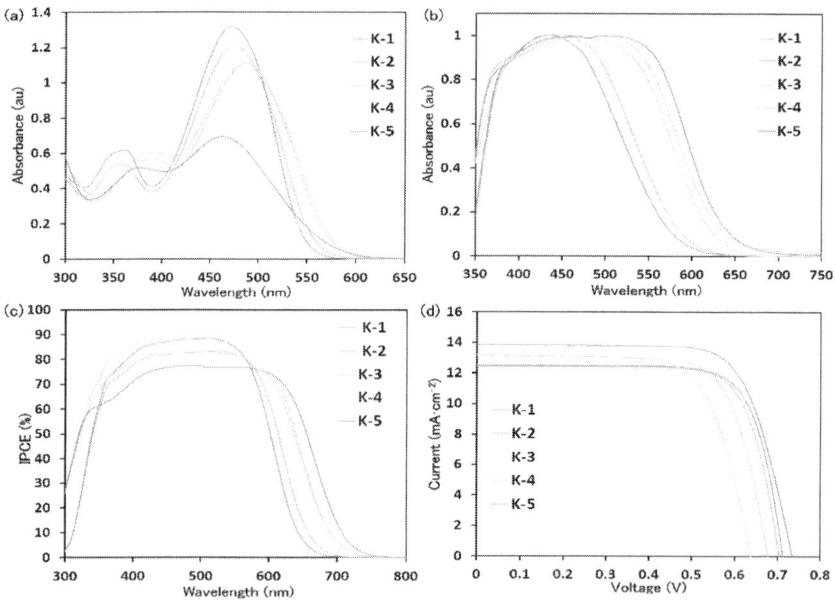


*Figure 9. Absorption spectra of K-dyes (a) in chloroform, and (b) on a transparent $TiO_2$ film. (c) IPCEs of nanocrystalline $TiO_2$ film sensitized by K-dyes. (d) I–V characteristics of K-dyes. Adapted with permission from reference (35). Copyright 2014 Elsevier. Adapted from reference (36) with permission from The Royal Society of Chemistry.*

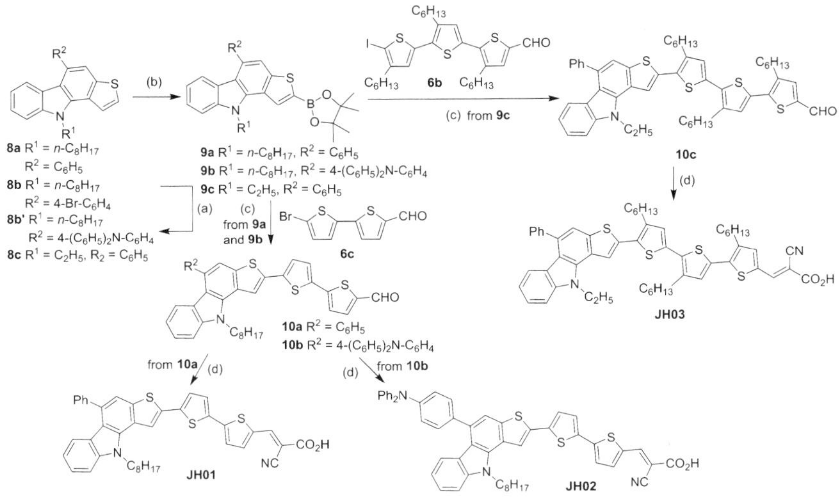

*Scheme 2. Synthesis of JH-dyes. (a) $Ph_2NH$, $Pd_2(dba)_3$ (1 mol %), $(t\text{-}Bu)_3P$ (4 mol %), $NaO^tBu$, toluene, 100 °C, 75 %. (b) n-BuLi, THF, -78 °C for 30 min followed by 2-isopropoxy-4,4,5,5-tetramethyl-1,3,2-dioxaborolane, THF, -78 °C to rt for 2 h. (c) **6b** or **6c**, $Pd(PPh_3)_4$ (10 mol %), $K_2CO_3$, DME/$H_2O$ (10/1), reflux; **10a**: 81 %, **10b**: 85 %, **10c**: 80 %. (d) $CNCH_2COOH$, $NH_4OAc$/HOAc, 110 °C; **JH01**: 66 %, **JH02**: 60 %, **JH03**: 89 %. Adapted with permission from reference (39). Copyright 2015 Elsevier.*

**Table 11. Absorption and electrochemical properties of JH-dyes. Reproduced with permission from reference (*39*). Copyright 2015 Elsevier.**

| *Dye* | $\lambda_{max}$ *(nm)* ($\varepsilon \times 10^4$ $M^{-1}cm^{-1}$)[a] | $\lambda_{max}$ *(nm) on* $TiO_2$[b] | *IP (eV)*[c] | $E_{0\text{-}0}$ *(eV)*[d] | *LUMO (eV)*[e] |
|---|---|---|---|---|---|
| **JH01** | 486 (3.7) | 539 | -5.46 | 2.03 | -3.35 |
| **JH02** | 492 (4.0) | 551 | -5.34 | 2.00 | -3.24 |
| **JH03** | 476 (3.4) | 558 | -5.30 | 2.01 | -3.26 |

[a] Absorption maxima, measured in chloroform at room temperature. [b] Absorption measured on a transparent 4 μm $TiO_2$ film. [c] Ionization potential (IP) of absorbed dyes on the nanocrystalline $TiO_2$ film was determined by using the photoemission yield spectrometer (Riken Keiki, AC-3E). [d] Optical bandgap ($E_{0\text{-}0}$) was determined from the onset of the absorption in chloroform. [e] The LUMO energy levels were calculated from the expression of LUMO = IP + $E_{0\text{-}0}$.

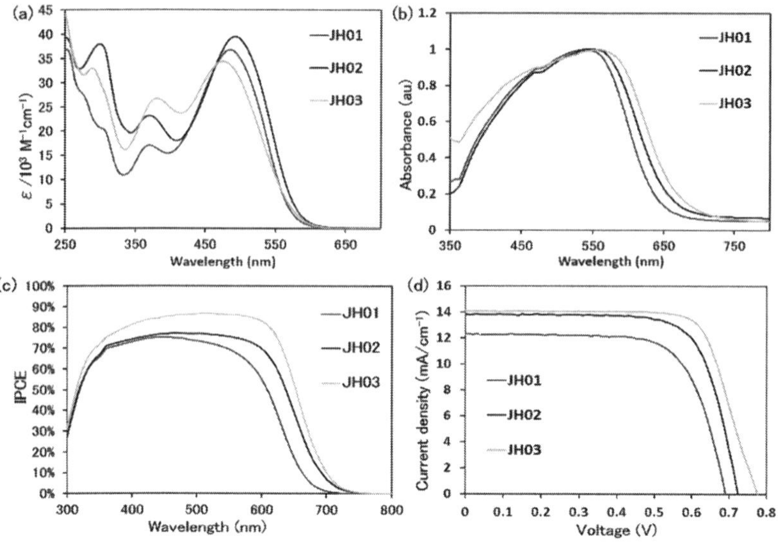


*Figure 10. Absorption spectra of JH-dyes (a) in chloroform, and (b) on a transparent $TiO_2$ film. (c) IPCEs of nanocrystalline $TiO_2$ film sensitized by JH-dyes. (d) I–V characteristics of JH-dyes. Reproduced with permission from reference (39). Copyright 2015 Elsevier.*

**Table 12. DSC performances of JH-dyes[a]. Reproduced with permission from reference (*39*). Copyright 2015 Elsevier.**

| *Dye* | $J_{sc}$ *[mA cm⁻¹]* | $V_{oc}$ *[V]* | *FF* | $\eta$ *[ %]* |
|---|---|---|---|---|
| **JH01** | 12.30 | 0.691 | 0.701 | 5.96 |
| **JH02** | 13.85 | 0.724 | 0.723 | 7.25 |
| **JH03** | 14.08 | 0.777 | 0.735 | 8.04 |

[a] Measurements were performed under AM 1.5 irradiation on the DSCs devices with 0.25 $cm^2$ active surface area defined by a metal mask.

## 1.3. Summary of DSSCs

We have presented some useful strategies to obtain highly efficient sensitizers by independently alternating the donor, π-linker, and acceptor moieties. **AK01** dye has successfully resulted in 6.2 % efficiency. The $J_{sc}$ was measured at 15.4 mA and the onset IPCE was close to 800 nm, which proves that **AK01** is potentially useful as panchromatic sensitizer. The naphthylamine donor based YF-dyes showed an interesting performance. The dyes **YF02** and **YF03**, containing amine donors showed higher $V_{oc}$ due to their twisted geometry. The increase in the output voltage indicates that the inclusion of naphthalene moiety in organic dyes is an auspicious process to prevent the molecular aggregation of dyes thus improving the performance of DSSCs. The designed molecules based on silyl linkers **YS01-03** showed better charge separation and increment in photocurrent, resulting in an impressive efficiency of 5.06 %. This work provided the significance of silyl moiety as linkers in D-π-A systems. The sensitizers **Ba-01-03** were low in molecular weight and yielded an efficiency of 6.38 %. Thienocarbazole based new dyes were also explored and the study suggested that the presence of thienocarbazole donor in K-dyes and JH-dyes significantly increased the molecular planarity and achieved the highest efficiency of 8.04 % for **JH03** dye by combination of thieno[3,2-*a*]carbazole donor with hexyl-substituted terthiophene π-linker. The efficient charge transport of the custom synthesized dyes has provided insights into designing and introducing precautious steps for newer organic molecules to be used in DSSCs.

## 2.1. Functional Fullerenes for Bulk Heterojunction Solar Cells

Since the first report in 1995 describing a bulk-heterojunction (BHJ) polymer solar cell (PSC) incorporating a methanofullerene (*4*), significant progress has been made in improving device performance and the scope of device research has broadened widely (*40, 41*). The most widely used active layer in BHJ device consists of a blend of electron-donating materials such as p-type conjugated polymers, and electron-accepting n-type fullerene derivatives (*42, 43*). In the BHJ structure, the blending of the polymer donor and the fullerene acceptor materials forms an interpenetrating network film, which is sandwiched between a transparent indium-tin oxide (ITO) positive electrode and a negative metal electrode (Figure 11). Upon illumination of the blend film through the transparent ITO, both donor and acceptor absorb photons to generate excitons, which diffuse to the donor and acceptor interfaces, where excitons dissociate to electrons and holes due to the proper offset of the LUMO energy levels of donor and acceptor (Figure 12). The dissociated electrons and holes will be drifted by the electric field and collected by corresponding negative and positive electrodes, respectively, to generate electricity. In principle, the power conversion efficiency (PCE) of BHJ devices is proportional to the related parameters, such as short-circuit current density ($J_{sc}$), open-circuit voltage ($V_{oc}$), and fill factor (*FF*), and hence simultaneously increasing both $J_{sc}$ and $V_{oc}$ is of important for achieving high PCEs (*44*). From the nature of semiconducting materials, $J_{sc}$ is generally correlated with the light absorption capability, exciton generation and dissociation, carrier mobility, and balanced charge transport ability of polymer or small molecule donors, and fullerene acceptors. $V_{oc}$ is mainly predominated by the energy level difference between the HOMO of the donor and the LUMO of the acceptor (Figure 12). For these reasons, various new donor and acceptor materials have been developed in the past two decades (*44–50*). Particularly, numerous new π-conjugated polymer donors have been synthesized with respect to the broad and strong absorption in the visible-near infrared region, and lower HOMO energy levels. To date, the state-of-the-art of BHJ devices have a PCE approaching 10 % based on blending the newly developed low bandgap donors and the most representative benchmark acceptor $PC_{61}BM$ or $PC_{71}BM$ (Figure 13) (*7–9*).

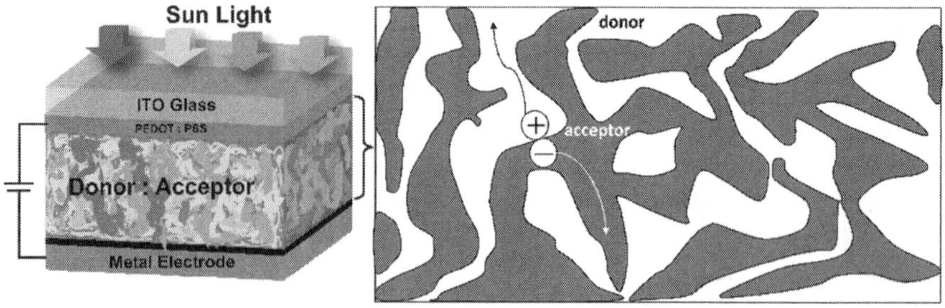


*Figure 11. BHJ solar cell configuration and blend film morphology.*

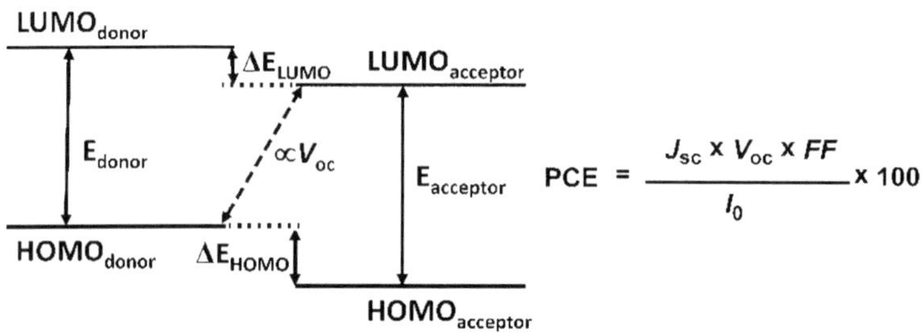


*Figure 12. Schematic drawing of donor and acceptor energy levels.*

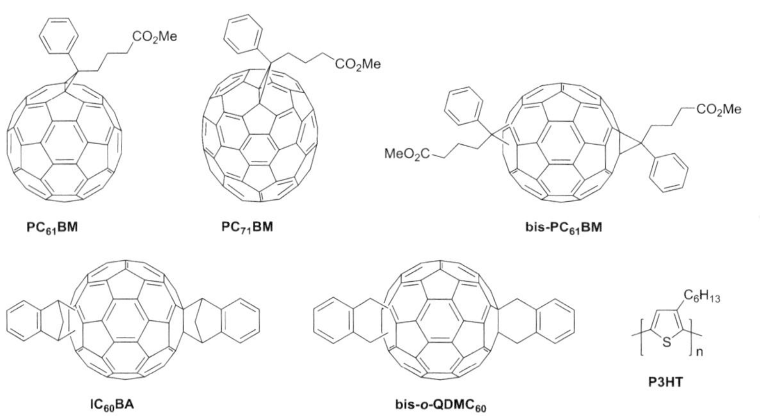


*Figure 13. Representative acceptor and donor materials.*

Functional fullerenes have been used widely as excellent n-type semiconductors in solution processable organic electronics (*51–54*). Especially, they serve as a unique electronic accepting material for BHJ solar cells due to their tunable solubility in organic solvents, a tunable energy level, and the superior packing structure in the solid state via versatile functional groups on $C_{60}$ core, while maintain the properties of pristine fullerenes, such as high electron mobility and high electron affinity (*48–50*). In addition to the most well-studied acceptor 1-(3-methoxycarbonyl)propyl-1-phenyl[6,6]$C_{61}$ (PCBM) (*55, 56*), various new functional fullerenes have been developed recently to improve the performance of BHJ solar cells, but only a limited number of new fullerene derivatives has been successfully employed. For example, the bisfunctional 56$\pi$-electron fullerenes with an up-shifted LUMO, such as indene-$C_{60}$-bisadduct ($IC_{60}BA$) (*57, 58*), bis$PC_{61}BM$ (*59*), and bis-*o*-quinodimethane $C_{60}$ (bis-*o*-$QDMC_{60}$) (*60, 61*), which are favorable for increasing $V_{oc}$ of BHJ devices, have been reported to show higher PCEs than that of PCBM when poly(3-hexylthiophene) (P3HT) was used as a donor (Figure 13). Nevertheless, as stated above, $PC_{61}BM$ and $PC_{71}BM$ are still by far the most commonly used n-type components in BHJ solar cells due to their high solubility, high electron transport ability, and suitable compatibility with a wide range of donors. However, those efficient acceptors including PCBM are still very expensive due to low yields, low selectivities, and harsh synthetic conditions. For example, $PC_{61}BM$ has been prepared by a one-pot reaction over two-steps in 58 % yield through the reaction of $C_{60}$ with methyl 4-benzoylbutyrate *p*-tosylhydrazone at 70 °C followed by isomerization of the resulting [5,6]$PC_{61}BM$ to $PC_{61}BM$ at 180 °C (*55*). Therefore, the development of new and practical fullerene functionalizations toward synthesizing new fullerene acceptors with high production yields is highly desirable for achieving high performance and low-cost BHJ solar cells, which may expedite the practical application of the BHJ solar cells in the next few years.

In this context, a variety of chemical functionalization of fullerenes have been developed during the last few decades, involving nucleophilic addition of organometallic reagents (*62–64*), cycloadditions (*48, 49, 54*), addition of radicals (*65*) in terms of the electron-accepting feature of the fullerene core. Among them, the nucleophilic addition of organometallic reagents such as organolithiums or Grignard reagents is one of the classical methods for the formation of monofunctionalized fullerenes. However, this method is always accompanied by the low conversion yields, low mono-selectivity, high loading of organometallic reagents, and limited functional group compatibility. In addition, the radical addition of fullerene proceeds rapidly, which often produces polysubstituted fullerenes. On the other hand, the transition-metal-catalyzed fullerene functionalization has been proved to exhibit high efficiency and selectivity under mild reaction conditions as well as a high compatibility with a wide range of functional groups, which offers much opportunity for creating novel functional fullerenes (*66–75*). For example, the selective monoarylation of [60]fullerene with organoboron reagents has been reported to proceed efficiently in the presence of Rh or Pd catalyst (*67, 68*). Recently, we have been interested in developing various catalytic functionalizations of fullerenes via generation of a fullerene radical (*71–77*). For example, we have reported a novel cobalt-catalyzed monoalkylation of [60]fullerene with various active alkyl bromides at ambient temperature to give monoalkylated hydrofullerenes in good to high yields (*71*). We have also demonstrated that the corresponding monoalkylated hydrofullerenes are important precursors for forming single-bonded fullerene dimers and 1,4-disubstitted fullerenes by activating the C-H bond of the fullerene core (*72–76*). Moreover, the corresponding monoalkylated fullerenes have been employed as a new class of acceptor materials in BHJ solar cells to show high photon-to-current efficiencies (*78*, *79*). This section mainly describes our recent achievements on the synthesis of various monosubstituted hydrofullerenes with cobalt catalyst and their optophysical properties as well as their application as acceptors in BHJ solar cells including the deuterium isotope effect.

## 2.2. Co-Catalyzed Selective Monoalkylation of [60]Fullerene for Synthesis of Hydrofullerenes and Deuteriofullerenes

Taking into account the facile radical addition of $C_{60}$ (*65*) and the transition-metal-catalyzed generation of organo radical species (*80*), we set out to develop a new transition-metal-catalyzed fullerene functionalization (*71*). Initially, various transition metal catalysts and additives have been screened for the reaction of fullerene ($C_{60}$) with benzyl bromide (**11a**, 3 equiv) in 1,2-dichlomethane (ODCB) with water ($H_2O$, 10 equiv) at room temperature under an argon atmosphere for 50 h (Table 13). On the basis of the activity of cobalt catalysts on radical transformation of alkenes with alky halides in the presence of Grignard reagents (*81*), we firstly examined various cobalt salts as catalysts. The reaction with 1,2-Bis(diphenylphosphino)ethanedichlorocobalt(II), 97% ($CoCl_2$dppe) catalyst alone did not produce any products (entry 1). To our delight, the use of $CoCl_2$dppe combined with manganese powder (Mn, 3 equiv) afforded the corresponding monobenzyl hydrofullerene **12** in 76 % isolated yield along with a small amount of multiadducts and recovered $C_{60}$ (entry 2). Among other metal additives tested, iron (Fe) additive could produce **12** in high yield with a prolonged reaction time, and zinc (Zn) additive gave a moderate yield of **12**, while copper (Cu) additive was totally inactive (entries 3-5). The varied activities should be attributed to their standard potentials. For example, the reduction potentials of Mn (-1.185), Zn (-0.7628), and Fe (-0.447) are able to reduce cobalt ($Co^{2+}$) to $Co^{o \text{ or } 1+}$ (-0.28), while Cu (0.521) cannot reduce $Co^{2+}$. We also examined other cobalt salts bearing different counteranions. $CoBr_2$dppe and $CoI_2$dppe were also active, while the yields of **12** were much lower than that of $CoCl_2$dppe (entries 6 and 7). The reaction with other cobalt catalysts having bidentate ligands, such as bis(diphenylphosphino)methane bis(cobalt(II) chloride) [$CoCl_2$dppm] and dichloro[1,1′-bis(diphenylphosphino)ferrocene]cobalt(II) [$CoCl_2$dppf], were active, giving **12** in 71 % and 75 % yields, respectively (entries 8 and 9). In sharp contrast, the cobalt catalysts having a monodentate ligand or without ligand were almost inactive (entries 10 and 11). In addition to cobalt catalysts, other transition metal catalysts, such as palladium (II) chloride dppe [$PdCl_2$dppe], chlorotris(triphenylphosphine)rhodium(I) [$RhCl(PPh_3)_3$], and nickel (II) chloride dppe [$NiCl_2$dppe] were used to be not effective to produce **12** (entries 12-14). Finally, it is worthy to note that the reaction without using water did not proceed to give any products, and the reaction efficiency was suppressed dramatically in the presence of oxygen.

**Table 13. Screening of catalysts and additives for reaction of $C_{60}$ with benzyl bromide[a]**

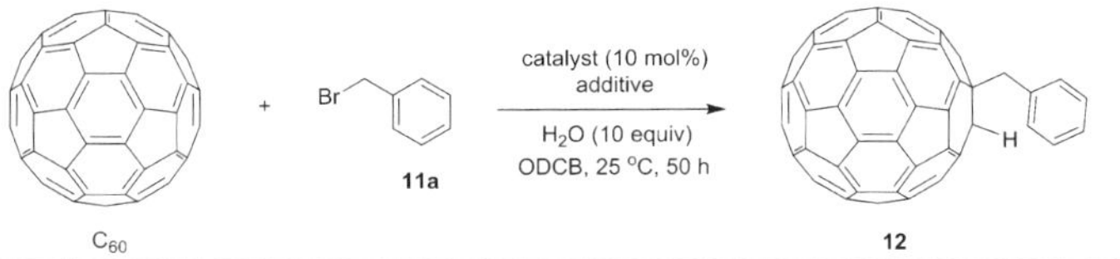


| entry | catalyst | additive (3.0 equiv) | yield of **12** ( %)[b] | recovery of $C_{60}$ ( %)[c] |
|---|---|---|---|---|
| 1 | $CoCl_2$dppe | none | 0 | 96 |
| 2 | $CoCl_2$dppe | Mn | 88 (76)[d] | (8)[d] |
| 3 | $CoCl_2$dppe | Fe | 80[e] | 12 |
| 4 | $CoCl_2$dppe | Zn | 56 | 25 |
| 5 | $CoCl_2$dppe | Cu | 0 | 97 |
| 6 | $CoBr_2$dppe | Mn | 40 | 39 |
| 7 | $CoI_2$dppe | Mn | 45 | 24 |
| 8 | $CoCl_2$dppm | Mn | 71 | 17 |
| 9 | $CoCl_2$dppf | Mn | 73 | 14 |
| 10 | $CoI_2(PPh_3)_2$ | Mn | 0 | 85 |
| 11 | $CoCl_2$ | Mn | 4 | 88 |
| 12 | $PdCl_2$dppe | Mn | 0 | 88 |
| 13 | $RhCl(PPh_3)_3$ | Mn | 29 | 58 |
| 14 | $NiCl_2$dppf | Mn | 20 | 68 |

[a] Reaction conditions: A 1,2-dichlorobenzene (ODCB, 4 mL) solution of catalyst (10 mol %) and additive (3 equiv) under Ar atmosphere was stirred for 1 hr at 25 °C. Next, $C_{60}$ (21.6 mg), $H_2O$ (10 equiv), and benzyl bromide (3 equiv) were added, and the resulting mixture was stirred for 50 hrs at 25 °C. [b] $^1H$ NMR yield determined using 2,4,6-trimethylpyridine as an internal standard. [c] HPLC yield determined using $C_{70}$ as an internal standard. [d] Isolated yield. [e] The reaction time is 5 days. Reproduced from reference 71.

Under the optimized conditions, $CoCl_2$dppe (10 mol %), Mn (3 equiv), $H_2O$ (10 equiv) in 1,2-dichlomethane (ODCB) at room temperature, we studied the substrate scope by using a wide range of active alkyl bromides (Table 14). The reactions were monitored by high performance liquid chromatography (HPLC) analysis and the products were purified by using silica gel chromatography. The corresponding products were determined unambiguously by using (proton) $^1H$ and (carbon) $^{13}C$ nuclear magnetic resosance (NMR) spectra as well as high resolution mass. The reactions with benzyl bromides **(11b-f)** bearing various functional groups on the phenyl ring, such as ester, cyano, and methoxy, afforded the corresponding hydrofullerenes **13-17** in good to high yields with a high monobenzyl selectivity regardless of the electronic properties of aromatic rings. It was noted that the monobenzyl hydrofullerenes (**13**, **14**, **16**, **17**) showed a high solubility in ODCB or toluene due to the ester and methoxy groups. The reaction was also compatible with other active alkyl bromides. For example, ally bromide (**11g**), propargyl bromide (**11h**), and 1-bromo-2-butyne (**11i**) were successfully employed as alkyl bromides to give the corresponding hydrofullerenes **18-20** in good yields. Methyl 2-bromoacetate (**11j**) showed a relatively lower reactivity, affording the corresponding hydrofullerene **21** in a moderate yield under a prolonged reaction time. It should be noted that in comparison with these active alkyl bromides, inactive alkyl bromides such as butyl bromide or bromomethylcyclopropane were used to be totally inactive under the present hydroalkylation conditions even at elevated temperatures.

The present hydroalkylation reaction was proved to be effective for the construction of potentially useful electronic materials of fullerene-bound macromolecules. For example, the reaction of $C_{60}$ with benzyl bromide **11k** bearing a zinc porphyrin (ZnP) group produced the corresponding ZnP-$C_{60}$ product **22** in 30 % yield with Mn reductant, while the yield of **22** could be increased to 40 % if Fe reductant was used instead of Mn powder (Scheme 3a). This method was also applied to the reaction of $C_{60}$ with the dendrimer [G-3]-Br **11l** having a benzyl bromide moiety at the focal point, affording the fullerene-bound dendrimer **23** in 48 % yield under the standard conditions (Scheme 3b). The product **23** exhibited a very high solubility in various organic solvents due to the dendrimer moiety. The double hydroalkylation of $C_{60}$ with 1,4-bis(bromomethyl)benzene **11m** having long alkyl chains on the phenyl ring also proceeded smoothly to give the fullerene dimer **24** connected by a benzyl spacer in 39 % yield, which showed a good solubility in ODCB or toluene (Scheme 3c).

The reaction mechanism is proposed in Scheme 4. Firstly, reduction of Co(II) complex by Mn reductant produces the electron-rich Co(0) complex (*82*), which undergoes a single electron transfer to benzyl bromide to generate a benzyl radical species **A** together with a Co(I) complex **B**. The facile addition of benzyl radical **A** to $C_{60}$ forms a fullerene monoradical **C**, which could be stabilized by radical delocalization on the $C_{60}$ core. Subsequently, the Co(I) complex **B** reacts with the fullerene monoradical **C** to form a fullerenyl-cobalt complex **D**, which is thought to suppress the further addition of radical species to form multiadducts. Finally, protonolysis of the complex **D** by the added water produces the corresponding product **12**, and the active Co(0) complex can be regenerated by the reduction of the Co(II) complex with Mn reductant.

In order to support our proposed mechanism, we carried out some control experiments (Scheme 5). For example, the reaction of $C_{60}$ with benzyl bromide **11b** in the presence of $D_2O$ instead of $H_2O$, the corresponding monobenzyl deuteriofullerene **25** was formed in 46 % yield (eq 1), indicating that the proton on the $C_{60}$ core was derived from the water additive. Moreover, when water was replaced by *n*-$Bu_3SnH$ in the reaction of $C_{60}$ with benzyl bromide **11a** under otherwise standard conditions, the corresponding product **12** was obtained in a slightly lower yield of 40 % (eq 2) (*83*), while the reaction seems to proceed more quickly probably due to the rapid trapping of the organotin hydride by the fullerene monoradical **C**.

**Table 14. Substrate scope for co-catalyzed monofunctionalization of $C_{60}$ with various active alkyl bromides[a,b]**

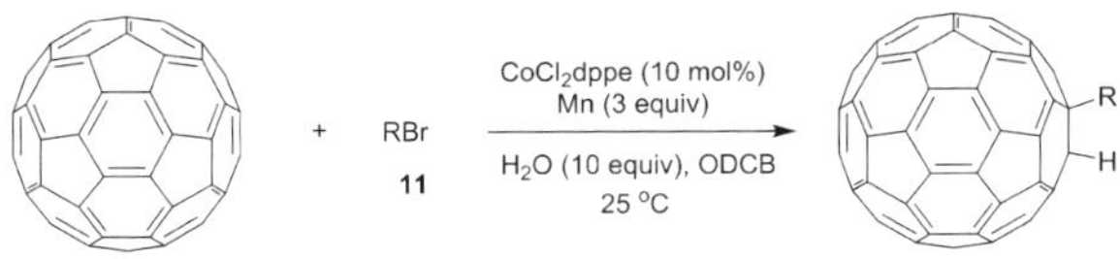


13-21

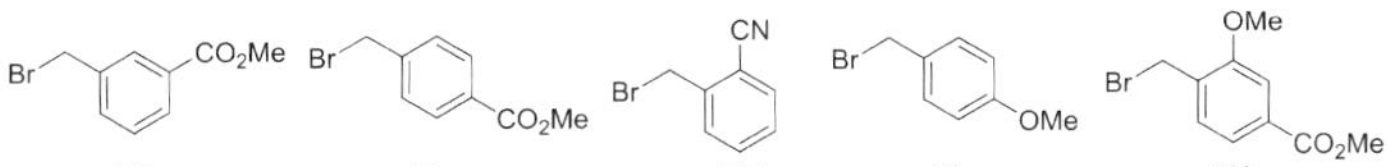


11b 11c 11d 11e 11f

11g 11h 11i 11j

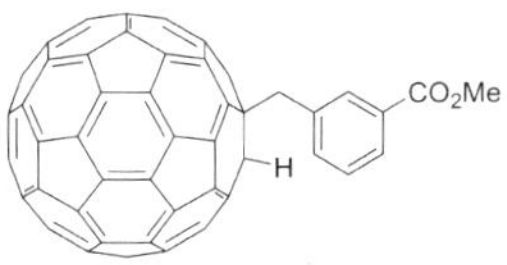


**13**, 58% (48 h)

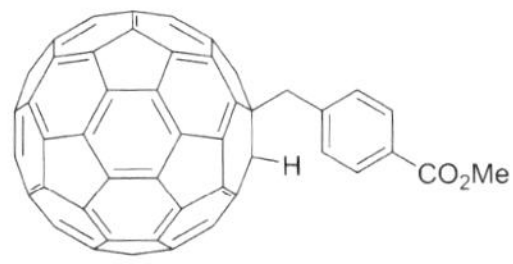


**14**, 66% (48 h)

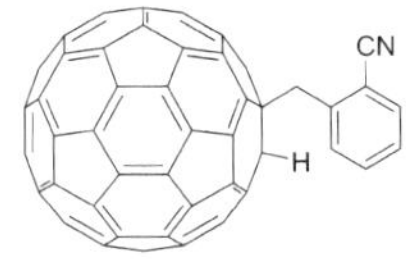


**15**, 51% (40 h)

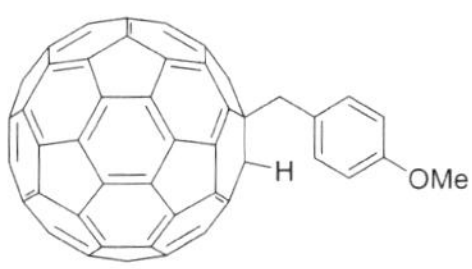


**16**, 66% (48 h)

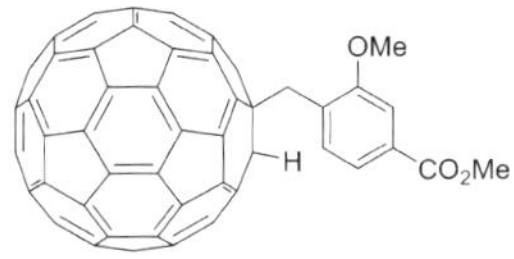


**17**, 60% (48 h)

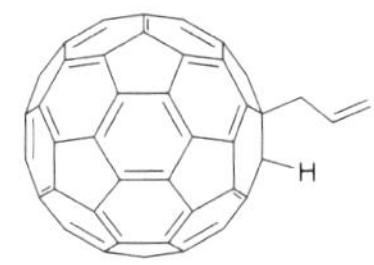


**18**, 74% (86 h)

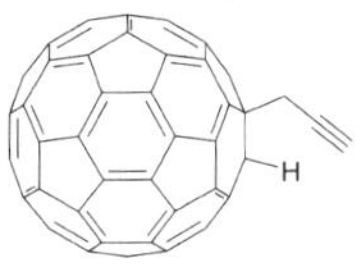


**19**, 75% (86 h)

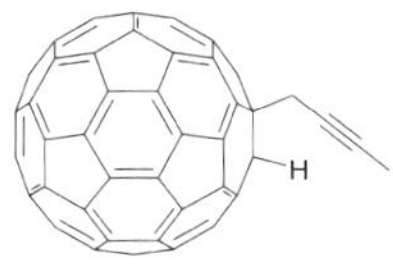


**20**, 42% (60 h)

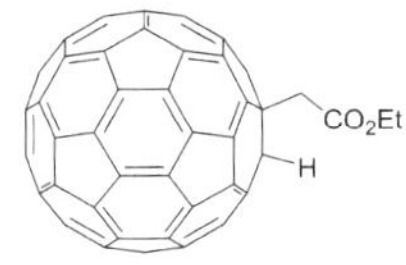


**21**, 36% (96 h)

[a] Reaction conditions: A 1,2-dichlorobenzene (ODCB, 4 mL) solution of $CoCl_2$dppe (10 mol %) and Mn (3 equiv) under Ar atmosphere was stirred for 1 hr at 25 °C. Next, $C_{60}$ (21.6 mg), $H_2O$ (10 equiv), and RBr (1.5-4.0 equiv) were added and the resulting mixture was stirred for 34-96 hrs at 25 °C. [b] Isolated yields. Adapted from reference 71. Adapted with permission from reference 78. Copyright 2013 Elsevier.

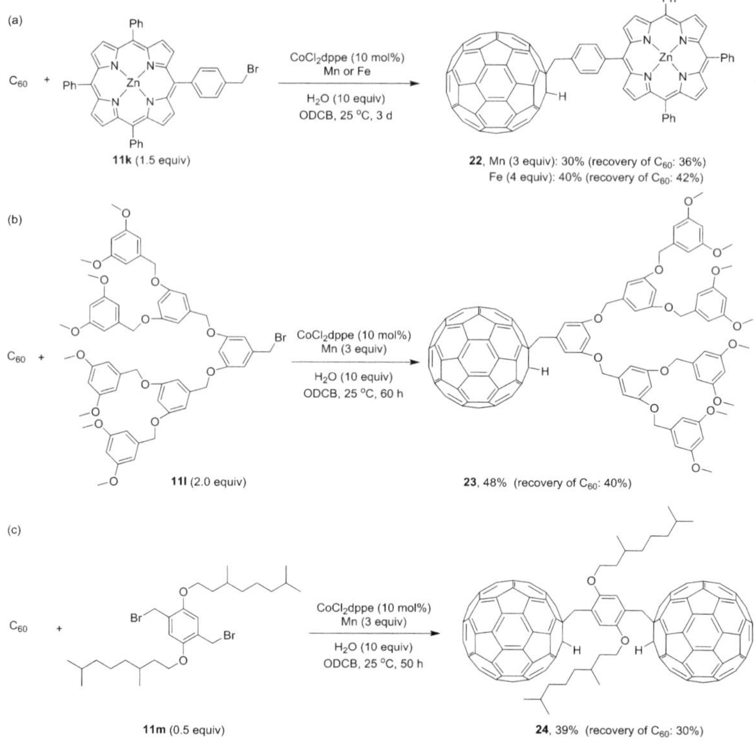


*Scheme 3. Co-catalyzed synthesis of fullerene-bound macromolecules. Reproduced from reference (71).*

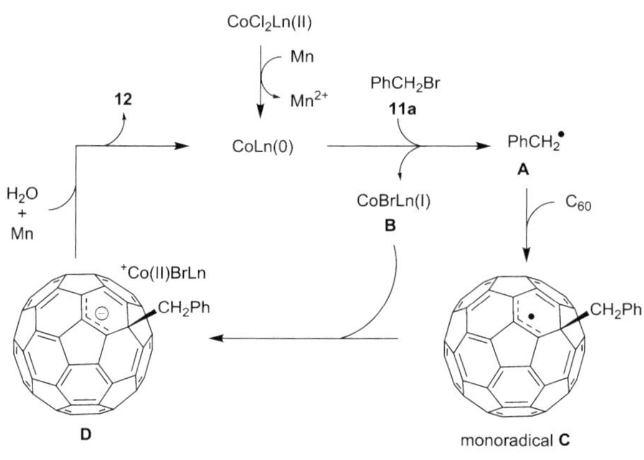


*Scheme 4. Proposed reaction mechanism. Reproduced from reference (71).*

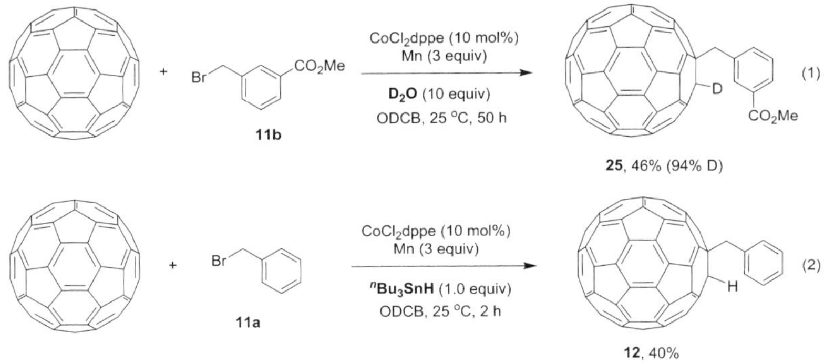

*Scheme 5. Control experiments. Reproduced from reference (71).*

On the basis of the reaction shown in eq 1, we synthesized a variety of monobenzyl deuteriofullerenes to investigate the deuterium isotope effect (DIE) on the performance of BHJ solar cells (Table 15) (*79*). It is noted that although the DIE has been reported in organic light emitting diodes to increase external quantum efficiency as compared to the protonated analogues (*84*), the DIE on BHJ solar cells has never been reported. Thus, the reaction of $C_{60}$ with the *meta*-ester-substituted benzyl bromide **11n** bearing a deuterium at the benzylic position under the standard conditions using deuterium oxide ($D_2O$, 10 equiv) instead of water gave the corresponding monobenzyl deuteriofullerene **26** in 47 % yield. Similarly, various *para*-methoxy-substituted benzyl bromides (**11e**, **11o-s**) with or without deuterium atoms on the benzyl moiety were tested. As a result, the corresponding monobenzyl deuteriofullerenes **27-32** were synthesized in good yields. All the products were purified by silica gel chromatography, and showed a comparable solubility with their protonated analogues in toluene or chloroform.

**Table 15. Co-Catalyzed monobenzylation of $C_{60}$ for synthesis of deuteriofullerenes[a,b]**

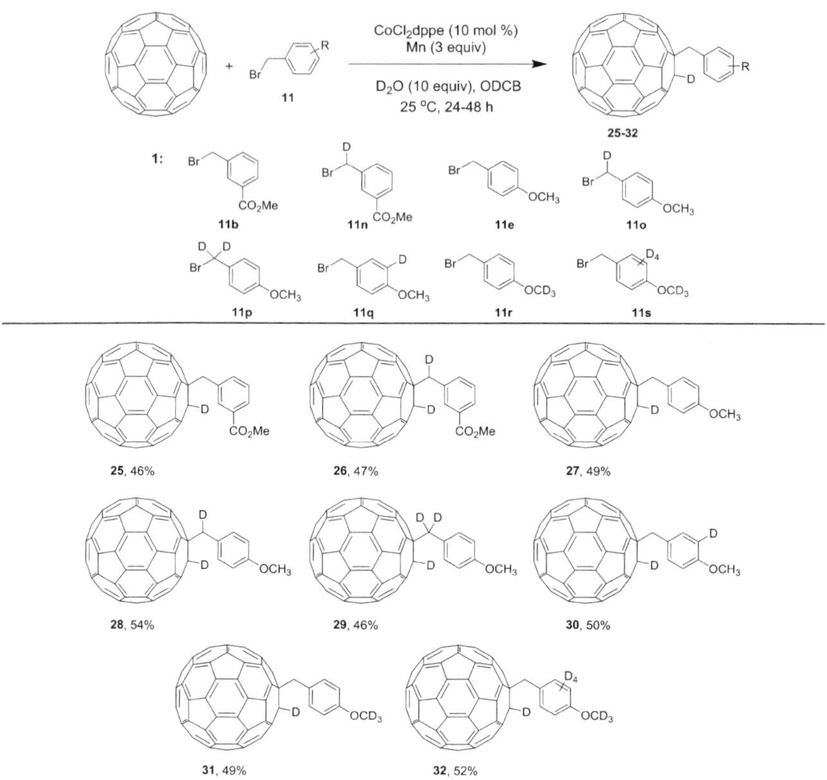


[a] Reaction conditions: A 1,2-dichlorobenzene (ODCB, 4 mL) solution of $CoCl_2$dppe (10 mol %) and Mn (3 equiv) under Ar atmosphere was stirred for 1 hr at 25 °C. Next, $C_{60}$ (21.6 mg), $D_2O$ (10 equiv), and RBr (1.5 equiv) were added and the resulting mixture was stirred for 24-48 hrs at 25 °C. [b] Isolated yields. Reproduced from reference 79.

## 2.3. Photophysical Properties of Monobenzyl Hydrofullerenes and Deuteriofullerenes

The UV-vis absorption spectra of hydrofullerenes and deuteriofullerenes measured in chloroform are shown in Figure 14 together with the reference acceptor $PC_{61}BM$ for comparison (*78*, *79*). The hydrofullerenes (**13**, **14**, **16**, **17**, and **23**) show a similar absorption band with $PC_{61}BM$ in the range of 250-280 nm despite of the varied functional groups on the phenyl ring (Figure 14a) and slightly enhanced absorption are observed in the 280-330 nm and 400-500 nm regions compared to that of $PC_{61}BM$ (Figure 14a, inset). The deuteriofullerenes **25-32** exhibit similar characteristic absorption spectra compared to that of hydrofullerenes (Figure 14b), indicating that the absorption of deuteriofullerenes is not influenced by the replacement of hydrogen atoms with deuterium atoms, as well as by the different number and position of deuterium atoms.

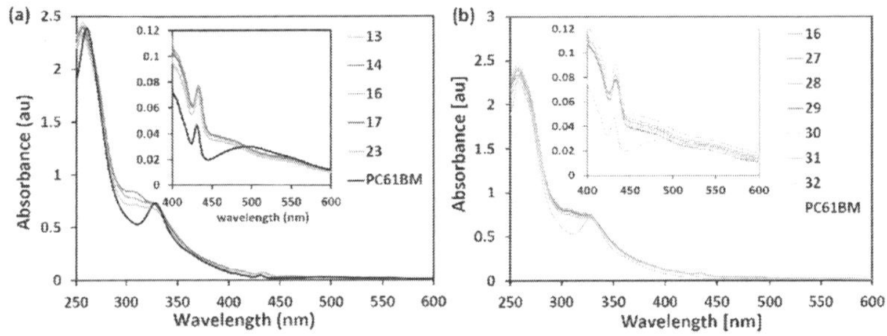


*Figure 14. UV-vis absorption of monobenzyl hydrofullerenes (a) (Reproduced with permission from reference (78). Copyright 2013 Elsevier.), and deuteriofullerenes (b) (Reproduced from reference (79)) in chloroform.*

The reduction potentials were measured by cyclic voltammetry (CV) using Ag/AgCl as a reference electrode, tetrafluoroborate (Bu4NBF4, 0.05 M) as a supporting electrolyte, Pt wire as a counter electrode, and glassy carbon as a working electrode in o-Dichlorobenzol (ODCB, Table 16). Both hydro- and deuteriofullerenes exhibit similar three pseudo-reversible reduction waves without being influenced by hydrogen or deuterium atoms and by the electron-donating or electron-withdrawing groups on the phenyl rings. Moreover, slightly negative shifts of the first reduction potentials have been observed for hydro- and deuteriofullerenes compared to the reference $PC_{61}BM$. Thus, the LUMO energy levels estimated by the first reduction potentials of hydro- and deuteriofullerenes are calculated to be in the range of -3.54 eV to -3.58 eV, which are similar or slightly higher than that of the reference $PC_{61}BM$ (-3.58 eV) under the same measurement conditions. We conclude that the relatively high LUMO energy levels of new fullerene derivatives must be attributed to the benzyl $sp^3$ carbon that blocks the conjugation systems between the aryl group and the $C_{60}$ core, which are favorable for achieving the high $V_{oc}$ of BHJ solar cells (*85*).

**Table 16. Electrochemical reduction potentials and LUMO energy levels of monobenzyl hydrofullerenes and deuteriofullerens, and $PC_{61}BM^{a,b}$. Adapted with permission from reference (*78*). Copyright 2013 Elsevier. Adapted from reference (*79*).**

| *compound* | $E_{1/2}^{1}/V$ | $E_{1/2}^{2}/V$ | $E_{1/2}^{3}/V$ | *LUMO/eV* |
|---|---|---|---|---|
| **13** | -0.61 | -0.98 | -1.46 | -3.55 |
| **14** | -0.59 | -0.98 | -1.48 | -3.57 |
| **16** | -0.61 | -1.01 | -1.56 | -3.55 |
| **17** | -0.62 | -1.01 | -1.56 | -3.54 |
| **23** | -0.60 | -0.98 | -1.49 | -3.56 |
| **25** | -0.60 | -0.99 | -1.45 | -3.56 |
| **26** | -0.60 | -0.99 | -1.54 | -3.56 |
| **27** | -0.60 | -0.99 | -1.56 | -3.56 |
| **28** | -0.59 | -0.99 | -1.54 | -3.57 |
| **29** | -0.58 | -0.98 | -1.53 | -3.58 |
| **30** | -0.59 | -0.98 | -1.52 | -3.57 |
| **31** | -0.59 | -0.98 | -1.55 | -3.57 |
| **32** | -0.59 | -0.99 | -1.55 | -3.57 |
| $PC_{61}BM$ | -0.58 | -0.98 | -1.48 | -3.58 |

[a] Potential values are versus Ag/AgCl reference electrode; reduction potential of ferrocene (0.64 V) is versus Ag/AgCl. [b] The LUMO energy levels were estimated from the first oxidation potentials.

## 2.4. BHJ Solar Cell Performances Using Hydro- and Deuteriofullerenes as Acceptors

The BHJ device was fabricated in the configuration indium tin oxide (ITO) and/or poly(3,4-ethylenedioxythiophene):poly(styrenesulfonate) (PEDOT:PSS)/ poly(3-hexylthiophene): fullerenes/lithium fluoride / aluminium [ITO/PEDOT:PSS/P3HT:fullerenes/LiF/Al] (*78*, *79*). An ODCB solution of a blend of P3HT and new fullerene (1:1, weight ratio) blend was spin-coated onto the PEDOT:PSS surface, forming an active layer with thicknesses ranging from 200 to 260 nm. The substrate with the active layer was dried at 110 °C for 10 min in the $N_2$ glovebox. Finally, LiF (1 nm) and Al (80 nm) were deposited onto the active layer by means of conventional thermal evaporation at a chamber pressure lower than $5\times10^{-4}$ Pa, resulting in the devices with an active area of $2\times2$ mm$^2$. The photovoltaic characterization of the hydrofullerene-based devices under illumination with 100 mW/cm$^2$ of AM 1.5 are summarized in Table 17, and their current density-voltage curves and incident photon-to-current conversion efficiency (IPCE) spectra are shown in Figure 15.

BHJ devices based on the hydrofullerenes **13**, **14**, and **16** having an ester or a methoxy group on the phenyl ring showed similar $J_{sc}$ of 8.24-8.76 mA/cm$^2$ and $V_{oc}$ of 0.60-0.61 V, resulting in the PCEs of 2.84 % for **13**, 2.80 % for **14**, and 2.39 % for **16**, respectively. The device based on **17** having both ester and methoxy groups on the phenyl ring exhibited an increased $J_{sc}$ of 9.66 mA/cm$^2$, $V_{oc}$ of 0.63 V, and $FF$ of 61.6 %, resulting in a high PCE of 3.75 % comparable to the reference $PC_{61}BM$ device (3.78 %). In comparison with the $PC_{61}BM$ device, the increased $V_{oc}$ of the device **17** is ascribed to the slightly higher LUMO energy level of **17** than that of $PC_{61}BM$. In contrast, the device based on **23** bearing a bulky dendrimer moiety showed a very low PCE of 0.13 % mainly due to the very low $J_{sc}$ and $FF$, indicating that the bulky G3 dendrimer group dramatically affects the packing arrangement of the $C_{60}$ cages, which is related to the charge transfer. Moreover, the highest $J_{sc}$ of the device **17** is in consistent with the highest IPCE value in the 350-650 nm region (Figure 15b).

**Table 17. BHJ Performances based on P3HT and hydrofullerene acceptors (w/w = 1/1)[a]. Reproduced with permission from reference (*78*). Copyright 2013 Elsevier.**

| *acceptor* | $J_{sc}$ [mA cm$^{-2}$] | $V_{oc}$ [V] | *FF* [ %] | *PCE* [ %] |
|---|---|---|---|---|
| **13** | 8.76 | 0.61 | 53.2 | 2.84 |
| **14** | 8.38 | 0.60 | 55.7 | 2.80 |
| **16** | 8.24 | 0.60 | 48.4 | 2.39 |
| **17** | 9.66 | 0.63 | 61.6 | 3.75 |
| **23** | 0.66 | 0.56 | 36.6 | 0.13 |
| $PC_{61}BM$ | 9.86 | 0.60 | 64.0 | 3.78 |

[a] Blend film was prepared using P3HT (20 mg) and hydrofullerenes (20 mg) in 1,2-dichlorobenzene (1 mL); annealing temperature is 110 °C (10 min).

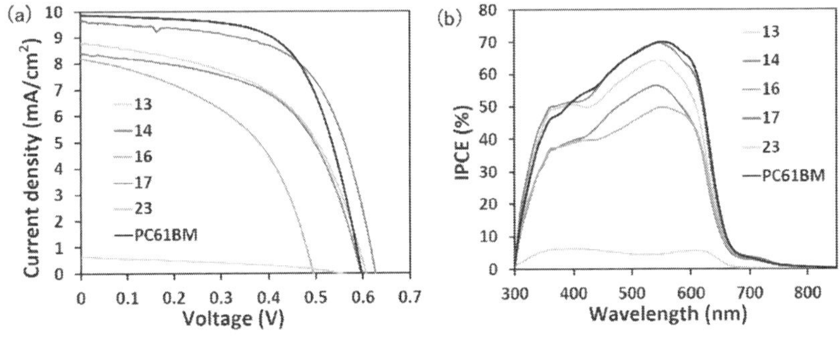


*Figure 15. BHJ solar cell performance based on monobenzyl hydrofullerenes. (a) Current density-voltage curves and (b) incident photon-to-current conversion efficiency. Reproduced with permission from reference (78). Copyright 2013 Elsevier.*

The surface morphologies of the active layers blending with P3HT donor were measured by means of atomic force microscopy (AFM). The homogeneous surfaces of the blend films showed obvious nanoscale phase separation and the changing of functional groups or positions on the phenyl ring influenced the film roughness. For example, the films of **13** and **14** having an ester group at the *meta*- or *para*-position of the phenyl ring displayed the root mean squares (rms) of 3.99 nm and 0.63 nm (Figures 16a and 16b), respectively, in which the hydrofullerene **13** showed a better solubility than that of the hydrofullerene **14** in ODCB or chloroform. In addition, the rms of the film **16** having a methoxy group was measured to be 5.25 nm (Figure 16c), which was higher than that of the film **14** having an ester group. The blend film of **17** having the highest photoelectric conversion efficiency (PCE) showed a very smooth surface with the rms of 0.54 (Figure 16d), implying the optimized interfacial contact between the P3HT donor and the acceptor **17**. It is noted that the blend film **23** bearing a bulky dendrimer on the $C_{60}$ cage showed a largest rms of 21.7 nm, which should be the reason of the low $J_{sc}$.

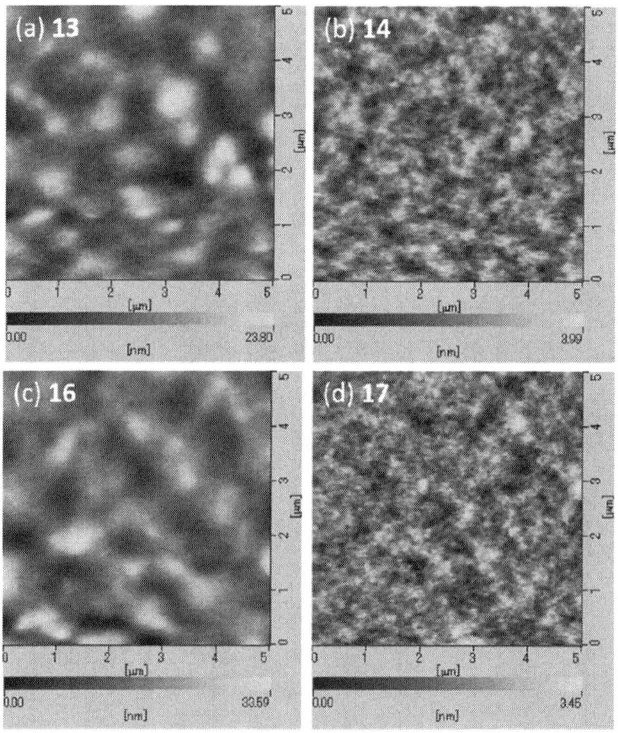


*Figure 16. AFM images of P3HT donor with different monobenzyl hydrofullerenes. Reproduced with permission from reference (78). Copyright 2013 Elsevier.*

Next, a series of BHJ devices were fabricated based on the deuteriofullerene acceptors and poly(3-hexylthiophene) (P3HT) donor under the same method to investigate the deuterium effect on the photovoltaic performances (*79*). Interestingly, when the deuteriofullerene acceptors **25** and **26** having an ester group and a deuterium atom at the $C_{60}$ cage or the benzylic position were blended with the P3HT donor, the PCEs of 3.75 % and 3.63 % were obtained, respectively (Table 18), which were much higher than that of their hydrofullerene analogue **13** (Table 17, 2.84 %). The $V_{oc}$ values of the devices **25** and **26** were almost similar to that of the device **13**, which was in consistent with their similar LUMO energy levels (Table 16). The higher PCEs of the devices **25** and **26** are attributed to their increased $J_{sc}$ and *FF* compared to that of the device **13**.

**Table 18. PSC Performances based on monobenzyl deuteriofullerene acceptors with P3HT donor (w/w=1/1)[a]. Reproduced from reference (*79*).**

| *acceptor* | $J_{sc}$ *[mA cm⁻²]* | $V_{oc}$ *[V]* | *FF [ %]* | *PCE [ %]* |
|---|---|---|---|---|
| **25** | 10.72 | 0.60 | 58.2 | 3.75 |
| **26** | 10.13 | 0.62 | 58.2 | 3.63 |
| **27** | 10.08 | 0.62 | 57.4 | 3.63 |
| **28** | 10.75 | 0.61 | 59.5 | 3.93 |
| **29** | 11.10 | 0.62 | 60.3 | 4.16 |
| **30** | 9.16 | 0.60 | 59.5 | 3.28 |
| **31** | 10.04 | 0.60 | 58.3 | 3.48 |
| **32** | 11.44 | 0.61 | 57.6 | 4.03 |
| 5D-$PC_{61}BM$ | 10.29 | 0.58 | 61.5 | 3.66 |

[a] Blend film was prepared using P3HT (20 mg) and deuteriofullerens (20 mg) in 1,2-dichlorobenzene (1 mL); annealing temperature is 110 °C (10 min).

We also examined other deuteriofullerenes **27-32** having a methoxy group on the phenyl ring as acceptors blending with the P3HT donor to observe the similar deuterium effect. In spite of the position and number of the deuterium atoms, the devices **27-32** exhibited much higher PCEs (3.28 % to 4.16 %, Table 18) compared to their hydrofullerene analogue **16** (2.39 %, Table 17). Particularly, the device **29** having a deuterium at the $C_{60}$ core and two deuterium atoms at the benzylic position showed the highest PCE of 4.16 % with $J_{sc}$ of 11.10 mA/cm², $V_{oc}$ of 0.62 V, and *FF* of 60.3 %, which was even higher than that of the reference $PC_{61}BM$ device (3.78 %, Table 17). The photovoltaic parameters in Table 18 clearly showed that the $J_{sc}$ and *FF* of the deuteriofullerene devices were much higher than that of their hydrofullerene analogue devices, which mainly contributed to the higher PCEs. Moreover, the deuteriofullerene devices showed much higher incident photon-to-current efficiency (IPCE) values than that of the hydrofullerene devices in the 350-650 nm region, and hence resulted in the higher $J_{sc}$. It should be noted that the every device was tested at least for 3 times under the same fabrication method and higher average PCEs were achieved in all the deuteriofullerene devices compared to the hydrofullerene devices, indicating the high reproducibility of the present device performances.

The atomic force microscopy (AFM) image of the representative **29**/P3HT blend film showed obvious nanoscale phase separation. The smaller rms of 0.78 nm for the deuteriofullerene **29** than that of the hydrofullerene analogue **16** (5.25 nm) indicated that the roughness of the blend film was influenced by the introduction of deuterium atoms (Figures 18a and 18b). Moreover, the transmission electron microscopy (TEM) and electron energy loss spectroscopy (EELS) mapping of the cross section showed that the morphology of the deuteriofullerene **29** device had a more clear interpenetrating network structure compared to that of the hydrofullerene **16** device across the film (Figures 18c and 18d), indicating the optimal interfacial contact between the deuteriofullerene acceptor and the P3HT donor. Taking into consideration the similar absorption and LUMO energy levels of both hydro- and deuteriofullerenes, we ascribe the increased PCEs and $J_{sc}$ for the deuteriofullerene devices to the optimal thin film morphology probably due to the higher stability of the deuteriofullerene thin films upon fabrication, which may be derived from the more stable carbon-deuterium bonds than carbon-hydrogen bonds. Finally, it is worthy to note that the deuterium-incorporated fullerene acceptors are not always effective for increasing the photovoltaic performances as we found that the $PC_{61}BM$ and 5D-$PC_{61}BM$ devices with P3HT donor exhibited almost same PCEs of 3.78 % and 3.66 %, respectively, under the present fabrication method.

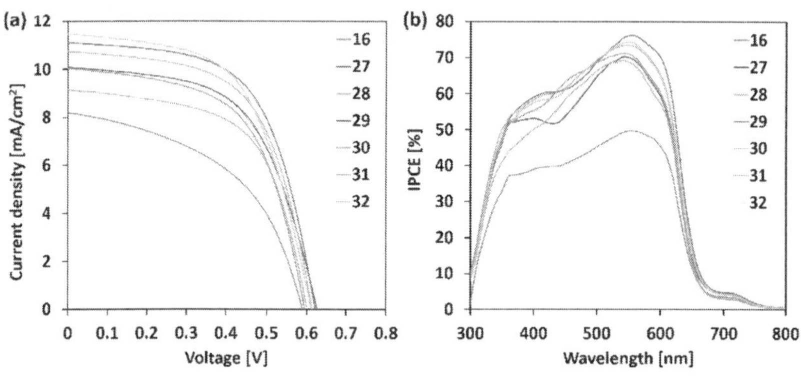


*Figure 17. BHJ solar cell performances of deuteriofullerene devices. (a) Current density-voltage (I-V) curves and (b) incident photon-to-current conversion efficiency (IPCE). Reproduced from reference (79).*

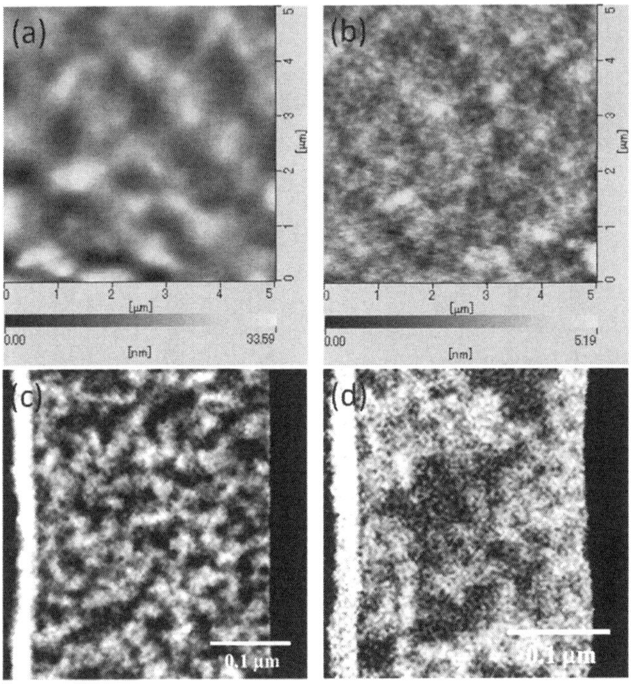


*Figure 18. AFM images of P3HT donor with (a) hydrofullerene (**16**), and (b) deuteriofullerene (**29**); TEM-EELS mappings of cross-section of (c) hydrofullerene (**16**), and (d) deuteriofullerene (**29**) devices. The bright and dark regions represent the P3HT-rich domains and fullerene-acceptor-rich domains, respectively. Reproduced from reference (79).*

## 2.5. Summary of BHJ Solar Cells

In this section, we have described a new Co-catalyzed highly selective and efficient monofunctionalization of [60]fullerene with active alkyl bromides toward synthesis of monoalkylated fullerene derivatives. The resulting monobenzyl hydro- and deuteriofullerenes were successfully employed as new electron acceptors with P3HT donor in BHJ solar cells to show high photovoltaic performances. In particular, the deuteriofullerene-based devices exhibited an unprecedented deuterium isotope effect, resulted in the higher PCEs as compared to their hydrofullerene analogue devices mainly due to the improved $J_{sc}$ and *FF*. The highest PCE was up to 4.16 %, which was higher than that of the reference $PC_{61}BM$ device under the same device conditions. This research demonstrates that partial replacement of a proton by a deuterium atom in the fullerene acceptors should be one of the efficient approaches for improving the photovoltaic performances.

## Conclusion

Regarding the research on DSSCs, some useful strategies to obtain higher power conversion efficiency have been presented. Molecular design and synthesis of donors, π-spacers, and acceptors of new organic dyes were carried out in our laboratories. Efficient charge transport of the synthesized dyes has provided insights into taking precaution against molecular design of molecules. So far, our dyes have not yet reached to very satisfactory level for practical application. We are still in pursue to increase the panchromatic response of dyes to have better absorption, which should lead to improved performance. Regarding the research on BHJ solar cells, we have developed new and practically useful methods for the synthesis of a wide range of functional fullerenes; the reaction proceeds in catalytic manner and makes a large scale production feasible. Some of such functional fullerenes were applied as acceptors of BHJ solar cells, and we could achieve higher power conversion efficiency than the well-known BHJ using $PC_{61}BM$ device. We are now ready to have a number of functional fullerenes and related carbon materials at hand, and therefore it is expected that we are able to achieve much higher photovoltaic performances in BHJ solar cells.

## Acknowledgments

This work was supported by a Scientific Research (B) from Japan Society for Promotion of Science (JSPS) (No. 25288043), and World Premier International Research Center Initiative (WPI), MEXT, Japan. Authors also extend gratitude of the Solar Energy Research Institute (SERI) to the University Kebangsaan Malaysia (UKM) through the Escience Fund research grant with code 03-01-02-SF1149 of the Ministry of Science, Technology and Innovation (MOSTI), Malaysia for their kind contribution as well.

## Authors' Contribution

Associate professor Tienan Jin wrote BHJ solar cells and helped very much to write DSSCs. Associate professor Md. Akhtaruzzaman wrote DSSCs. Professor emeritus Y. Yamamoto put together both sections and reviewed them.

## References

1. Jackson, P.; Hariskos, D.; Lotter, E.; Paetel, S.; Wuerz, R.; Menner, R.; Wischmann, W.; Powalla, M. New world record efficiency for Cu (In,Ga)$Se_2$ thin-film solar cells beyond 20 %. *Prog. Photovolt: Res. Appl.* **2011**, *19*, 894–897.
2. Green, M. A.; Emery, K.; Hishikawa, Y.; Warta, W.; Dunlop, E. D. Solar cell efficiency tables (Version 45). *Prog. Photovolt: Res. Appl.* **2015**, *23*, 1–9.
3. O'Regan, B.; Grätzel, M. A low-cost, high-efficiency solar cell based on dye-sensitized colloidal $TiO_2$ films. *Nature* **1991**, *353*, 737–740.

4. Yu, G.; Gao, J.; Hummelen, J. C.; Wudl, F.; Heeger, A. J. Polymer photovoltaic cells: Enhanced efficiencies via a network of internal donor-acceptor heterojunctions. *Science* **1995**, *270*, 1789–1791.
5. Yella, A.; Lee, H. W.; Tsao, H. N.; Yi, C.; Chandiran, A. K.; Nazeeruddin, M. K.; Diau, E. W.-G; Yeh, C.-Y.; Zakeeruddin, S. D.; Grätzel, M. Porphyrin-sensitized Solar cells with cobalt(II/III)–based redox electrolyte exceed 12 percent efficiency. *Science* **2011**, *334*, 629–633.
6. Han, L.; Islam, A.; Chen, H.; Malapaka, C.; Chiranjeevi, B.; Zhang, S.; Yang, X.; Yanagida, M. High-efficiency dye-sensitized solar cell with a novel co-adsorbent. *Energy Environ. Sci.* **2012**, *5*, 6057–6060.
7. Liu, Y.; Chen, C.-C.; Hong, Z.; Gao, J.; Yang, Y. (M.); Zhou, H.; Dou, L.; Li, G.; Yang, Y. Solution-processed small-molecule solar cells: Breaking the 10 % power conversion efficiency. *Sci. Rep.* **2013**, *3*, 3356.
8. You, J.; Dou, L.; Yoshimura, K.; Kato, T.; Ohya, K.; Moriarty, T.; Emery, K.; Chen, C.-C.; Gao, J.; Li, G.; Yang, Y. A polymer tandem solar cell with 10.6 % power conversion efficiency. *Nat. Commun.* **2013**, *4*, 1446.
9. Kan, B.; Zhang, Q.; Li, M.; Wan, X.; Ni, W.; Long, G.; Wang, Y.; Yang, X.; Feng, H.; Chen, Y. Solution-processed organic solar cells based on dialkylthiol-substituted benzodithiophene unit with efficiency near 10 %. *J. Am. Chem. Soc.* **2014**, *136*, 15529–15532.
10. Grätzel, M. Photoelectrochemical cells. *Nature* **2001**, *414*, 338–344.
11. Robertson, N. Optimizing dyes for dye-sensitized solar cells. *Angew. Chem., Int. Ed.* **2006**, *45*, 2338–2345.
12. Mishra, A.; Fischer, M. K. R.; Bäuerle, P. Metal-free organic dyes for dye-sensitized solar cells: From structure: Property relationships to design rules. *Angew. Chem., Int. Ed.* **2009**, *48*, 2474–2499.
13. Hagfeldt, A.; Boschloo, G.; Sun, L.; Kloo, L.; Pettersson, H. Dye-sensitized solar cells. *Chem. Rev.* **2010**, *110*, 6595–6663.
14. Dennler, G.; Scharber, M. C.; Brabec, C. J. Polymer-fullerene bulk-heterojunction solar cells. *Adv. Mater.* **2009**, *21*, 1323–1338.
15. Krebs, F. C. Fabrication and processing of polymer solar cells: A review of printing and coating techniques. *Sol. Energy Mater. Sol. Cells* **2009**, *93*, 394–412.
16. Hagfeldt, A.; Grätzel, M. Molecular photovoltaics. *Acc. Chem. Res.* **2000**, *33*, 269–277.
17. Grätzel, M. Conversion of sunlight to electric power by nanocrystalline dye-sensitized solar cells. *J. Photochem. Photobiol. A* **2004**, *164*, 3–14.
18. Zeng, W.; Cao, Y.; Bai, Y.; Wang, Y.; Shi, Y.; Zhang, M.; Wang, F.; Pan, C.; Wang, P. Efficient dye-sensitized solar cells with an organic photosensitizer featuring orderly conjugated ethylenedioxythiophene and dithienosilole blocks. *Chem. Mater.* **2010**, *22*, 1915–1925.
19. Ning, Z. J.; Fu, N.; Tian, H. Improvement of dye-sensitized solar cells: What we know and what we need to know. *Energy Environ. Sci.* **2010**, *3*, 1170–1181.
20. Clifford, J. N.; Martinez-Ferrero, E.; Viterisi, A.; Palomares, E. Sensitizer molecular structure-device efficiency relationship in dye sensitized solar cells. *Chem. Soc. Rev.* **2011**, *40*, 1635–1646.

21. Chen, Y. S.; Chou, H. H.; Chen, Y. C.; Hsu, C. Y.; Lin, J. T. Recent developments in molecule-based organic materials for dye-sensitized solar cells. *J. Mater. Chem.* **2012**, *22*, 8734–8747.
22. Liang, M.; Chen, J. Arylamine organic dyes for dye-sensitized solar cells. *Chem. Soc. Rev.* **2013**, *42*, 3453–3488.
23. Wu, Y.; Zhu, W. Organic sensitizers from D–π–A to D–A–π–A: effect of the internal electron-withdrawing units on molecular absorption, energy levels and photovoltaic performances. *Chem. Soc. Rev.* **2013**, *42*, 2039–2058.
24. Ambrosio, F.; Martsinovich, N.; Troisi, A. What is the best anchoring group for a dye in a dye-sensitized solar cell? *J. Phys. Chem. Lett.* **2012**, *3*, 1531–1535.
25. Ardo, S.; Meyer, G. J. Photodriven heterogeneous charge transfer with transition-metal compounds anchored to $TiO_2$ semiconductor surfaces. *Chem. Soc. Rev.* **2009**, *38*, 115–164.
26. Chiba, Y.; Islam, A.; Watanabe, Y.; Komiya, R.; Koide, N.; Han, L. Dye-sensitized solar cells with conversion efficiency of 11.1 %. *Jpn. J. Appl. Phys.* **2006**, *45*, L638.
27. Hwang, S.; Lee, J. H.; Park, C.; Lee, H.; Kim, C.; Park, C.; Lee, M.-H.; Lee, W.; Park, J.; Kim, K.; Park, N.-G.; Kim, C. A highly efficient organic sensitizer for dye-sensitized solar cells. *Chem. Commun.* **2007**, 4887–4889.
28. Akhtaruzzaman, M.; Islam, A.; Yang, F.; Asao, N.; Kwon, E.; Singh, S. P.; Han, L.; Yamamoto, Y. A novel metal-free panchromatic $TiO_2$ sensitizer based on a phenylenevinylene-conjugated unit and an indoline derivative for highly efficient dye-sensitized solar cells. *Chem. Commun.* **2011**, *47*, 12400–12402.
29. Akhtaruzzaman, M.; Menggenbateer; Islam, A.; El-Shafei, A.; Asao, N.; Jin, T.; Han, L.; Alamry, K. A.; Kosa, S. A.; Asiri, A. M.; Yamamoto, Y. Structure–property relationship of different electron donors: novel organic sensitizers based on fused dithienothiophene π-conjugated linker for high efficiency dye-sensitized solar cells. *Tetrahedron* **2013**, *69*, 3444–3450.
30. Akhtaruzzaman, M.; Seya, Y.; Asao, N.; Islam, A.; Kwon, E.; El-Shafei, A.; Han, L.; Yamamoto, Y. Donor–acceptor dyes incorporating a stable dibenzosilole π-conjugated spacer for dye-sensitized solar cells. *J. Mater. Chem.* **2012**, *22*, 10771–10778.
31. Usta, H.; Lu, G.; Facchetti, A.; Marks, T. J. Dithienosilole-and dibenzosilole-thiophene copolymers as semiconductors for organic thin-film transistors. *J. Am. Chem. Soc.* **2006**, *128*, 9034–9035.
32. Yabusaki, Y.; Ohshima, N.; Kondo, H.; Kusamoto, T.; Yamanoi, Y.; Nishihara, H. Versatile synthesis of blue luminescent siloles and germoles and hydrogen-bond-assisted color alteration. *Chem.-Eur. J.* **2010**, *16*, 5581–5585.
33. Horiuchi, T.; Miura, H.; Sumioka, K.; Uchida, S. High efficiency of dye-sensitized solar cells based on metal-free indoline dyes. *J. Am. Chem. Soc.* **2004**, *126*, 12218–12219.
34. Yang, F.; Akhtaruzzaman, M.; Islam, A.; Jin, T.; El-Shafei, A.; Qin, C.; Han, L.; Alamry, K. A.; Kosa, S. A.; Hussein, M. A.; Yamamoto, Y. Structure–property relationship of naphthalene based donor–π–acceptor

organic dyes for dye-sensitized solar cells: remarkable improvement of open-circuit photovoltage. *J. Mater. Chem.* **2012**, *22*, 22550–22557.

35. Zhao, J.; Jin, T.; Islam, A.; Kwon, E.; Akhtaruzzaman, M.; Asao, N.; Han, L.; Alamry, K. A.; Kosa, S. A.; Asiri, A. M.; Yamamoto, Y. Thieno [2, 3-*a*] carbazole-based donor–π–acceptor organic dyes for efficient dye-sensitized solar cells. *Tetrahedron* **2014**, *70*, 6211–6216.
36. Zhao, J.; Oniwa, K.; Islam, A.; Qin, C.; Asao, N.; Han, L.; Yamamoto, Y.; Jin, T. Thieno[2,3,*a*]carbazole donor-based organic dyes for high efficient dye-sensitized solar cells. *Org. Chem. Front.* **2015**, *2*, 253–258.
37. Ferrara, G.; Jin, T.; Oniwa, K.; Zhao, J.; Asiri, A. M.; Yamamoto, Y. Synthesis of new donor-acceptor-donor materials via Au-catalyzed double cascade cyclization. *Tetrahedron Lett.* **2012**, *53*, 914–918.
38. Nishida, J.-i.; Masuko, T.; Cui, Y.; Hara, K.; Shibuya, H.; Ihara, M.; Hosoyama, T.; Goto, R.; Mori, S.; Yamashita, Y. Molecular design of organic dye toward retardation of charge recombination at semiconductor/dye/electrolyte interface: introduction of twisted π-linker. *J. Phys. Chem. C* **2010**, *114*, 17920–17925.
39. Jiang, H.; Oniwa, K.; Islam, A.; Zhao, J.; Han, L.; Sun, Y.-J.; Bao, M.; Asao, N.; Yamamoto, Y.; Jin, T. Efficient thieno[3,2-*a*]carbazole-based organic dyes for dye-sensitized solar cells. *Tetrahedron* **2015**DOI10.1016/j.tet.2015.04.018.
40. Dennler, G.; Scharber, M. C.; Brabec, C. J. Polymer-fullerene bulk-heterojunction solar cells. *Adv. Mater.* **2009**, *21*, 1323–1338.
41. Krebs, F. C. Fabrication and processing of polymer solar cells: A review of printing and coating techniques. *Sol. Energy Mater. Sol. Cells* **2009**, *93*, 394–412.
42. Ma, W.; Yang, C.; Gong, X.; Lee, K.; Heeger, A. J. Thermally stable, efficient polymer solar cells with nanoscale control of the interpenetrating network morphology. *Adv. Funct. Mater.* **2005**, *15*, 1617–1622.
43. Li, G.; Shrotrita, V.; Huang, J. S.; Yao, Y.; Moriarty, T.; Emery, K.; Yang, Y. High-efficiency solution processable polymer photovoltaic cells by self-organization of polymer blends. *Nat. Mater.* **2005**, *4*, 864–968.
44. Yip, H.-L.; Jen, A. K. Y. Recent advances in solution-processed interfacial materials for efficient and stable polymer solar cells. *Energy Environ. Sci.* **2012**, *5*, 5994–6011.
45. Ma, H.; Yip, H. L.; Huang, F.; Jen, A. K. Y. Interface engineering for organic electronics. *Adv. Funct. Mater.* **2010**, *20*, 1371–1388.
46. Cheng, Y. J.; Yang, S. H.; Hsu, C. S. Synthesis of conjugated polymers for organic solar cell applications. *Chem. Rev.* **2009**, *109*, 5868–5923.
47. Chen, J. W.; Cao, Y. Development of novel conjugated donor polymers for high-efficiency bulk heterojunction photovoltaic devices. *Acc. Chem. Res.* **2009**, *42*, 1709–1718.
48. He, Y. J.; Li, Y. F. Fullerene derivative acceptors for high performance polymer solar cells. *Phys. Chem. Chem. Phys.* **2011**, *13*, 1970–1983.
49. Li, C.-Z.; Yip, H.-L.; Jen, A. K. Y. Functional fullerenes for organic photovoltaics. *J. Mater. Chem.* **2012**, *22*, 4161–4177.

50. Matsuo, Y. Design concept for high-LUMO-level fullerene electron-acceptors for organic solar cells. *Chem. Lett.* **2012**, *41*, 754–759.
51. Anthony, J. E.; Facchetti, A.; Heeney, M.; Marder, S. R.; Zhan, X. n-Type organic semiconductors in organic electronics. *Adv. Mater.* **2010**, *2*, 3876–3892.
52. Beaujuge, P. M.; Fréchet, J. M. J. Molecular design and ordering effects in π-functional materials for transistor and solar cell applications. *J. Am. Chem. Soc.* **2011**, *133*, 20009–20029.
53. Thompson, B. C.; Fréchet, J. M. J. Polymer–fullerene composite solar cells. *Angew. Chem., Int. Ed.* **2008**, *47*, 58–77.
54. Martín, N. New challenges in fullerene chemistry. *Chem. Commun.* **2006**, 2093–2104.
55. Hummelen, J. C.; Knight, B. W.; LePeq, F.; Wudl, F. Preparation and characterization of fulleroid and methanofullerene derivatives. *J. Org. Chem.* **1995**, *60*, 532–538.
56. Wienk, M. M.; Kroon, J. M.; Verhees, W. J. H.; Knol, J.; Hummelen, J. C.; van Hal, P. A.; Janssen, R. A. J. Efficient methano[70]fullerene/MDMO-PPV bulk heterojunction photovoltaic cells. *Angew. Chem., Int. Ed.* **2003**, *42*, 3371–3375.
57. He, Y.; Chen, H.-Y.; Hou, J.; Li, Y. Indene-C60 bisadduct: A new acceptor for high-performance polymer solar cells. *J. Am. Chem. Soc.* **2010**, *132*, 1377–1382.
58. He, Y.; Zhao, G.; Peng, B.; Li, Y. High-yield synthesis and electrochemical and photovoltaic properties of indene-C70 bisadduct. *Adv. Funct. Mater.* **2010**, *20*, 3383–3389.
59. Lenes, M.; Wetzelaer, G.-J. A. H.; Kooistra, F. B.; Veenstra, S. C.; Hummelen, J. C.; Blom, P. W. M. Fullerene bisadducts for enhanced open-circuit voltages and efficiencies in polymer solar cells. *Adv. Mater.* **2008**, *20*, 2116–2119.
60. Voroshazi, E.; Vasseur, K.; Aernouts, T.; Heremans, P.; Baumann, A.; Deibel, C.; Xue, X.; Herring, A. J.; Athans, A. J.; Lada, T. A.; Richtere, H.; Rand, B. P. Novel bis-$C_{60}$ derivative compared to other fullerene bis-adducts in high efficiency polymer photovoltaic cells. *J. Mater. Chem.* **2011**, *21*, 17345–17352.
61. Kim, K.-H.; Kang, H.; Nam, S. Y.; Jung, J.; Kim, P. S.; Cho, C.-H.; Lee, C.; Yoon, S. C.; Kim, B. J. Facile snthesis of *o*-xylenyl fullerene multiadducts for high open circuit voltage and efficient polymer solar cells. *Chem. Mater.* **2011**, *23*, 5090–5095.
62. Hirsch, A.; Soi, A.; Karfunkel, H. R. Titration of $C_{60}$: A method for the synthesis of organofullerenes. *Angew. Chem., Int. Ed. Engl.* **1992**, *31*, 766–768.
63. Matsuo, Y.; Iwashita, A.; Abe, Y.; Li, C.-Z.; Matsuo, K.; Hashiguchi, M.; Nakamura, E. Regioselective synthesis of 1,4-di(organo)[60]fullerenes through DMF-assisted monoaddition of silylmethyl grignard reagents and subsequent alkylation reaction. *J. Am. Chem. Soc.* **2008**, *130*, 15429–15436.
64. Matsuo, Y.; Nakamura, E. Selective multiaddition of organocopper reagents to fullerenes. *Chem. Rev.* **2008**, *108*, 3016–3028.

65. Tzirakis, M. D.; Orfanopoulos, M. Radical reactions of fullerenes: from synthetic organic chemistry to materials science and biology. *Chem. Rev.* **2013**, *113*, 5262–5321.
66. Martín, N.; Altable, M.; Filippone, S.; Martín-Domenech, A.; Echegoyen, L.; Cardona, C. M. Retro-cycloaddition reaction of pyrrolidinofullerenes. *Angew. Chem., Int. Ed.* **2006**, *45*, 110–114.
67. Nambo, M.; Noyori, R.; Itami, K. Rh-catalyzed arylation and alkenylation of $C_{60}$ using organoboron compounds. *J. Am. Chem. Soc.* **2007**, *129*, 8080–8081.
68. Mori, S.; Nambo, M.; Chi, L.-C.; Bouffard, J.; Itami, K. A bench-stable Pd catalyst for the hydroarylation of fullerene with boronic acids. *Org. Lett.* **2008**, *10*, 4609–4612.
69. Filippone, S.; Maroto, E. E.; Martín-Domenech, A.; Suarez, M.; Martín, N. An efficient approach to chiral fullerene derivatives by catalytic enantioselective 1,3-dipolar cycloadditions. *Nat. Chem.* **2009**, *1*, 578–582.
70. Xiao, Z.; Matsuo, Y.; Nakamura, E. Copper-catalyzed formal [4 + 2] annulation between alkyne and fullerene bromide. *J. Am. Chem. Soc.* **2010**, *132*, 12234–12236.
71. Lu, S.; Jin, T.; Bao, M.; Yamamoto, Y. Cobalt-catalyzed hydroalkylation of [60]fullerene with active alkyl bromides: Selective synthesis of monoalkylated fullerenes. *J. Am. Chem. Soc.* **2011**, *133*, 12842–12848.
72. Lu, S.; Jin, T.; Kwon, E.; Bao, M.; Yamamoto, Y. Highly efficient $Cu(OAc)_2$-catalyzed dimerization of monofunctionalized hydrofullerenes leading to single-bonded [60]fullerene dimers. *Angew. Chem., Int. Ed.* **2012**, *51*, 802–806.
73. Lu, S.; Si, W.; Bao, M.; Yamamoto, Y.; Jin, T. Co-catalyzed radical cycloaddition of [60]fullerene with active dibromides: Selective synthesis of carbocycle-fused fullerene monoadducts. *Org. Lett.* **2013**, *15*, 4030–4033.
74. Si, W.; Lu, S.; Bao, M.; Asao, N.; Yamamoto, Y.; Jin, T. Cu-catalyzed C–H amination of hydrofullerenes leading to 1,4-difunctionalized fullerenes. *Org. Lett.* **2014**, *16*, 620–623.
75. Si, W.; Zhang, X.; Asao, N.; Yamamoto, Y.; Jin, T. Ni-catalyzed direct 1,4-difunctionalization of [60]fullerene with benzyl bromides. *Chem. Commun.* **2015**, *51*, 6392–6394.
76. Lu, S.; Jin, T.; Bao, M.; Yamamoto, Y. NaOH-catalyzed dimerization of monofunctionalized hydrofullerenes: Transition-metal-free, general, and efficient synthesis of single-bonded [60]fullerene dimers. *Org. Lett.* **2012**, *14*, 3466–3469.
77. Si, W.; Lu, S.; Asao, N.; Bao, M.; Yamamoto, Y.; Jin, T. NBS-promoted oxidation of fullerene monoradicals leading to regioselective 1,4-difunctional fullerenes. *Chem. Commun.* **2014**, *50*, 15730–15732.
78. Lu, S.; Jin, T.; Yasuda, T.; Ashraful, I.; Akhtaruzzaman, M.; Han, L.; Alamry, K. A.; Kosa, S. A.; Asiri, A. M.; Yamamoto, Y. Functional 2-benzyl-1,2-dihydro[60]fullerenes as acceptors for organic photovoltaics: Facile synthesis and high photovoltaic performances. *Tetrahedron* **2013**, *69*, 1302–1306.

79. Lu, S.; Jin, T.; Yasuda, T.; Si, W.; Oniwa, K.; Alamry, K. A.; Kosa, S. A.; Asiri, A. M.; Han, L.; Yamamoto, Y. Deuterium isotope effect on bulk heterojunction solar cells. Enhancement of organic photovoltaic performances using monobenzyl substituted deuteriofullerene acceptors. *Org. Lett.* **2013**, *15*, 5674–5677.
80. Ford, L.; Jahn, U. Radicals and transition-metal catalysis: An alliance par excellence to increase reactivity and selectivity in organic chemistry. *Angew. Chem., Int. Ed.* **2009**, *48*, 6386–6389.
81. Wakabayashi, K.; Yorimitsu, H.; Oshima, K. Cobalt-catalyzed tandem radical cyclization and cross-coupling reaction:? Its application to benzyl-substituted heterocycles. *J. Am. Chem. Soc.* **2001**, *123*, 5374–5375.
82. Amatore, M.; Gosmini, C. Direct method for carbon–carbon bond formation: The functional group tolerant cobalt-catalyzed alkylation of aryl halides. *Chem.-Eur. J.* **2010**, *16*, 5848–5852.
83. Fagan, P. J.; Krusic, P. J.; Evans, D. H.; Lerke, S. A.; Johnston, E. Synthesis, chemistry, and properties of a monoalkylated buckminsterfullerene derivative, tert-Bu$C_{60}$ anion. *J. Am. Chem. Soc.* **1992**, *114*, 9697–9699.
84. Tong, C. C.; Hwang, K. C. Enhancement of OLED efficiencies and high-voltage stabilities of light-emitting materials by deuteration. *J. Phys. Chem. C* **2007**, *111*, 3490–3494.
85. Brabec, C. J.; Cravino, A.; Meissner, D.; Sariciftci, N. S.; Fromherz, T.; Rispens, M. T.; Sanchez, L.; Hummelen, J. C. Origin of the open circuit voltage of plastic solar cells. *Adv. Funct. Mater.* **2001**, *11*, 374–380.

# Section IV

Chapter 10

# Biosafety Evaluation of Nanoscaled Porous Energy Materials

**S. Bashir,[1] Z. Luo,[2] B. Martinez,[1] U. Okakpu,[1] and J. Liu[*,1,3]**

[1]Department of Chemistry, Texas A&M University-Kingsville, Kingsville, Texas 78363, United States
[2]Department of Chemistry and Physics, Fayetteville State University, Fayetteville, North Carolina 28301, United States
[3]Department of Chemistry, Texas A&M University, College Station, Texas 77843, United States
[*]E-mail: jingbo.liu@tamuk.edu, jingbo.liu@tamu.edu

The broad concerns related to energy materials are end-of-life effects in the environment and human health due to a possible seepage of catalytic components into the water table. The biosafety aspects of catalysts or other components, such as noble metal nanoparticles (NPs), metal oxides NPs and metal-organic frameworks (MOFs), were investigated to determine performance benchmarks. Toxicity study on microorganism indicated that metal-oxide ($TiO_2$, Ag-$TiO_2$) and MOFs are more toxic than noble metal (Ag) NPs. Additionally, study on human cells of porous materials indicated possible toxicity. The mechanism of microbial cell death was due to degradation of macromolecules and leakage of cellular potassium ions. In contrast, human cells were damaged by oxidative stress of the plasma membrane and depolarization of the mitochondrial membrane potential.

## Preamble

This standalone chapter addresses a relatively new concern of biosafety of discarded energy devices, storage components, or accidental release of the catalytic elements in the environment, related to water safety. This is tackled in this chapter in three parts. The first part related to gas capture of methane using select metal-organic frameworks (MOFs), and *in vitro* cell culture assay on human retinal pigment epithelium (RPE), where we demonstrate that certain MOFs agents are toxic. The second part deals with the relationship between MOFs and zeolites since both structures have structural commonalities and differences that are compared and contrasted. The last section demonstrates the biosafety of MOFs. A whole series of MOFs were explored to cover the wide spectrum of likely candidates documented in the literature with different organic linkers and metal coordination sites. Not all MOFs were evaluated for gas storage, against RPE cells or microorganisms found in water, but were distributed across this parameter space using random walk methodology. The results further indicate that selected MOFs will kill microorganism such as *Escherichia coli* in water.

## 1. Gas Storage

This study focused on development of porous materials (metal-organic frameworks, MOFs) to improve the uptake of gases, such as storage of methane derived from the shale gas. The shale gas is one type of unconventional natural gases with lower permeability than 1 millidarcy. It was found trapped within shale formations. The chemical components and associate characteristics in typical shale gas are listed in Table 1. At the beginning of this century, shale gas has become one of the important sources of natural gas in the USA and the rest of the world. Study indicated that the shale gas will expand into a world energy supply (*1*). Therefore, production and storage of major components ($CH_4$) are critical for scientists to explore.

**Table 1. Mole fraction, molar mass, and collision diameter δ for each component in a shale gas**

| *Gas Component* | *Mole Percent (%)* | *Collision Diameter (Nm)* | *Molar Mass (G/Mol)* |
|---|---|---|---|
| $Ch_4$ | 87.4 | 0.4 | 16.0 |
| $C_2H_6$ | 0.12 | 0.52 | 30.0 |
| $Co_2$ | 12.48 | 0.45 | 44.0 |
| Average | N/A | 0.41 | 19.5 |

### 1.1. Metal-Organic Frameworks

MOFs are also known as porous coordination polymers (PCPs) and become an emerging type of porous materials, which are formed by the self-assembling of metallic centers and bridging organic linkers. The metal ions or clusters reacted to form secondary building units (SBUs) (*2*). MOFs are extensively acknowledged as an important family of compounds with prospective applications in gas storage (*3*), separation (*4*), catalysis (*5*), drug delivery (*6*), and molecular sensing (*7*). MOFs are characterized by their tunable pore sizes, topologies, and functionality and their inherent flexibility (*8*) arising from their inorganic–organic hybrid nature, leading to their classification as soft porous crystals. Such flexibility is not generally observed in more classic carbon- or oxide-based porous materials (*9*). Framework flexibility usually occurs as a guest-induced transformation between two or more structures that occurs upon desolvation and/or guest re-adsorption (*10*). Of great importance is the design and synthesis of organic linkers which are pivotal to the realization of MOFs with required structures and properties (*11*) and evaluated in the results section (*12–15*).

### 1.2. Water Safety

Energy storage and conversion materials can be released into water table. Some components in fuel cells and solar photovoltaics such as the electrodes may have catalytic activity in biological systems. These energy materials can be used as disinfection or cellular proliferation. The effects of accidental release of select metal organic-frameworks or silver or silver-titania was investigated with respect to water disinfection, due to the critical requirement of water to sustain life. The United Nations have designated 2005-2015 as the 'Water for Life' decade (*16*), recognizing the criticality of water to human health, particularly under global warming conditions (*17*). Irrespective of the merits or shortcomings of global warming models, generation of clean water has been, is, and will be of paramount importance (*18*). The most common routes toward water purification are water filtration (*19*), boiling (*20*) and chemical treatment (*21*). These approaches provide affordable and effective strategies to combat pathogenic microbes such as *Escherichia coli* (*E. coli*). *E. coli* is a Gram-negative microbe, responsible for causing ~20,000 cases of food-related illnesses in the United States annually (*22*). The microbe was recently identified as a potential cause for deaths in Germany in 2011 (*23*).

Water is routinely disinfected using bleach (5.25-6.15 mass % sodium hypochlorite), which exhibits a broad spectrum of antimicrobial activity and can be used to rapidly disinfect water (*24*). However, bleach exhibits limited stability in light, has low activity at high pH, and can generate toxic by-products such as trihalomethane (*25*). The application of nanomaterials as an alternative to bleach presents a different route to disinfection (*26*). Nanomaterials can be made a using a variety of approaches, are affordable, and are designed to be effective against a wide range of pathogens for an extended period without generating toxic by-products (*27*). The objectives of this stud are to compare and contrast different type of nanosystems and bleach in terms of disinfection and persistance and to compaere most likely method of inhibition and where appropriate to compare to other similar systems such as zeolites. With this framework, three generations of nanodisinfectants were developed: core-shelled silver (Ag) nanoparticles (NPs: 1st generation); Ag-titania ($Ag/TiO_2$: 2nd generation); and nanoscaled metal-organic frameworks (MOFs: 3rd generation). The first generation NPs (Ag) can be readily prepared using natural products as reducing agents (*28*). The rationale for using nanomaterials is that their mode of action is different from bleach and may be applicable in circumstances where microbes have developed a tolerance, particularly when low concentrations of bleach are used for disinfection (*29*). MOFs are zeolite-like structures in which the ligands form the network around a central atom. In this manner, extensive networked three-dimensional structures can be prepared and have only recently been used in disinfection or cancer therapy (*30*). Comparison with bleach is included since bleach is one of the earliest compounds used in disinfection. Other generations of disinfectants have also been used and described in the literature, starting with silver and titania. The advantage of composites (Ag-$TiO_2$) over silver (Ag) are lower cost (less amount of silver) and greater effectiveness under visible light conditions (unlike titania, which is effective under ultraviolet conditions). MOFs take this advantage one-step further, due to their ultrahigh surface area and tunability in design (ability to use different scaffolds, compared to metal colloids). In this regard, MOFs are another step forward in design and effectiveness in water disinfection. Presently, the advantages and disadvantages of each system and a comparison between MOFs and zeolites were discussed, in terms of effective strateergies towards disinfection of potable water.

# 2. Experimental

## 2.1. Methods and Materials

All chemicals were from Sigma-Aldrich (St. Louis, MO) unless otherwise stated and doubly distilled and sterilized water was used to make solutions.

## 2.2. MOFs Synthesis

In this study, the carboxylate ligands were used to product MOFs. The carboxylate units are important due to their preference to stabilize the MOFs through *in situ* formed secondary building units (SBUs) such as metal carboxylates [$M_2(COO)_4$] and zinc acid anhydride or carboxylate [$Zn_4O(COO)_6$] (*12*). Thus, carboxylate-containing units with aromatic backbones could be used as linkers to construct porous MOFs. The structures, framework topologies, pore/cage sizes and porosities can be functionalized (*11*). Use of m-benzenedicarboxylate (BTC) unit, copper-benzenedicarboxylate [$Cu_3(BTC)_2$], Hong Kong University of Science and Technology-1, HKUST-1), displaying the highest methane adsorption has been synthesized (*13*). It was found that zirconium (Zr)-based MOFs confer superior stability compared to common zirconium/copper (Zn/Cu) based MOFs (*14*). Therefore, the terphenyl-4,4-dicarboxylic acid with the two carboxylates was used as the primary building. Totally, eight formulations of MOFs were prepared by solvothermal chemistry under moderate conditions. Specifically, the mixture of inorganic ionic compounds iron (III) nitrate [($Fe(NO_3)_3$], and zinc (II) nitrate [$Zn(NO_3)_2$] zirconium(IV) chloride [$ZrCl_4$]) and ligands provided by Dr. Zhou's group (Figure 1), solvent and benzoic acid were placed in a glass bottle, which was tightly capped. The above solution was then ultrasonically dissolved and heated at 100 °C for 2 days, accordingly. The final product was cooled down to ambient condition, naturally.

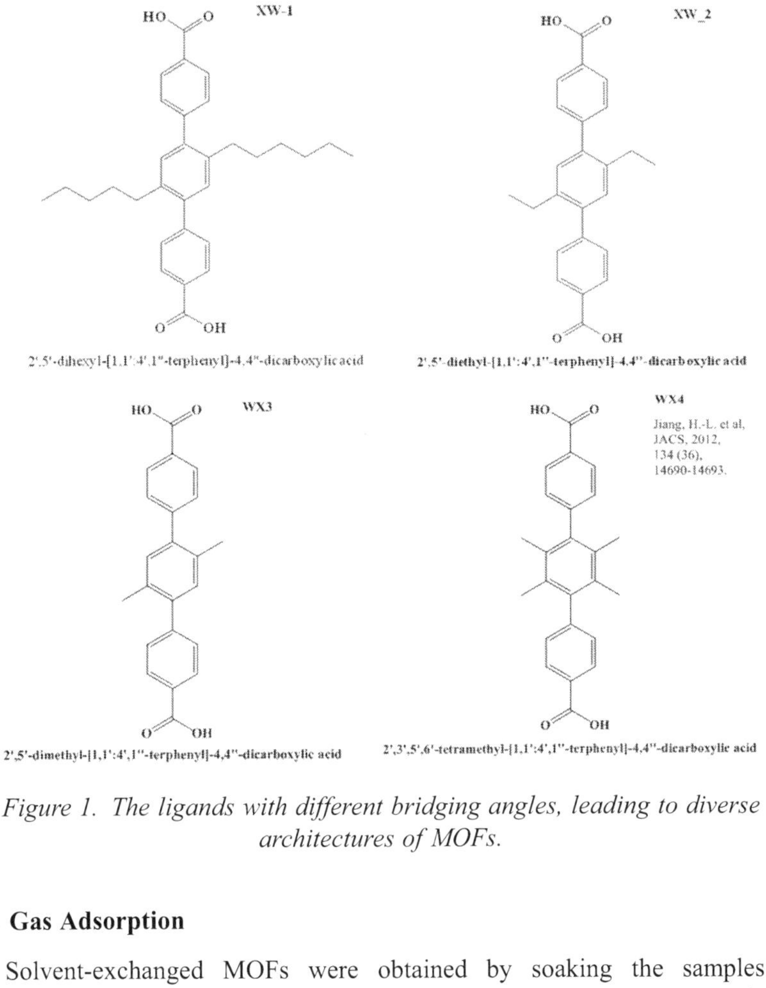


*Figure 1. The ligands with different bridging angles, leading to diverse architectures of MOFs.*

## 2.3. Gas Adsorption

Solvent-exchanged MOFs were obtained by soaking the samples in acetone/dichloromethane (DCM) for 3 days, refreshing every 5 h. The completely desolvated MOFs would be afforded by heating the solvent-exchanged bulk at 393 K under a vacuum overnight. The samples were further activated by using the degassing port in the surface area analyzer for 10 h at 393 K to measure the gas uptake. Low-pressure methane ($CH_4$) sorption experiments were recorded on a Quantachrome IQ2 system under different conditions. The Micrometrics ASAP 2020 Physisorption Surface Area and Pore (ASAP) Analyzer was used to determine methane adsorption isotherms at 195 K. The ASAP 2020 measures pressure and then computes volume adsorbed as a result of pressure changes.

## 2.4. In Vitro Materials Safety Evaluation

It is also important to ensure the MOFs and gas-absorbed MOFs will not cause health concerns due to their incidental release into the environment. In this study, we conducted the degree of apoptosis (cell death) in retinal pigment epithelium (RPE) cells (*31*). The cytotoxicity bioassay was based on total fluorescence yield of 4-amino-5-methylamino-2′,7′-difluorofluorescein diacetate (DAF FM DA, as measure of NO), which was a ratio between treated and untreated cells in PBS (pH 7.4) solution. The NO was measure at 10 minute time point versus control RPE ($120\times10^3$ cells/well) whole cells [WC] with the following treatments: 0.5 μL (1mg/mL of MOFs). Other control agents were also added for comparison. These were: 25 μL (10 mg/mL of sodium cyanide) [CN]; 25 μL (1 mg/mL of 7-Ethyl-10-hydroxy-camptothecin)[SN-38], 5 μL (100 μg/mL of rotenone) [Rot]; and 5 μL (100 μg/mL of carbonyl cyanide 4-(trifluoromethoxy)phenylhydrazone) [FCCP].

## 2.5. Synthesis of NPs

Nanoparticles reduced by natural product (citrate, ascorbic acid) were synthesized following the previously established procedure (*32*). Briefly, the reducing agents were prepared in stoichiometric amounts in ultrapure water and dispersed with titanium (IV) butoxide (0.014 mol) mechanically (< 20 revolutions-per-minute). Butanol, acetic acid and ammonia were added drop-wise to control the hydrolysis rate of $TiO_2$. A 1:20 molar ratio of titania to silver nitrate (0.0071 mol in ultrapure water) was also used. Arabic gum (3 % Wt) was subsequently added and the mixture was stirred for 30 min at 60 °C. The solution was cooled and Ti-dispersion added drop-wise (to the reduced silver Arabic gum solution) under vigorous agitation with addition of reducing agent (volume equivalent to 0.40 M) for 2 hours. This dispersion was subject to pre-heat treatment and vacuum filtration to obtain filtrate, which was heated at various temperatures.

In addition to the discussion in *2.2. MOFs Synthesis*, a series of MOFs were also synthesized accordingly to a previous procedure (*33*). The MOFs structure can be tuned according to the ligand structure and bond angle (Figure 2). Briefly, single crystal(s) with the formula of [$Metal_x(H_2O)_2(ligand)(H_2O)_y$] (abbreviated as M-L) were prepared using n-topic carboxylic terminating or amine-terminating ligands. Three metals ions (such as Cu and cobalt (Co)) and four carboxylic-based ligands (oxalic acid, terephthalic acid, benzene-1,3,5-tricarboxylic acid, and 4,4'4"-(1,3,5-triazine-2,4,6-triyl)tribenzoic acid) were used in a combinatorial manner. The carboxylic ligand ($6.3 \times 10^{-5}$ mol) was added to the appropriate metal salt ($10^{-4}$ mol) in N-dimethylformamide to which nitric acid (2 M) was added drop-wise to control the overall acidity of the mixture. The ligand-metal mixture was agitated mechanically and transferred to an air-tight vial, which was heated to 85°C for 24 h. A number of formulations were prepared in this manner, summarised in Table 2-3.

The synthesized nanoparticles (NPs) were characterized using electron microscopy before and after addition to microbes in water. The morphology of the microbes was also examined before and after addition of the nanodisinfectants and based on the changes in microbial morphology, the mode of inhibition inferred, correlated with literature precedence and our observations.

| Nomenclature of ligands | Two dimensional structure of ligands | Three dimensional structure of ligands |
|---|---|---|
| Oxalic acid (OA) | 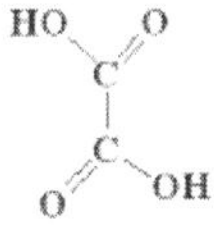  oxalic acid | 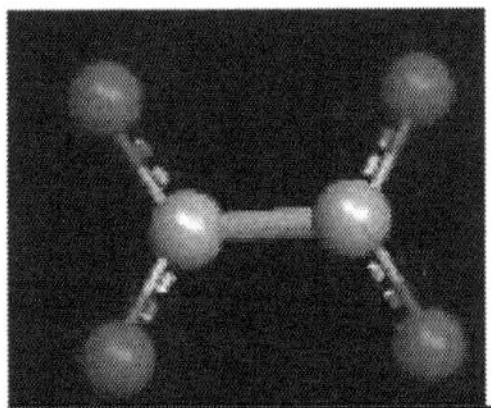 |
| Terephthalic acid (BDC), | 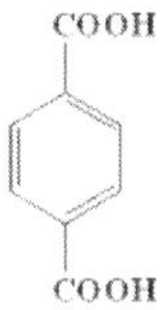  terephthalic acid | 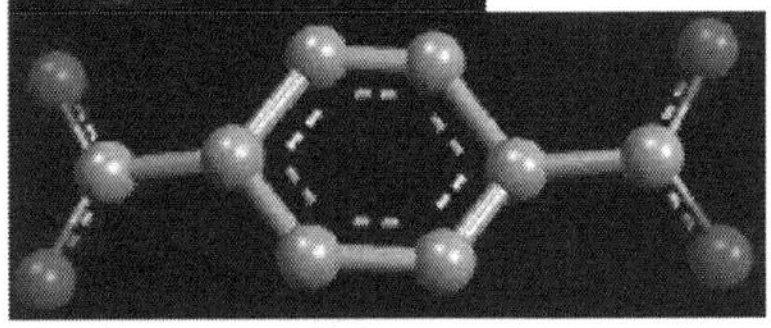 |
| Benzene-1,3,5-tricarboxylic acid (BTC) | 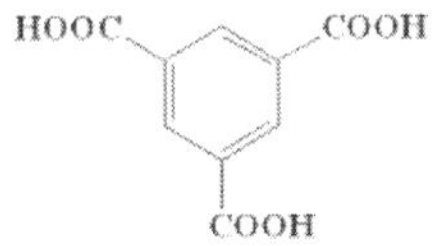  benzene-1,3,5-tricarboxylic acid | 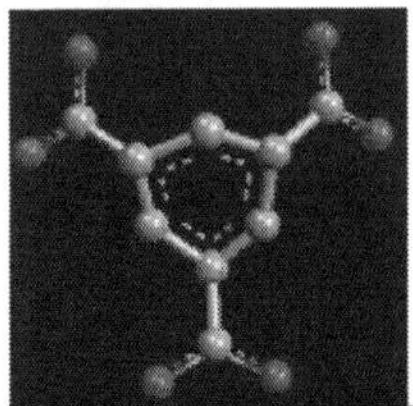 |
| 4,4'4''-(1,3,5-triazine-2,4,6-triyl)tribenzoic acid (TATB). | 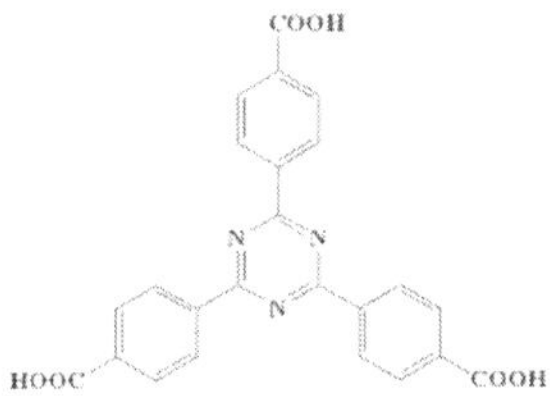  4,4',4''-(1,3,5-triazine-2,4,6-triyl)tribenzoic acid | 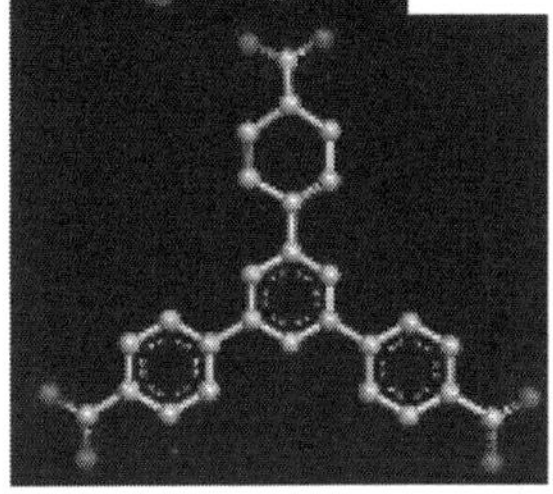 |

*Figure 2. Different ligand provides the different binding angles to tune the porosity and pore size. The optimized or semi-optimized geometry is also shown (red color for oxygen atom, gray color for carbon, hydrogen omitted for clarity. ArgusLab 4.0.1 (Mark Thompson and Planaria Software LLC, geometry optimization module using PM3 Hamiltonian function, with convergence set over 100 steps at a gradient of 10 kcal/mol/Ang).*

**Table 2. Summary of ligands and cations used to produce series of MOFs. All values were calculated except crystal space group, which was determined using X-ray diffraction (XRD) at TAMU College Station, Department of Chemistry.**

| **Ligands → Cation ↓** | **Oxalic acid (OA)** | **Terephtalic acid or 1,4-benzenedicarboxylic acid (BDC)** | **Trimesic Acid or 1,3,5-benzenetricarboxylic acid (BTC)** | **4-[bis(4-carboxyphenyl)-1,3,5-triazin-2-yl]benzoic acid (TATB)** |
|---|---|---|---|---|
| $Co^{2+}$ | Cobalt (II) oxalate | Cobalt (II) teraphtalate | Cobalt (II) 1,3,5-benzenetricarboxylate | Cobalt (II) 4-[bis(4-carboxyphenyl)-1,3,5-triazin-2-yl] benzoate |
| $Cu^{2+}$ | Copper (II) oxalate | Copper (II) teraphtalate | Copper (II) 1,3,5-benzenetricarboxylate | Copper (II) 4-[bis(4-carboxyphenyl)-1,3,5-triazin-2-yl] benzoate |
| $Zn^{2+}$ | Zinc (II) oxalate | Zinc (II) teraphtalate | Zinc (II) 1,3,5-benzenetricarboxylate | Zinc (II) 4-[bis(4-carboxyphenyl)-1,3,5-triazin-2-yl]benzoate |
| Information for addition MOFs is also listed. The MOF was too small to generate a single crystal, however was used as a suspension in the antimicrobial study. | | | | |
| $Ag^{+}$ | Silver (I) oxalate | Silver (I) teraphtalate | Silver (I) 1,3,5-benzenetricarboxylate | Silver (I) 4-[bis(4-carboxyphenyl)-1,3,5-triazin-2-yl]benzoate |

**Table 3. Summary of properties of MOFs beginning with ligand type and metal with resulting space group geometry and selected unit cell dimensions determined from small single crystals. Other physical values or physical chemical data either calculated or simulated is also given. Finally small three-dimensional representations of similar space group structures from the crystallography open database are also shown for comparison. The [Metal$_x$($H_2O$)$_2$(ligand)($H_2O$)$_y$] configuration is heterogeneous in isomorphism, as well as n-topic depending on inclusion of single crystals or polycrystals.**

| **MOF nomenclature →**<br>**MOF properties ↓** | **Metal Oxalate** | **Metal 1,4-benzenedi-carboxylate** | **Metal 1,3,5-benzenetri-carboxylate** | **Metal 4-[bis(4-carboxyphenyl)-1,3,5-triazin-2-yl]benzate** |
|---|---|---|---|---|
| Space Group | Monoclinic P | Hexagonal P | Triclinic P | Hexagonal P |
| Unit Cell determination | a=5.19Å, a=8.27 Å and b=13.23 Å, M=Cu/Co α=90°, β=107.47° γ=90°, and V=542 $Å^3$ | a=16.58Å, b=16.58Å, and c=16.58Å, M=Cu α=90°, β=90° γ=120°, and V=3342 $Å^3$ | a=23.04Å, b=24.37Å and c=27.46Å, M=Cu α=67.64°, β=67.87° γ=74.96°, and V=3342 $Å^3$ | a=33.89Å, b=9.88Å and c=18.73Å , M=Zn α=90°, β=90.74° γ=90°, and V=6271 $Å^3$ |
| Pore size of each MOF (nm) | *ca.* 0.5 | > 1 | > 2 | > 2.6 |
| Volume of each MOF ($Å^3$) | 542 | 3135 | 13000 | 18190 |
| Formulae of ligands | $C_2H_2O_4$ | $C_8H_6O_4$ | $C_9H_6O_6$ | $C_{24}H_{15}N_3O_6$ |
| Molar mass of ligands (g/mol) | 90.0349 | 166.1308 | 210.1403 | 441.3924 |
| Atom Count | 8 | 18 | 21 | 48 |
| Calculated Log P (*, mol/l) | -0.26 | 1.29 | 0.95 | 5.68 |
| Calculated Polarizability (*, $Å^3$) | 5.87 | 15.08 | 17.58 | 45.47 |
| Calculated 2D Polar Surface Area (*, Å) | 74.6 | 74.6 | 111.9 | 150.57 |
| Calculated 3D Van der Waals Surface Area (*, Å) | 104.04 | 210.49 | 247.49 | 549.07 |
| Calculated Refractivity (*, $m^3.mol^{-1}$) | 14.44 | 40.57 | 47.83 | 150.52 |
| Self-consistent field (SCF) Energy (a.u.) | -46.7885 | -71.0942 | -83.3718 | -243.3037 |
| Metal ions include: **$Co^{2+}$, $Cu^{2+}$ and $Zn^{2+}$** | | | | |
| Reported single crystal structure from Crystallography Open Database [file PDF #] http://www.crystallography.net | 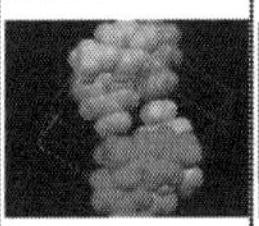 [2221171] |  [1008710] |  [4338016] |  [7029663] |

## 2.6. Characterization of MOFs and NPs

Nanoparticles (NPs) and MOFs were characterized using transmission electron microscopy (TEM) from which system dimensions, pore size and surface profile were obtained. Briefly, The Tecnai G2 F20 TEM (FEI Company, Hillsboro, OH) equipped with energy dispersive (EDS), energy electron loss spectroscopy (EELS), and post-column Gatan Image Filter was used to evaluate the morphology of the nanomaterials. The analytes were dispersed in absolute ethanol and deposited onto the carbon-coated copper grid to prepare TEM specimen. The scanning TEM (STEM) mode was used to obtain Z-contrast STEM images using a high-angle angular dark-field detector. The scanning electron microcope (FEI, 650 FEG Quanta SEM, Hillsboro, Oregon) was operated in energy dispersive spectrometer mode with an operating voltage of 20 kV. Lastly, ultraviolet-visible spectroscopies were used to characterize the samples before and after disinfection with the microorganism. The data were able to demonstrate inculsion of metal into the structured systems. In addition, no leaching of metal occurred within the time-frame of the treatment period.

## 2.7. Minimum Bactericidal Concentration Test

The minimum bactericidal concentrations (MBC) of different nanomaterials were tested as a benchmark of bactericidal performance of MOFs and NPs. The details of the assay have been previously described (*34*). *E. coli* (a Gram-negative biosafety level 1 microorganism, ATCC #10798) was cultured in nutrient broth (NB) overnight in a shaker (5 rpm, 37 °C, dark). Briefly, $1.5 \times 10^7$ colony forming units per millilitre (cfu/mL) were cultured and diluted until 2.5 $10^5$ cfu (*35*, *36*). The microbes were examined using optical microscopy and were further diluted with NB were spread onto a nutrient agar plate and the plate was incubated at 37 °C for 16 h. The number of colonies was counted (~ 250 colonies/plate) and a correlation between cfu and optical density (OD) measurement at 600 nm estabilished. Subsequent experiments used OD as a measure of disinfection ($0.3 <$ O.D $< 0.6$). The preliminary results demonstrated that the molarity of NPs could be controlled through stoichiometry and that negligible amount of NPs was 'lost' or adhered to the microbes or tube surfaces, when evaluated within 5 minutes of incubation (*37*). In addition, OD measurements were found to correlate with plate counting of colonies, as well as a good predictor of nanosystem toxicity (Table 4 for zeolite MBCs). Data were also consistent with other assays for toxicity in the literature such as the β-galactosidase assay (*35*) and were used for subsequent assays.

**Table 4. Summary of major studies using zeolites as matrices for disinfection, as well as studies with MOFs, indicating strain of microbe, the zeolite or system, biological effectiveness, and appropriate citation**

| Microorganism/ (ATCC # or Strain) [a] | Zeolite Name and amount [b] | Percent killed [c] | [d] |
|---|---|---|---|
| *Escherichia. coli* (*E. coli*, PTA-852) | Ag-MOF, 0.6 ppm, 2 h | 100% MBC | cs |
| *E. coli* | Co-TDM, 15 ppm, 0.5-2h | 100% MBC | [71] |
| *E. coli*, *Staphylococcus aureus (S. aureus)* | 0.37 mg Ag in MOF, 24 h | $MIC_{upper\_limit}$ 308 µg/mL | [72] |
| *E. coli, S. aureus, Pseudomonas. aeruginosa (P. aeruginosa), Candida albicans (C. albicans)* | 34 mg Ag in PTA=S MOF (compound 1), 24 h | $MIC_{upper\ limit}$ 40 µg/mL | [73] |
| *E. coli* (CSH7) | [Zeomic AG10D]-Ag, 100 ppm, 20 h | 100% MBC | [74] |
| *Bacillus cereus* (*B. cereus*, WN129), *B. subtilis* (WN131), *B. anthracis* Sterne (WN742) | [AgION]-(Ag/Zn) 25,000/140,000 ppm 5-24 h | 99.9% MBC | [75] |
| *E. coli* (NIHJ JC2) | [Na-Zeo, 10g]-Ag (1 M), 5 min | LOG 7 | [76] |
| *E. coli*, *S. aureus* | [Na-Y]-Ag (9.7%), time=not stated, but generally is 24 h | $MIC_{upper\ limit}$ 0.3 mg/mL | [77] |
| *E. coli* | [Montmorillonite clay (B550T), 3g]-Ag (3.36%) 24h | $MIC_{upper\_limit}$ 2.5 mg/mL | [78] |
| *Methicillin-resistant S. aureus* | Ti-[zeotype LTA]-Ag (2.3 wt %), 24 h | LOG 3 | [79] |
| *S. aureus (29213,) Streptococcus mutans (25175), Streptococcus mutans Ingbritt, Actinomyces viscosus (A. viscosus, 15987), Lactobacillu casei (L. casei, 393), S. aureus (29213), C. albicans (90028)* | ZnOw AT-83 2.5 mg/mL, 24 h. [Other agents tested were: Basic magnesium hypochlorite, Basic magnesium hypochlorite, Longbei inorganic antibiotic powder, IONPURE-H, and ZnOw AT-88] | 312.5 µg/ml for *S. aureus* (1250 mg/mL MBC) | [80] |
| *E. coli (K12W-T), P. aeruginosa (NCIMB8295), S. aureus (NCIMB6571)* | zeolite X (5.8 wt/wt %)-$Ag^+$ (2.0 Wt %), dose=1g/L, 2h | LOG 2.5 for *E. coli* | [81] |
| *S. aureus (9213)* | [ZSM-5]-Ag (0.2 Wt%) 100 mg/mL, 2-6 h [Also found order of increased potency as: $AgNO_3$ > Ag-ZSM-5> $Ag_2O$] | LOG 2.5. | [82] |
| *E. coli (XL-Blue)* | Alumina supported-Y-zeolite films(2–3 µm, 0.005M $Ag^+$), 2 h | LOG 3 | [83] |
| *E. coli, P. aeruginosa* | [Clinoptilolite Clay, 0.2 g]-Ag (0.1 N), 24 h, Zone Diameter (ZD) using 8 mm disk | $ZD_{max}$=20 mm (*P. aeruginosa* ) | [84] |
| *E. coli, Bacillus subtilis Saccharomyces cerevisiae (S. cerevisia), C. albicans* | [Na-Y and Na-X zeolites, 2.5 g] – Ag (9.7 wt % Na-Y and 9.8 wt % Na-X) 24h [48 h for *S. cerevisiae]*. Ag Loading= 0.01 to 1.00 mg/mL | $MIC_{max}$0.3 mg/ ml (*E. coli*); $MIC_{max}$ 1.0 mg/mL (*S. cerevisia, C. albicans)* | [85] |

Note: [a] A= *Actinomyces*, C= *Candida*, L=*Lactobacillu*, S= *Staphylococcus* or *Saccharomyces*, or *Streptococcus* depending upon context.

[b] 1 parts-per-million (ppm) = 1mg/L = 1mg /mL = 1000mg/L = 1µg/mL. Wt percentage ×10, 000 = ppm.

The MW of an ion species is equal to its concentration in ppm at $10^{-3}$ M (e.g. Ag MW is 107.9 Da, therefore a $10^{-3}$ M solution is equal to 107.9 ppm; Log Reduction ($log_{10}$) is the difference in cell population between initial population and final population, therefore $Log_{10}$ 6 reduction would correspond to a decrease in the population of 999999 cells or 99.9999% inactivation.

[c] The incubation time was listed in hours (h) unless otherwise specified in the study, ZD= Zone Diameter on plate, and [d] cs= current study [Refs: listed in main manuscript].

### 2.8. Statistical Analysis

The MBC for 100% lethality was estimated using two-methods: culture plate counting and, OD measurements. One-way analysis of variance (ANOVA) with ad-hoc test was carried out between control and experimental samples (*38*).

## 3. Results and Discussion

### 3.1. MOFs Preparation

Different formulations of MOFs were prepared by solvo-thermal synthesis. Figure 1 showed selected ligands (provided by Dr. Zhou's group) used in MOF preparation.

### 3.2. Gas Adsorption of $CH_4$

Low pressure $CH_4$ uptake was measured using the micrometrics ASAP 2020 instrument at 195 K. The methane uptake of zirconium-4-[4-(4-carboxyphenyl)-2,5-diethylphenyl]benzoic acid [Zr-WX2] is shown in Figure 3. It is estimated that the isoreticular MOFs in the **WX** series will show higher uptake and higher surface area with the presence of the extended side aliphatic chains (further study will be carried out).

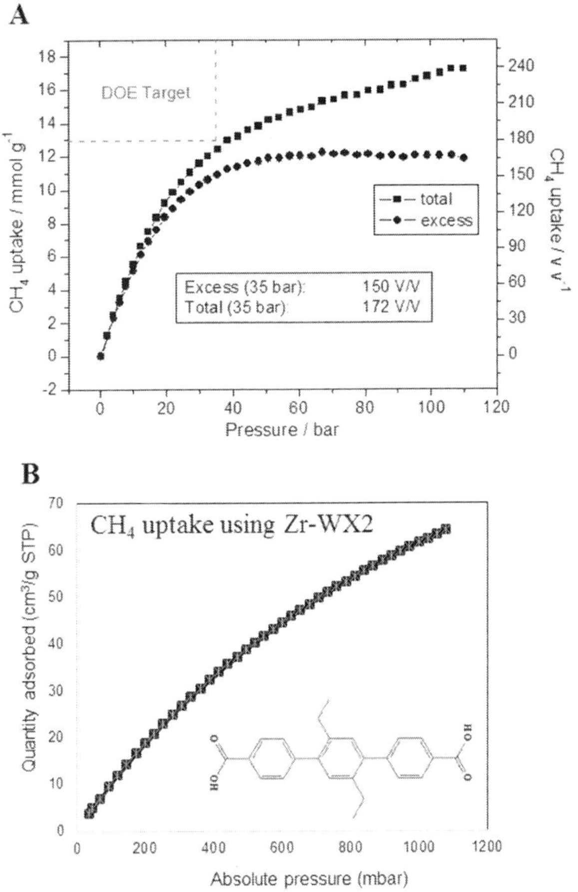


*Figure 3. The $CH_4$ adsorption isotherm: **A**: $CH_4$ (courtesy for Dr. Zhou); **B**: adsorption of Zr-WX2.*

### 3.3. Toxicity Study

Comparison of measured NO from cells treated with the MOFs and other chemical agents whose mode of inhibition or stimulation in NO generation was conducted. Through comparison, we can see that some MOFs are not toxic to RPE cells, whereas others exhibited toxicity (Figure 4). We have shown that the bioassay using NO as a diagnostic molecule is a broad responsive bioassay to examine the toxicity of MOFs. The results demonstrate the NO is a viable test assay and that careful synthesis tune-up to derive highly porous crystalline frameworks is not harmful to human health through release of NO as a protective agent against oxidative stress.

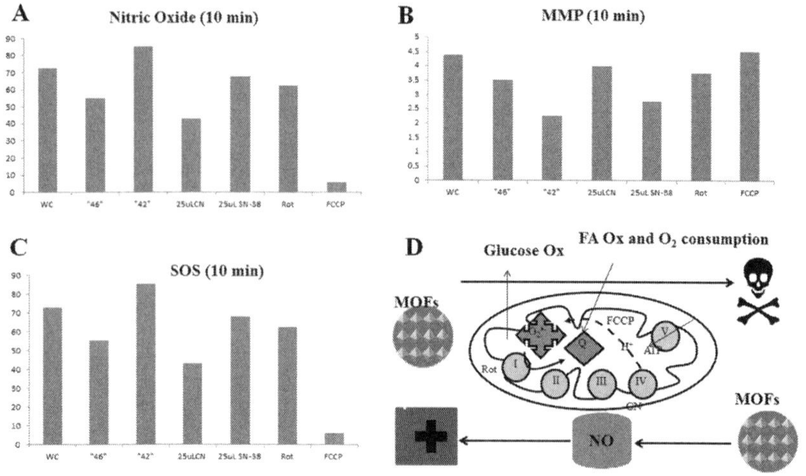


*Figure 4. Toxicity study of MOFs (two MOFs were chosen), **A**: Fluorescence of DAF FM DA, as measure of NO, **B**: Fluorescence of tetramethylrhodamine ethyl ester as measure of mitochondrial membrane potential; **C**: Fluorescence of singlet oxygen sensor green; **D**: Proposed mechanism for increased RPE cell death induced by MOFs.*

### 3.4. Biosafety Evaluation of MOFs

The selected MOF which may be used as electrode components in fuel cells were evaluated for bio toxicity against human retinal pigment epithelium. The results demonstrate that these materials pose no toxicity or may be toxic depending on their degree of crystallinity and pore size. The degree of toxicity was related to ability to sequester and release nitric oxide (NO). If NO was released in parts-per-million (ppm) concentrations, the cell was able to tolerate oxidative stress and the MOF was non-toxic, if the MOF did not sequester NO or released NO at high doses, the cell was unable to tolerate oxidative stress and under these conditions the MOF was toxic.

The results suggested accidental release of fuel cell electrode catalysts or other components; structurally similar to the evaluated MOFs may be toxic to humans. The most plausible route would release into the environment and environment to water table and to human consumption of contaminated water supply. The propensity of metal-organic frameworks bio-toxicity to microorganisms in water is evaluated in terms of disinfection quotient. The bioassay is compared with two silver nanoparticles (NPs) and silver-titania composite NPs (*32–35*) and bleach. The bactericidal effectiveness of each system is described, followed by general mechanism of disinfection.

### 3.5. Rationale of Using MOFs

This work focused on: (1) designing nanosystems using facile chemistry approach and providing a synthetically reproducible and optimal formulation method; (2) testing effectiveness of bactericidal performance of nanosystems on *Escherichia coli* (*E. coli*); and (3) understanding the mechanism of action of MOFs, NPs and bleach on disinfection. Extensive research has been done, regarding the antimicrobial effects of silver-containing materials. However, MOF based antimicrobial agents have only been introduced recently with little comparison to other systems, particularly zeolites, to which they are similar. The new aspects of this study is to utilize the MOF architecture, using benign starting materials, which yields two advantages: (a) the tunable surface area and porosity of the MOFs are retained; (b) the synthesis process has been simplified without the necessity for sonication (to generate hydrogen radicals) or microwaves (thermal and electron assisted reduction). We also include a comparison among the various systems, such as selected generations of disinfectants. A description of the inactivation mode amongst the disinfectants is also included.

### 3.6. Comparison of Bactericidal Performance

Three generated systems, Ag NPs (*39*), Ag-$TiO_2$ NPs (*40*), and MOFs were evaluated as disinfectants, which were compared with bleach (*41*). The minimum bactericidal concentrations (MBCs) were carried out over a two-week period. The MBCs of Ag NPs were determined to be 2.5 ppm (Figure 5a); while Ag-$TiO_2$ NPs to be 0.6 ppm (Figure 5b) for 100% inactivation of *E. coli*. Within this time-period, no bacterial growth was observed, suggesting that nanodisinfection persists at least for two weeks. Similar results were observed for *Staphylococcus aureus* (*S. aureus*), a Gram-positive microorganism, however higher doses (~ × 4) were required when compared to *E. coli* (*33*, *42*). One drawback of Ag NPs is that the time needed to disinfect the water of infectious agents is ~ 2 hr at ~10 ppm and ~6 hr at < 1 ppm. A major difference between Ag and Ag-$TiO_2$ NPs is that Ag-$TiO_2$ NPs are cheaper than Ag and more effective under visible light conditions (unlike $TiO_2$ alone) (*43*). Previously, disinfection using culture methods and optical density for the strains was carried out and good agreement found between both studies, in addition to literature reports (*44*). Our observations suggested that the cells were in exponential growth phase. In this manner, the microorganism maintained the same cells' physiological state as long as the optical density was carefully monitored to ensure utilizable carbon was not the limiting step, which tends to become limiting at optical densities (O.D) > 0.6, which has also been reported in the literature (*45*).

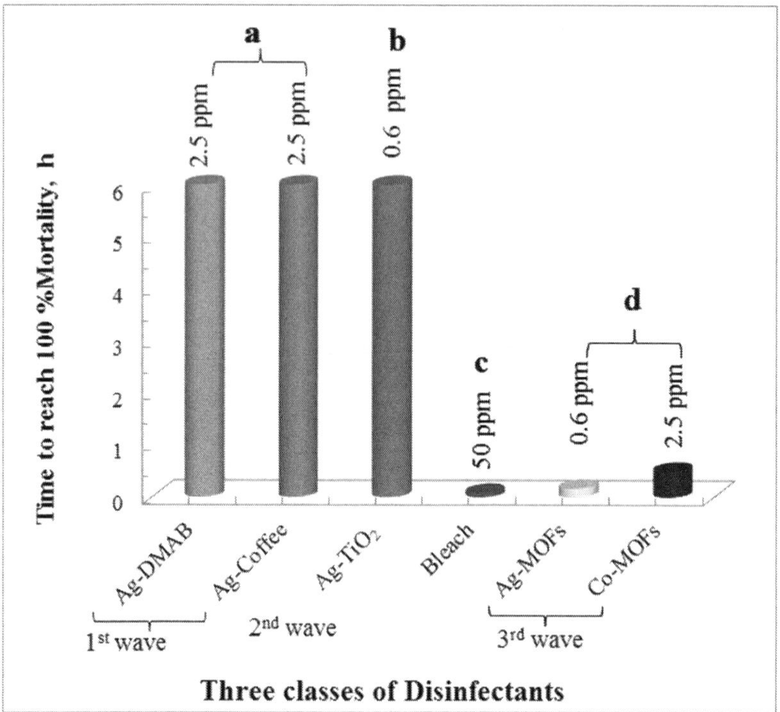


*Figure 5. Plot of time required to reach 100 % mortality versus disinfectant,* ***a****: the first-generation (Ag NPs, MBC = 2.5 ppm at 6 h);* ***b****: the second-generation (Ag-titania NPs, 0.6 ppm at 6 h);* ***c****: the zero-generation bleach (50 ppm at 10 min);* ***d****: third-generation ($Ag_3BTC$ MOF: 0.6 ppm at 10 min and $Co_4TDM$ MOF: 2.5 ppm at 30 min. microorganism at $2.5 \times 10^5$ cfu/mL).*

By comparison, bleach acts rapidly with a MBC of 50 ppm (Figure 5c) for up to 3 days in sunlight (*30*). To maintain disinfection, an additional dose of bleach is required periodically. At 50 ppm of bleach, disinfection was achieved through total microbial disintegration, indicating no significant residues were left to image. Data also indicated Ag and Ag-$TiO_2$ NPs destroy microbes (Figure 5d), but do not obliterate the cells as bleach does. One advantage of these NP disinfectants compared to bleach is that their MBCs were considerably lower and persistence was greater (*46*). The inactivation mode resulting in microbial destruction by bleach is through cellular amino acid modification and membrane lipid oxidation (*47*). These modifications result in loss of nucleic acid and protein production and repair. On the other hand, the NPs do not lead to total microbial disintegration (Figure 6) (*42*). The above results suggest that titania-silver nanocomposites would yield similar antimicrobial potency to pure silver, in terms of dose and duration of disinfection (persistence), as well as their slow speed of inhibition, compared with bleach, although at much lower concentrations that what bleach.

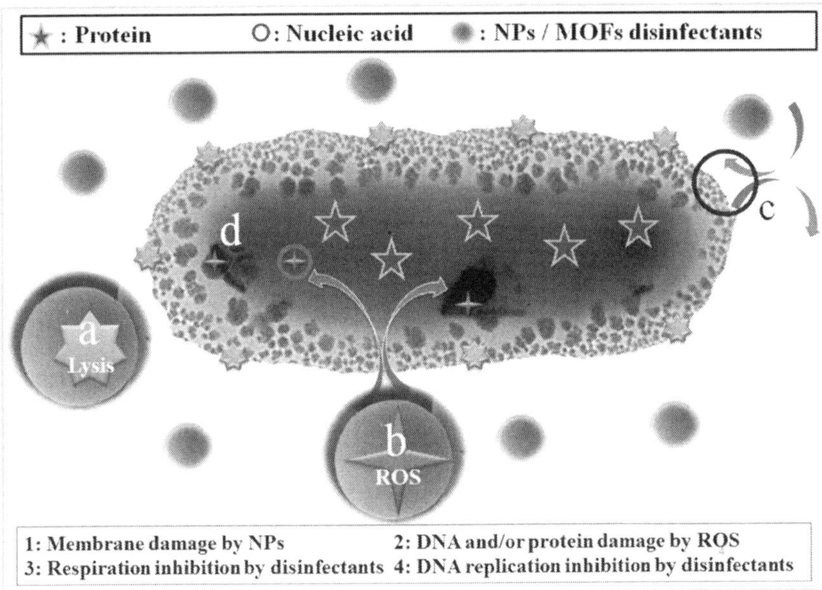


*Figure 6. An overview of the four major processes involved in microbial inactivation, **a**: membrane lysis; **b**: reactive oxygen species-induced damage of deoxyribonucleic acids (DNA) and proteins; **c**: inhibition/modification of critical proteins such as respiratory proteins; **d**: inhibition or denaturation of DNA.*

### 3.7. Bactericidal Performance of NPs

In the present study, bactericidal activities of Ag and Ag-$TiO_2$ NPs were evaluated for comparision. The MBCs for Ag NPs were found to be 2.5 ppm within 6 hrs. Ag-$TiO_2$ NPs showed greater potency with MBCs at 0.6 ppm within 6 hrs. Both NPs demonstrated a broad spectrum of activity, long term persistence, negligible toxic by-product generation, and effective activity under visible light. However the NPs disinfect and inactivate microbial activity over a prolonged time period (two weeks). Microscopic analyses of *E. coli* (Figure 7) suggest bacteria undergo protein (*48*) (Figure 7a), and DNA (*49*) degradation (Figure 7b), and cellular potassium ion ($K^+$) leakage (Figure 7c) (*50*). This inference was made on the basis of a comparison of morphological changes and EELS mapping. Since DNA macromolecules lack sulphur and the location was not in the plasma membrane (PM), it was inferred these were from DNA. Phase contrast microscopy showed elongation and fragmentation of the region containing the DNA strand, observations consistent with literature observations on effect of silver nanoparticles on microorganisms. In a similar manner, protein degradation was inferred through mapping of nitrogen/phosphorous, sulphur. It was also found bacteria cell death resulted from ROS-induced outer membrane degradation, by comparing lesions in the cytoplasm and cell membrane cross-referenced with known studies (*51*).

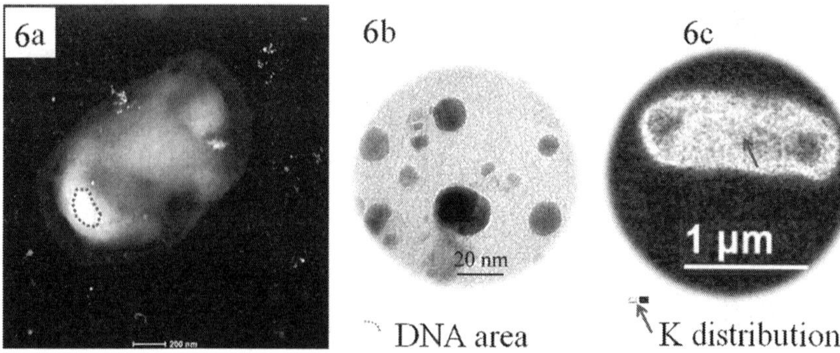


*Figure 7. Characterization of damaged Escherichia coli (E. coli, at 2.5 × $10^5$ cfu/mL) microbe and nanoparticles (Ag-$TiO_2$: 6.5-20 nm) as disinfectants.* ***a***: *Scanning transmission electron microscope (STEM) analysis of (E. coli showing DNA denaturation*,* ***b***: *TEM of nanoparticle morphology and size distribution (6.5-20 nm);* ***c***: *potassium ions ($K^+$) mapped by electron energy loss spectroscopy (EELS) to demonstrate membrane depolarization.*

Denaturation can be assessed through analysis of phase contrast of the cell, by either optical or electron microscopy and is consistent with other published studies which have measured fragmentation. Thus morphological changes obtained by STEM and EELS are excellent predictors of biochemical changes with respect to ion leakage and denaturation. The EELS mapping showed nitrogen, sulphur that are common building blocks of proteins and nitrogen and phosphorous that are common building blocks of nucleic acids. From these EELS maps and visual inspection of DNA strand un-zipping, it was concluded, MOF-DNA unzipping had occurred, or protein degradation, phenomena that have also been observed and reported in the literature with Ag NPs. Variation in size distribution was less than 3 nm.

### 3.8. Bactericidal Performance of MOFs

In the present section, a series of Ag-MOFs, Co-MOFs, and Cu-MOFs were developed (Figure 3 and Table 2-3). These MOFs exhibit rapid and effective *E. coli* inactivation (0.6-2.5 ppm, < 30 min) for an extended duration (Figure 5d). It was found that MOFs have higher potency than zeolites (Figure 8 and Table 4). The rationale for comparing MOFs and zeolites as disinfection was that both systems are highly porous, have high surface area and Ag-zeolites has previously been used towards disinfection. A literature review of all major publications involving zeolites as disinfectants (Table 4) indicates that MOFs possess superior potency, due to lower MBCs and greater speed of disinfection. MOFs also exhibit similar advantages to NPs, such as a broad spectrum of activity, long term persistence, negligible toxic by-product generation, and effective activity under visible light. Additionally, MOFs, like bleach, rapidly disinfect and inactivate microbial activity within minutes. Microscopic analyses of *E. coli* (Figure 9) suggest bacteria also undergo a number of biochemical and morphological change. Analyses of Figure 5 indicate extensive protein degradation (Figure 9a, light and dark regions) (*48*). In the figure, numerous vesicle-like formations can also be observed, indicating cell wall disassembly. Loss of wall plasticity is achieved through disruption of cross-wall linking and weakening of the phosphor-backbone (confirmed through EELS). Figure 9a also indicates degradation of DNA (Figure 9a, dark aggregation) (*51*). The morphological change observed by phase contrast is through the formation of aggregation bodies, suggesting inhibition of DNA replication, followed by DNA migration and partial unzipping. The described morphological changes were observed within minutes using the MOFs (Figure 9b). Although the MOFs are highly crystalline structures in micrometre scale, their secondary building units are nanoscale in dimension and promote rapid membrane peroxidation and cytosis. EELS mapping data (Figure 9c) for potassium ion ($K^+$) confirmed membrane damage (*50*). Potassium ion leakages were a result of disruption to the plasma membrane potential and a marker for cell death, since the cell is unable to maintain a potential gradient for active transport. The mechanisms using MOFs and NPs (Figure 6) for membrane depolarization, cell wall disassembly, inhibition of replication and partial lysis are similar.

It was also found bacteria cell death resulted from ROS-induced outer membrane degradation (*51*), as concluded when compared to untreated (Figure 10a) and heat-treated (Figure 10b) controls, respectively. The control microbe has the full complement of flagella for movement, intact cell wall and plasma membrane. When the cell is heated to imitate heat shock, cell wall lysis was observed, including DNA aggregation (gray area) and loss of water soluble nutrients. This was conducted to compare heat shock with oxidative stress, which is one mechanism of cell damage promoted by MOFs and NPs. The effectiveness is the degree of "dissolution" of silver relative to degree of immobilization (sulfidation) and deactivation (photoreduction of thiol- or protein-bound silver) or encapsulation (by nanoscale film formation). Under the present conditions, the nanosystems were not transformed to lesser toxic precursors, which may account for the long term persistence of disinfection; toxicity not observed in higher animals due to potential biotransformation.

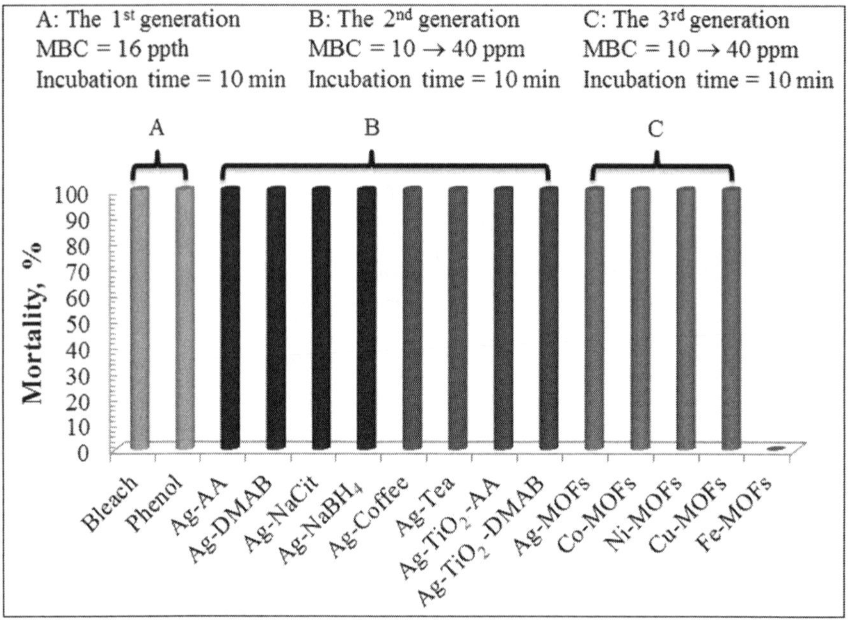


*Figure 8. Summary of each generation of disinfectant, common disinfectants (bleach, phenol), first-wave (silver, silver-titania) disinfectants reduced with various reducing agents (ascorbic acid [AA], dimethylamine borane [DMAB], sodium citrate [NaCit], sodium borohydride [$NaBH_4$], coffee, tea,] and third wave metal organic frameworks with different metals [Ag, cobalt (Co), nickel (Ni), Cu and iron (Fe)], including minimum bacterial concentration (MBC), Incubation time and percent cell mortality (for 100% Inhibition) . The MBCs were determined using OD measurements that were previously calibrated to colony forming units/mL. All the samples were characterized using spectroscopy and microscopy, described above and NPs had a particle size averaging 20 nm (± 5 nm) [Top row]. (Reprinted with permission from (Liu, J., Chamakura, K., Perez-Ballestero, R., and Bashir, S. In Nanomaterials for Biomedicine. Nagarajan, R. (ed). American Chemical Society, Washington, D.C). Copyright (2013) American Chemical Society.). [Also see Chamakura, K., Perez-Ballestero, R., Luo, Z., Bashir, S., and Liu, J. (2011). Comparison of bactericidal activities of silver nanoparticles with common chemical disinfectants. Colloids and Surfaces B: Biointerfaces, 84(1): 88-96]. Items bracketed in Part C are listed in order of most effective (Ag) to least effective (Fe), although all except Fe was able to disinfect the microorganism tested.*

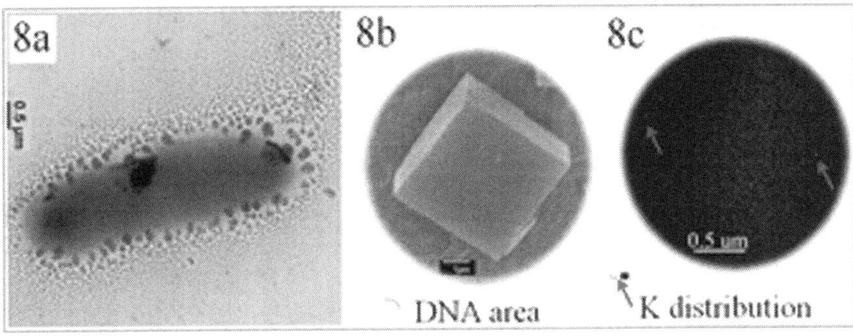


*Figure 9. Morphological analysis of the damaged E. coli microbe and metal-organic frameworks as disinfectants, **a**: TEM image of E. coli showing extensive damage including cell wall and membrane disintegration, **b**: SEM image of MOF (Cu-BTC) crystals (ca. μ5 m length for the singly crystal); **c**: EELS map showing $K^+$ signal. Microorganism at $2.5 \times 10^5$ cfu/mL.*

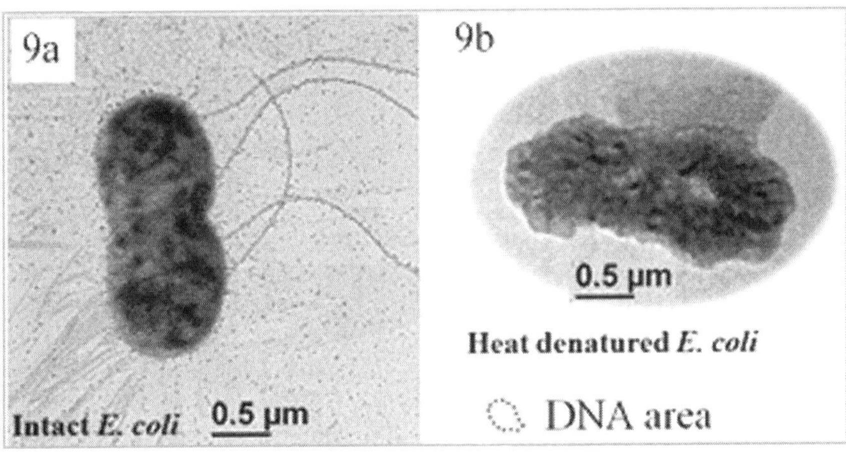


*Figure 10. TEM image of an E. coli microbe, **a**: intact control and **b**: heat-denatured positive control, showing plasma membrane peeling from the cell wall, DNA migration and fragmentation, and lysis releasing cytoplasm to the environment.*

A comparison of disinfection potency using MBCs ranks as: MOFs ~ Ag-$TiO_2$ < Ag, ~ zeolites < bleach, with MOFs as the most potent and bleach as the least potent. A literature comparison with zeolites indicated a wide fluctuation in MBCs when converted to the ppm scale. The lowest calculated ppm values (most potent) were slightly higher than the current MBCs for MOFs (summarised in Figure 8). Data also showed that MOFs (Figure 5d) display similar MBCs to NPs (Figures 5a and 5b), but with a superior dose-response curve (act within minutes) and did more damage to the cell wall. Numerous vesicle-like nodules were observed around the cell wall, phenomenologically similar nodules were observed for bleach, suggesting that both MOFs and bleach exert their biological effectiveness through generation of a localised oxidizing environment, but unlike bleach, MOFs redistribute this to DNA and protein fragmentation, whereas for bleach (at 50 ppm) the lysis of the cell wall continues until the cell is totally disintegrates into a few cytoplasmic fragments. Although the MOFs format micrometre-length single crystals or polycrystals (Figure 9b), in solution phase their actions appears to resemble that of colloidal silver-titania. In addition, like Ag-$TiO_2$ (and unlike bleach), MOFs also depolarise the plasma membrane, leading to loss of potassium ion into the extracellular matrix. The loss of potassium is diagnostic for microbial inactivation, since no microorganisms have been observed to grow in media after potassium ion leakage. This suggests either a state of inactivation or indefinite stasis (bacteriostatic or vegetative-state). The evidence from the literature coupled with extensive cell wall damage would support inactivation (*52*).

### 3.9. Proposed Inactivation Mechanism

The inactivation mechanism of NPs and MOFs is through generation of reactive oxygen species (ROS) (*53*) at the double layer between the solvent (water) and NPs/MOFs (*54*). The ROS oxidizes microbial outer membranes and phospholipids; and weaken cell-wall cross-links to induce cell lysis (*55*). A comparison of the micrographs would support deoxyribonucleic acid (DNA) base pair disruption (*56*), inhibition of DNA replication (*57*) and protein degradation (*58*). Analysis of cell culture and EDS would support inhibition of microbial cell growth (*59*), membrane depolarization (*54*) respectively. Similar to bleach, MOFs and NPs disrupt critical cellular functions, leading to cell growth inhibition and cell death (*34*).

MOFs represent a technical breakthrough that may cause a re-evaluation of the status quo is the use of MOFs (third generation) as disinfectants. MOFs are crystalline compounds that consist of metal ions coordinated with organic linkers to form structured frameworks with specific properties. The indicated MOFs are designed to oxidize and depolarize outer microbial membranes and inhibit protein synthesis through MOFs' metal atom. Using appropriate design, MOFs can be generated to target specific components involved in water purification, such as the removal of pathogens, organics, salts and metal ions (*60*). Due to the ultrahigh internal surface area of the MOFs, small molecules can be encapsulated and simply released by a change in environmental pH. Therefore, optimal effectiveness can be chemically engineered into MOFs (Figure 3). In this manner,

MOFs can be used at low concentrations for extended duration for maximum effectiveness. Importantly, these nanomaterials can serve as model systems to develop advanced water disinfection technologies, which may be extended to use in other cell types (*61*). For example, MOF have been demonstrated to be effective nano-disinfectants against prokaryotic microbes (Figure 8). It is also noted that the MOF platform can also be engineered into a number of different configurations. The tailoring is related towards specific end-points. Examples include drug delivery (*62*), imaging (*63*), diagnostics (*64*) and therapeutics (*65*).

These cited applications (*39–65*) and our demonstration of disinfection of water can be used to address needs relevant to the 'Water for Life' decade to improve human access to potable clean water at reduced cost (*66*) with minimal environmental footprint, with other practical applications in the field of human health, such as cancer diagnostics, therapy and drug delivery.

## 4.0. Modelling of Ion-Release

In this section, the ion-release from the MOFs and zeolites was studied to understand disinfection kinetics or processes. The impact of the disinfecting agent on the environment has to be assessed. For example, it is well-established that one environmental impact of accidental bleach release is the formation of toxic- by-products (*67*). Although silver and silver-titania do not generate toxic by-products they can complex with other minerals in soil, air or water and exert an influence (*68*). Silver will complex to form zero-valent silver or silver sulfide, which are less toxic than silver ions (*69*). The end-fate of these NPs is heavily influenced on the charge density, and ligands (*70*). Zeolites and MOFs are slightly different to NPs in that they are always coordinated within a ligand framework and as such as more susceptible to environmental modification than metal NPs. A brief review serves three useful purposes: (a) An estimation of the speed of disinfection can be obtained by considering leach rate from within the framework; this in turn will assist in understanding (b) stability and (c) likely environmental impact of accidental release into the environment. A number of assumptions are made in generation of the zeolite kinetic model, summarized below.

Review of zeolites (*71*) kinetics using expressions (1)–(7) gives rise to two points of interest arise. The first point arises from the assumption that zeolites disinfect from leaching of silver from silver-impregnated zeolites into the bulk phase. The rate can be modelled using pseudo-first order kinetics by assuming that silver ions within the zeolite ($ZAg^+$) leach out into the bulk media and facilitate a redox type reaction described by equations (1), (2) and modelled by (3) from which the half-life can be derived (4):

$$ZAg^+(s) + X^\bullet \rightarrow Ag + X + m\bar{e} \text{ (and } Ag \rightarrow Ag^+ + \bar{e}) \quad (1)$$

$$m(Ag^+) \rightarrow (Ag^+)_m \text{ and } ZAg^- \leftrightarrow zAg^+ \ mAg^+ + Ag^0 \quad (2)$$

$$\text{In } ([Ab^+]_{post}, t = t/[Ab^+]_{pre}, t=0) = -kt \quad (3)$$

$$t_{1/2} = \mathrm{Ln}\ 2/k \quad (4)$$

If the concentration of silver ions is known and converted to natural logarithmic scale, then the half-life of the reaction (4) can be estimated, where In [$Ag^+$] is the concentration of the silver ions (M) at pre-exposure (t=0) or post-exposure at a specific time (t, sec), with $t_{1/2}$ being the half-life of decay (sec); n is a constant ($M^{-1}s^{-1}$). From the above, the first point of interest is that there is a limit to total silver loading, with < 0.1 percent weight (% Wt.) content favouring formation of smaller silver nanoparticles on the outer zeolite and higher deforming (> 3 % Wt) the zeolite structure; the second attribute can be seen from equations (5)-(7). For example, the diffusion of an ion (e.g. species A) out of the zeolite to the vicinity of the microorganism depends on the zeolite mass ($Z_a$, g), the post-exposure times of the ion with the microorganism ($t_{post}$, sec) and the corresponding ion concentration in molarity and normality respectively as shown in (5).

$$A_i = ZaMa_{post}/N \quad (5)$$

The activity within the zeolite ($A_i$, a.u.), in turn is related to mass of ion (e.g. Ai) within the zeolite ($Z_{Ai}$, g), its pre- and post-concentration ($M_{Aipre}$, M; $M_{Aipost}$, M), where M is moles/liter; volume of the ion (Ai) in solution ($V_{Ai}$, L) and the zeolite exchange capacity (ZE, meq/m), summarized in (6)

$$Ai_Z = Z_{Ai}[M_{Aipre} - M_{Apost}]V_{Ai}/(ZE) \quad (6)$$

For most systems this activity can be approximated as unity, if the movement within and out of the system is similar.

The actual activity can be estimated as the ratio of the activity in the bound-state ($A_{ib}$, a.u.) and at equilibrium ($A_{iequilibrium}$, *a.u.*), summarized in (7).

$$\mathrm{Activity\ (Ai)} = A_b/A_{equilibrium} \quad (7)$$

The second point of interest for zeolites is the parameters that influence degree of disinfection *greatest* are: solvent used in the synthesis of the zeolite, solvent pH, power of reductant, percent incorporation of silver in the zeolite and the factor with the *least* influence on disinfection is size of incorporated nanoparticle on the zeolite surface.

In contrast to the kinetic analysis for zeolites, MOFs are stable and does not leach metal ions, unless the framework is designed to collapse. The environmental impact of zeolites is limited to ion leakage, with similar biological impact to release of silver ions. At present, there is no data relating to environmental impact of MOFs due to their recent introduction and limited application; however their usage is expected to increase over time. Unlike zeolites, the biological effectiveness of the MOFs is through the application of the entire metal-organic frameworks system and not related to ion leakage. MOFs are not expected to yield any ion movement out of the frameworks, since the metal is coordinated within an organic framework, resulting in highly stabilized system. Hypothetically, if ion leakage was observed (which was not the case with our MOFs), this leakage would suggest a highly unstable MOF structure, unless the MOF framework was designed to collapse.

Under these circumstances, the environmental impact of accidental release of a collapsible MOF impact would be similar to organically coated NPs, which form sulfides, carboxylates, nitrates, depending on the metal and ligand used.

# 4. Summary

Eight formulations of MOFs were synthesized and methane gas uptake was measured. The MOFs are estimated to show improved volumetric and gravimetric capacities. The ligands consisted mainly of four dicarboxylic acids with an aromatic backbone. The biochemical processes through which MOFs exhibits toxicity was evaluated and shown to be consistent with that mitochondrial oxidative stress and associated cellular dysfunction that play key roles in disrupting oxygen consumption, energy production and generation of NO. In addition other MOFs were shown to be potent disinfectants, suggesting that accidental release of materials in energy storage or fuel cells into the water table may pose health hazards suggesting disposal and end-of-life policy directive ought to be issued alongside performance parameter requirements.

# 5. Brief Future Prospects and Guidelines

The current study strongly suggests that physio-chemical means of inactivation are more important than MOF geometry (inferred between comparisons of MOFs with different ligands, yielding similar MBCs). Specifically, MOFs catalyse the formation of vesicle-like structures (assessed by microscopy) in the cell wall, resulting in extensive cell lysis (*86*). MOFs represent the third generation, of disinfectants exhibiting the same multi-fold advantages as nanomaterials but with the speed of inactivation similar to that observed for bleach, hence representing a possible "first choice" future class of disinfectants, however as materials used in gas storage devices, these disinfection properties may well be a beneficial side-effect of disposal if they can be segregated from accidental ingestion by humans, which may pose unaccounted health hazards.

We have demonstrated that MOF are a viable and effective alternative to chemical disinfectants and zeolites. They have MBCs in the low ppm range, have longer persistence than bleach and attach the microbe through different physical and biochemical processes. This observation strongly suggesting that antibiotic-resistant microorganism would be susceptible to MOF bactericidal activity. The ultralow doses and fast interaction time are a consequence of the ultrahigh surface area, porosity and small dimensions by the MOF, which extend the applications from carbon dioxide capture, hydrogen storage to cancer diagnosis and water purification. The end-of-life evaluation ought to be integrated alongside performance and stability parameters in the design of these energy devices.

## Acknowledgments

Acknowledgment is made to the Donors of the American Chemical Society Petroleum Research Fund for partial support of this research. The National Science Foundation (CBET-0930079), Graduate Scholarship from the College of Arts and Sciences, Texas A&M University-Kingsville (TAMUK) University Research Award (URA, 160323), R. Welch Foundation Departmental Grant (AC006) and TAMUK supported by the Department of Education PPHOA (501006) grant are duly acknowledged for their financial support. Dr. H.-C. Zhou and his group from Texas A&M University for providing ligands, Dr. Enrique Massa from TAMUK, Department of Biological and Health Sciences for allowing us to conduct the bioassay; and Dr. Jeffrey Wigle for providing the RPE assay and access to the Triservice laboratory (Fort Sam Houston, AirForce, EPA-07-029-HE-00-EPA) are acknowledged. The technical support from the TAMUK and the use of TAMU Center of Microscopy Imaging and Materials Characterization Facility are also duly acknowledged. Drs. T. Hays and P. Cox (TAMUK) are acknowledged for their comments and revision on this manuscript. Graduate students (Tingting Wang, Sandhya Koppaka, Shivashankar Thati and Joseph Medina) were acknowledged to participiate in the MOFs synthesis and bactericidal study. Xuan Wang (a PhD candiate at TAMU) is also acknowledged for her contribution to supervise undergraduate student (U. Okakpu) to prepare MOFs and measure the gas uptake capability.


## Authors' Contribution

Dr. S. Bashir conceived biolgoical research activities and oversaw the chemical and biological safety. He also collected and analyzed the toxicity data on RPE cells. He wrote the first draft and co-edited the second. Dr. Z. Luo conducted electron microscopic analysis on bactericidal activites. Two students (B. Martinez and U. Okakpu) were trained to prepare MOFs and exposed into the research field of porous materials used for gas storage. Dr. Liu conceived this project and carried out all the experimental procedures (except electron microscopy) and data analysis. She revised the drafts, as well as oversaw the research progress, manuscript submission and student training.

## References


1. (a) Technically Recoverable Shale Oil and Shale Gas Resources: An Assessment of 137 Shale Formations in 41 Countries Outside the United States. U.S. Energy Information Administration. http://www.eia.gov/analysis/studies/worldshalegas/archive/2013/pdf/fullreport_2013.pdf (accessed November 2015). (b) Modern Shale Gas. Development in the United States: A Primer. U.S. Department of Energy. http://energy.gov/sites/prod/files/2013/03/f0/ShaleGasPrimer_Online_4-2009.pdf (accessed November 2015).
2. Kitagawa, S.; Kitaura, R.; Noro, S. *Angew. Chem., Int. Ed.* **2004**, *43*, 2334–2375.

3. Murray, L. J.; Dincă, M.; Long, J. R. *Chem. Soc. Rev.* **2009**, *38*, 1294–1314.
4. Li, J.-R.; Kuppler, R. J.; Zhou, H.-C. *Chem. Soc. Rev.* **2009**, *38*, 1477–1504.
5. Lee, J.; Farha, O. K.; Roberts, J.; Scheidt, K. A.; Nguyen, S. T.; Hupp, J. T. *Chem. Soc. Rev.* **2009**, *38*, 1450–1459.
6. Horcajada, P.; Serre, C.; Maurin, G.; Ramsahye, N. A.; Balas, F.; Vallet-Regí, M.; Sebban, M.; Taulelle, F.; Férey, G. J. *J. Am. Chem. Soc.* **2008**, *130*, 6774–6780.
7. Allendorf, M. D.; Bauer, C. A.; Bhakta, R. K.; Houk, R. J. T. *Chem. Soc. Rev.* **2009**, *38*, 1330–1352.
8. Horike, S.; Shimomura, S.; Kitagawa, S. *Nat. Chem.* **2009**, *1*, 695–704.
9. Stylianou, K. C.; Rabone, J.; Chong, S. Y.; Heck, R.; Armstrong, J.; Wiper, P. V.; Jelfs, K. E.; Zlatogorsky, S.; Bacsa, J.; McLennan, A. G.; Ireland, C. P.; Khimyak, Y. Z. K.; Thomas, M.; Bradshaw, D.; Rosseinsky, M. J. *J. Am. Chem. Soc.* **2012**, *134* (50), 20466–20478.
10. Rosseinsky, M. J. *Microporous Mesoporous Mater.* **2004**, *73*, 15–30.
11. He, Y.-B.; Li, B.; O'Keeffe, M.; Chen, B.-L. *Chem. Soc. Rev.* **2014**, *43*, 5618–5656.
12. He, Y.; Zhou, W.; Qian, G.; Chen, B. *Chem. Soc. Rev.* **2014**, *43* (16), 5657–5678.
13. Mason, J. A.; Veenstra, M.; Long, J. R. *Chem. Sci.* **2014**, *5*, 32–51.
14. Cavka, J. H.; Jakobsen, S.; Olsbye, U.; Guillou, N.; Lamberti, C.; Bordiga, S.; Lillerud, K. P. *J. Am. Chem. Soc.* **2008**, *130*, 13850–13851.
15. Gupta, S. *Int. J. Oncol* **2003**, *22* (1), 15–20.
16. Rheingans, R.; Dreibelbis, R.; Freeman, M. C. *Global Public Health* **2006**, *1*, 31–48.
17. Dai, A. *Climatic Change* **2011**, *2*, 45–65.
18. Parmesan, C.; Yohe, G. *Nature* **2003**, *421* (6918), 37–42.
19. Fewtrell, L.; Kaufmann, R. B.; Kay, D.; Enanoria, W.; Haller, L.; Colford, J. M. *Lancet* **2005**, *5* (1), 42–52.
20. Gilman, R. H.; Skillicorn, P. *Bull. World Health Org.* **1985**, *63* (1), 157–63.
21. Scott, J. P.; Ollis, D. F. *Environ. Prog.* **2006**, *14* (2), 88–103.
22. Riley, L. W.; Remis, R. S.; Helgerson, S. D.; McGee, H. B.; Wells, J. G.; Davis, B. R.; Cohen, M. L. *N. Engl. J. Med.* **1983**, *308* (12), 681–685.
23. Altmann, M.; Wadl, M.; Altmann, D.; Benzler, J.; Eckmanns, T.; Krause, G.; An der Heiden, M. *Emerging Infect. Dis.* **2011**, *17* (10), 1906–1909.
24. Rutala, W. A.; Weber, D. J. *Clin. Microbiol. Rev.* **1997**, *10* (4), 597–610.
25. Meier, J. R. *Mutat. Res. Rev.* **1988**, *196* (3), 21–245.
26. Li, Q.; Mahendra, S.; Lyon, D. Y.; Brunet, L.; Liga, M. V.; Li, D.; Alvarez, P. J. *Water. Res.* **2008**, *42* (18), 4591–4602.
27. Savage, N.; Diallo, M. S. *J. Nano. Res.* **2005**, *7* (4), 331–342.
28. Medina-Ramirez, I.; Luo, Z.; Bashir, S.; Mernaugh, R.; Liu, J. L. *Dalton Trans.* **2010**, *40* (5), 1047–1054.
29. Adams, L. K.; Lyon, D. Y.; Alvarez, P. J. *Water Res.* **2006**, *40* (19), 3527–3532.
30. Weber, D. J.; Sickbert-Bennett, E.; Gergen, M. F.; Rutala, W. A. *J. Am. Med. Inform. Assoc.* **2003**, *289* (10), 1274–1277.

31. Sökmen, M.; Candan, F.; Sümer, Z. *J. Photochem. Photobiol., A* **2001**, *143* (2), 241–244.
32. Medina-Ramirez, I.; Bashir, S.; Luo, Z.; Liu, J. L. *Colloids Surf., B* **2009**, *73* (2), 185–191.
33. Bashir, S.; Dinn, J.; Liu, J. *MRS Proc.* **2013**, *1498*, 91–96.
34. Liu, J.; Wang, Z.; Luo, Z.; Bashir, S. *Dalton Trans.* **2013**, *42*, 2158–2166.
35. Tennen, R.; Setlow, B.; Davis, K. L.; Loshon, C. A.; Setlow, P. *J. Appl. Microbiol.* **2000**, *89* (2), 330–338.
36. Buckingham-Meyer, K.; Goeres, D. M.; Hamilton, M. A. *J. Microbiol. Methods* **2007**, *70* (2), 236–244.
37. Lara, H. H.; Garza-Treviño, E. N.; Ixtepan-Turrent, L.; Singh, D. K. *J. Nano Biotechnol.* **2011**, *9* (1), 30–31.
38. Tolaymat, T. M.; El Badawy, A. M.; Genaidy, A.; Scheckel, K. G.; Luxton, T. P.; Suidan, M. *Sci. Total Environ.* **2010**, *408* (5), 999–1006.
39. Eckhardt, S.; Brunetto, P. S.; Gagnon, J.; Priebe, M.; Giese, B.; Fromm, K. *Chem. Rev.* **2013**, *113* (7), 4708–4754.
40. Younas, H.; Qazi, I. A.; Hashmi, I.; Awan, M. A.; Mahmood, A.; Qayyum, H. A. *Environ. Sci. Pollut. Res.* **2014**, *21* (1), 740–752.
41. Lantagne, D.; Person, B.; Smith, N.; Mayer, A.; Preston, K.; Blanton, E.; Jellison, K. *Environ. Sci. Technol.* **2014**, *48* (9), 5093–5100.
42. Liu, J.; Chamakura, K. R.; Ballestero, R. P.; Bashir, S. Historical Overview of the First Two Waves of Bactericidal Agents and Development of the Third Wave of Potent Disinfectants. In *Nanomaterials for Biomedicine*; Nagarajan, R., Ed.; ACS Symposium Series 1119; American Chemical Society: Washington, DC, 2012; Chapter 6, pp 129−154.
43. Liu, J.; Luo, Z.; Bashir, S. *Biomater. Sci.* **2013**, *1*, 194–201.
44. Domínguez, M. C.; de la Rosa, M.; Borobio, M. V. *J. Antimicrob. Chemother.* **2001**, *47* (4), 391–398.
45. Sezonov, G. D.; Joseleau-Petit, D.; D'Ari, R. *J. Bacteriol.* **2007**, *189* (23), 8746–8749.
46. Petit, M.; George, I.; Servais, P. *Can. J. Microbiol.* **2000**, *46* (7), 679–684.
47. Ryter, S. W.; Kim, H. P.; Hoetzel, A.; Park, J. W.; Nakahira, K.; Wang, X.; Choi, A. M. *Antioxid. Redox Signaling* **2007**, *9*, 49–89.
48. Panáček, A.; Kvitek, L.; Prucek, R.; Kolar, M.; Vecerova, R.; Pizurova, N.; Sharma, V. K.; Nevečná, T.; Zboril, R. *J. Phys. Chem. B* **2006**, *110* (33), 16248–16253.
49. Damm, C.; Münstedt, H.; Rösch, A. *Mater. Chem. Phys.* **2008**, *108* (1), 61–66.
50. Kawahara, K.; Tsuruda, K.; Morishita, M.; Uchida, M. *Dent. Mater.* **2000**, *16* (6), 452–455.
51. Pokhrel, L. R.; Silva, T.; Dubey, B.; El Badawy, A. M.; Tolaymat, T. M.; Scheuerman, P. R. *Sci. Total Environ.* **2012**, *426* (2012), 414–422.
52. Feng, Q. L.; Wu, J.; Chen, G. Q.; Cui, F. Z.; Kim, T. N.; Kim, J. O. *J. Biomed. Mater. Res.* **2000**, *52* (4), 662–668.
53. Kim, J. S.; Kuk, E.; Yu, K. N.; Kim, J. H.; Park, S. J.; Lee, H. J.; Kim, S. H.; Park, Y. K.; Park, Y. H.; Hwang, C.-Y.; Kim, Y.-K.; Lee, Y.-S.; Jeong, D. H.; Cho, M. H. *Nano Med.* **2007**, *3*, 95–101.

54. Choi, O.; Hu, Z. *Environ. Sci. Technol.* **2008**, *42* (12), 4583–4588.
55. Page, K.; Palgrave, R. G.; Parkin, I. P.; Wilson, M.; Savin, S. L.; Chadwick, A. V. *J. Mater. Chem.* **2007**, *17* (1), 95–104.
56. Thurman, R. B.; Gerba, C. P.; Bitton, G. *Crit. Rev. Environ. Sci. Technol.* **1989**, *18* (4), 295–315.
57. Li, K.; Zhao, X. K.; Hammer, B.; Du, S.; Chen, Y. *ACS Nano.* **2013**, *7* (11), 9664–9674.
58. Akhavan, O.; Choobtashani, M.; Ghaderi, E. *J. Phys. Chem. C* **2012**, *116* (17), 9653–9659.
59. Choi, O.; Deng, K. K.; Kim, N. J.; Rose, J., Jr.; Surampalli, R. Y.; Hu, Z. *Water Res.* **2008**, *42* (12), 3066–3074.
60. Chui, S. Y.; Lo, S. M. F.; Charmant, J. P.; Orpen, A. G.; Williams, I. D. *Science* **1999**, *283* (5405), 1148–1150.
61. Horcajada, P.; Serre, C.; Vallet-Regí, M.; Sebban, M.; Taulelle, F.; Férey, G. *Angew. Chem.* **2006**, *118* (36), 6120–6124.
62. Horcajada, P.; Chalati, T.; Serre, C.; Gillet, B.; Sebrie, C.; Baati, T.; Eubank, J. F.; Heurtaux, D.; Clayette, P.; Kreuz, C.; Chang, J.-S.; Hwang, Y. K.; Marsaud, V.; Bories, P.-N.; Cynober, L.; Gil, S.; Férey, G.; Couvreur, P. *Nat. Mater.* **2009**, *9* (2), 172–178.
63. Rowe, M. D.; Thamm, D. H.; Kraft, S. L.; Boyes, S. G. *Biomacromology* **2009**, *10* (4), 983–993.
64. McKinlay, A. C.; Morris, R. E.; Horcajada, P.; Férey, G.; Gref, R.; Couvreur, P.; Serre, C. *Angew. Chem., Int. Ed.* **2010**, *49* (36), 6260–6266.
65. Theron, J.; Walker, J. A.; Cloete, T. E. *Crit. Rev. Microbiol.* **2008**, *34* (1), 43–69.
66. Chong, M. N.; Jin, B.; Chow, C. W.; Saint, C. *Water Res.* **2010**, *44* (10), 2997–3027.
67. Ahlborg, U. G.; Thunberg, T. M.; Spencer, H. C. *CRC Crit. Rev. Toxicol.* **1980**, *7* (1), 1–35.
68. Gottschalk, F.; Sonderer, T.; Scholz, R. W.; Nowack, B. *Environ. Sci. Technol.* **2009**, *43* (24), 9216–9222.
69. Kaegi, R.; Sinnet, B.; Zuleeg, S.; Hagendorfer, H.; Mueller, E.; Vonbank, R.; Boller, M.; Burkhardt, M. *Environ. Pollut.* **2010**, *158* (9), 2900–2905.
70. Tolaymat, T. M.; El Badawy, A. M.; Genaidy, A.; Scheckel, K. G.; Luxton, T. P.; Suidan, M. *Sci. Total Environ.* **2010**, *408* (5), 999–1006.
71. Zhuang, W.; Yuan, D.; Li, J.-R.; Luo, Z.; Zhou, H.-C.; Bashir, S.; Liu, J. *Adv. Healthcare Mater.* **2012**, *1*, 225–238.
72. Liu, Y.; Xu, X.; Xia, Q.; Yuan, G.; He, Q.; Cui, Y. *Chem. Commun.* **2010**, *46*, 2608–2610.
73. Jaros, S. W.; Smoleriski, P.; Guedes da Silva, M. F. C.; Florek, M.; Krol, J.; Staroniewicz, Z.; Prombeiro, A. J. L.; Kirillov, A. M. *Cry. Eng Comm* **2013**, *15*, 8060–8064.
74. Matsumura, Y.; Yoshikata, K.; Kunisaki, S. I.; Tsuchido, T. *Appl. Environ. Microbiol.* **2003**, *69*, 4278–4281.
75. Galeano, B.; Korff, E.; Nicholson, W. L. *Appl. Environ. Microbiol.* **2003**, *69*, 4329–4331.

76. Inoue, Y.; Hoshino, M.; Takahashi, H.; Noguchi, T.; Murata, T.; Kanzaki, Y.; Hamashima, H.; Sasatsu, M. *J. Inorg. Biochem.* **2002**, *92*, 37–42.
77. Yang, F. C.; Wu, K. H.; Huang, J. W.; Horng, D. N.; Liang, C. F.; Hu, M. K. *Mater. Sci. Eng., C* **2012**, *32* (5), 1062–1067.
78. Magaña, S.; Quintana, P.; Aguilar, D.; Toledo, J.; Chávez, C.; Cortéz, M.; Léon, L.; Pelegrín, Y.; Lopéz, T.; Sánchez, R. *J. Mol. Catal. A: Chem.* **2008**, *281*, 192–199.
79. Wang, J.; Wang, Z.; Guo, S.; Zhang, J.; Song, Y.; Dong, X.; Wang, X.; Yu, J. *Microporous Mesoporous Mater.* **2011**, *146*, 216–222.
80. Fang, M.; Chen, J.; Xu, X.; Yang, P.; Hildebrand, H. F. *Int. J. Antimicrob. Agents* **2006**, *27*, 513–517.
81. Kwakye-Awuah, B.; Williams, C.; Kenward, M. A.; Radecka, I. *J. Appl. Microbiol.* **2008**, *104*, 1516–1524.
82. Lalueza, P.; Monzón, M.; Arruebo, M.; Santamaría, J. *Mater. Res. Bull.* **2011**, *46*, 2070–2076.
83. Sabbani, S.; Gallego-Perez, D.; Nagy, A.; Waldman, W. J.; Hansford, D.; Dutt, P. K. *Microporous Mesoporous Mater.* **2010**, *135*, 131–136.
84. Top, A.; Ülku, S. *Appl. Clay Sci.* **2004**, *27*, 13–19.
85. Ferreira, L.; Fonseca, A. M.; Botelho, G.; Almeida- Aguiar, C.; Neves, I. C. *Microporous Mesoporous Mater.* **2012**, *160*, 126–132.
86. Hrenovic, J.; Milenkovic, J.; Ivankovic, T.; Rajic, N. *J. Hazard. Mater.* **2012**, *201*, 260–264.

## Chapter 11

# Postface: Nanomaterials for Energy, A Look Forward

**Stephen Creager***

**Hunter Laboratory, Clemson University, Clemson, South Carolina 29634, United States**
***E-mail: screage@clemson.edu**

Structural control of materials on the nanoscale is essential to achieve the properties needed for the efficient collection, transduction, storage and consumption of energy. Many recent advances in energy science and technology have been enabled by advances in synthesis and characterization of nanomaterials. This chapter will consider various scenarios for how such advances might continue in the future.

## Introduction

From the burning of wood or coal for heat to the refining of petroleum to power automobile engines to the generation of electricity from renewable sources such as wind or solar, the ability to collect, transduce, store and consume energy effectively has been fundamental to modern life since the dawn of civilization. Just as the light bulb transformed society by enabling us to work/play at night, green energy is the next revolution in decentralized, smart-grid, clean energy for the society of today and tomorrow. Energy-related issues continue to be important in the public arena; one need only think about how global geopolitics is affected by factors such as global climate change, the changing cost and availability of fossil fuels such as coal, petroleum, and natural gas, and the emergence of renewable energy sources such as wind and solar, to understand the importance of energy to modern society.

Against this backdrop of significance, the importance of nanomaterials for energy technology has only recently come to be appreciated. Nanomaterials harness time and space dimensions that are not normally accessed in bulk materials and because of this fact they often exchibit chemical, electrical, optical properties that are unmatched in their bulk counterparts. Of course, nanomaterials are not new; many materials that have been used for centuries in energy harvesting, storage and consumption are quite complex at the nanoscale. Rather, the recent focus on nanomaterials is due to a lack of appreciation until relatively recently that the properties of many materials including materials used in energy-related technologies are dominated by structure and dynamics on the nanoscale. This realization regarding the new properties of nanomaterials, combined with the relatively recent availability of tools for imaging and studying structure and dynamics on the nanoscale, has greatly deepened out understanding of nanomaterials for energy and in many other areas of science and technology.

This book on Nanomaterials for Sustainability has focused on select recent developments for topics related in some way to nanomaterials for energy sustainability. The book is divided into thematic chapters, where each chapter describes a recent development in a specific area. There are chapters on fuel cells including proton exchange membrane fuel cells (PEMFC) and solid oxide fuel cells (SOFC) that are the engines driving new generations of electrochemical power sources. There are chapters on the electrocatalysts that operate within fuel cells, catalyzing critical electrode reactions such as the electrochemical oxygen reduction reaction (ORR). There are also chapters on electrode modeling, electrode characterization techniques, and lastly but equally important energy policy. In this final "postface" chapter, I will attempt to look forward to where I see continued research and development efforts in five areas having the most impact. In doing so, I am venturing my opinion and as such will keep in mind the timeless proverb, attributed to many wise men including Yogi Berra, Mark Twain, Niels Bohr and others, which says something to the effect that "*It's tough to make predictions, especially about the future*." These are indeed wise words.

## Nanomaterials for Electrochemical Energy Conversion: Fuel Cells and Electrolysis Cells

Research activity on new materials for fuel cells and electrolysis cells, including catalysts, electrolytes and supporting structures, has steady increased over the past several decades and seems likely to at least remain steady. Two technology platforms receiving much attention are proton exchange membrane fuel cells (PEMFC) and solid oxide fuel cells (SOFC). These platforms differ principally in the temperature range in which they operate; PEMFC devices typically operate between 80-100 °C whereas SOFC devices require much higher temperatures, typically between 500 and 1000 °C. Because of these widely differing operating temperature regimes, nanomaterials research follows very different paths for these two platforms. In the PEMFC realm, there is a pressing need for oxygen reduction reaction (ORR) catalysts having higher activity, lower cost, and greater durability. Research has focused mostly on catalysts containing nanostructured platinum-group metals having high specific surface area that resist sintering. Recent reports on platinum-nickel nanoframe structures for ORR catalysis are promising and further developments seem likely. Catalysts that contain no platinum group metals (non-PGM catalysts) are also quite promising and are becoming much better understood, and it seems likely that further developments will be realized on these materials. The successful development of non-PGM ORR catalysts having activity and durability approaching that of the best PGM catalysts would be a breakthrough development, driving down cost and increasing availability of PEMFC devices.

SOFC platforms also benefit from improved ORR catalysis but not to the same degree as PEMFCs. Because of the higher operating temperature for SOFC devices, electrode reaction proceed relatively rapidly even without benefit of catalysis. Even so, higher activity and improved selectivity and durability are always desired in catalysts and will continue to be sought.

Most PEMFCs, and to a lesser extent SOFCs, utilize hydrogen gas as fuel. Hydrogen is convenient because it is lightweight and its electrochemical oxidation reaction (HOR) is catalyzed relatively easily with high activity and selectivity by small amounts of a platinum catalyst. Hydrogen is also inconvenient because there is no hydrogen-mine or natural tap to draw it like coal or petroleum, thus hydrogen would need to be generated, for example from water. Here, improved HOR catalysts are critical and are always being sought but the more difficult catalysis problem is electrooxidation of other fuels, usually fuels containing carbon. Another disadvantage of hydrogen is that it remains a gas under all except the most extreme conditions (low temperatures and high pressures). For many applications, particularly involving automotive transportation, it is desirable for fuels to be liquids under ambient conditions. Unfortunately, electro-oxidation of most liquid organic fuels is mechanistically complex with many microscopic steps and reaction intermediates. Even the best catalysts have low activity with the possibility of multiple reaction pathways. A pressing need exists for electrocatalysts having higher activity and greater selectivity for electrooxidation of liquid organic fuels. Many paths are possible for developing such catalysts, and it is not yet clear how much improvement may be realistically expected. Even so, it seems very likely that whatever catalysts are developed, their properties on the nanoscale will be intimately involved in determining catalyst activity. Catalyst development for electrochemical energy conversion is certain to involve a significant amount of nanomaterials research.

Electrocatalysis is important to accomplish fuel oxidation and oxygen reduction at electrodes, but equally important is the electrolyte that transports ions to and from the electrocatalyst active sites and between the anode and cathode in the electrochemical cell. PEMFC devices typically use a polyelectrolyte material, often a fluoropolymer, as a proton-conducting electrolyte, whereas SOFC devices typically use ceramic materials that conduct via oxide ions as electrolyte. Materials design considerations are very different for these two device classes, yet some things remain constant. For example, nanoscale integration of electrolyte with electrode / electrocatalyst is always important to provide a means for electrons and ions to move to and from the catalyst active sites, as they must for oxidation / reduction to occur. This tight integration of catalyst and electrolyte on the nanoscale is essential to achieving high electrocatalyst activity.

For PEMFC devices the principal drivers for improved electrolytes are increased ionic conductivity at temperatures below (possibly down to -40 °C) and above (possibly up to 120 °C) the usual operating range of 80-100 °C, and also increased conductivity under relatively dry conditions (e.g. less than 50% relative humidity). Most current PEM materials are poor ion conductors at these extremes of temperature and humidity, and research efforts will continue to provide materials having improved performance at these limits. There is also much interest in creating polymer electrolytes in which hydroxide anions are the charge carriers. Oxygen reduction under strongly alkaline conditions may be catalyzed using different and less-expensive and more earth-abundant materials from those commonly used under acid conditions. Many organic materials have poor long-term stability when exposed to hot and concentrated alkali solutions, and further research is needed to create new alkaline exchange materials suitable for use in fuel-cell devices. For SOFC devices, there is much interest in creating new ceramic / solid-state electrolytes that could enable fuel-cell operation at temperatures in the 300 – 500 °C range. Mixed conductors, e.g. materials that conduct both ions and electrons, are of interest for electrodes, as are catalyst layers that allow for use of chemically complex fuels, e.g. diesel fuel, and/or dirty fuel streams, e.g. fuels containing carbon monoxide, carbon dioxide, or other impurities / poisons. Mixed conductors are also potentially useful in applications involving gas purification.

Another area receiving high attention is that of electrolysis cells, which may be thought of simply as fuel cells run in reverse. Electrolysis technology is already well developed however the emergence of intermittent electrical power generation technologies from sources such as wind or solar has created a need for storing the electrical energy for later use. One way of storing electrical energy is via water electrolysis, which converts water into separate streams of hydrogen and oxygen. Much of the research on improved PEMFC and SOFC devices is also applicable to electrolyzers and has driven strong improvements in electrolyzer performance and cost over the past decade. Further improvements in performance, cost and durability are likely in response to the emerging high need for ways to accomplish large-scale energy storage using electrical energy produced by renewable but intermittent power sources that are already being planned for introduction on a 5-10 year timetable.

## Nanomaterials for Electrochemical Energy Storage

The past twenty years have seen remarkable improvements in the basic understanding and engineering performance of batteries for storing and providing electrical energy. Lithium batteries are of great interest due to their high cell voltage and low mass, which provide for high energy and power density. Most battery materials are inherently nanomaterials, and battery charge/discharge reactions nearly always involve electrochemical conversions on nanometric length scales. Much of the recent improvement in battery performance has come about due to improved understanding of nanoscale chemistry in battery materials and electrodes.

Further incremental improvements in lithium battery performance are likely from ongoing research for which a comprehensive review is not possible in this short chapter. Significant advances are also possible from several emerging battery chemistries and technologies that may be loosely grouped together as "Beyond Lithium Ion", which is also the name of an annual battery meeting sponsored by the US Department of Energy. Emerging chemistries in this category include lithium-sulfur and lithium-air, sodium and magnesium ion intercalation compounds, and others. Flow batteries are also of high interest particularly for applications in stationary energy storage, e.g. on the electric grid. Flow batteries are attractive in part because they are inherently scalable and tend to have very low capacity fade and be highly energy efficient, because all the reactions occur in a concentrated liquid phase rather than a solid phase. Work on traditional vanadium flow battery chemistry is still of interest and is being supplemented by emerging electrode chemistries including bromine/bromide, chlorine/chloride, and sulfur/sulfide. A particularly interesting approach involves fluids containing suspended micro/nanoparticles of redox-active solids, for example intercalation materials from traditional lithium-ion batteries. This approach to flow battery development combines the high energy density of a solid-state electrode material with the desirability of having an electrode that is a fluid. Nanoparticulate-based redox fluids are very interesting nanomaterials and I think that further developments are likely in this novel but promising approach to flow-battery development.

## Nanomaterials Characterization Techniques

Nanomaterials are traditionally characterized using electron and proximal probe microscopy techniques, often coupled with X-ray, electron and neutron diffraction techniques and techniques such as X-ray photoelectron spectroscopy (XPS), dynamic light scattering, zeta-potential measurements, nuclear magnetic resonance (liquid and solid state), gas porosimetry, and others. All of these techniques have undergone significant advancement over the past twenty years which has helped to drive research on a wide range of nanoscale phenomena and materials across many technology areas including energy. Incremental advancements in these (and nearly all) materials characterization techniques will continue to drive nanomaterials research over the coming decade. An emerging trend in the characterization area is an increased emphasis on *in-situ* characterization in high-pressure gas or liquid environments, possibly to include studies within operating devices such as batteries, fuel-cells or electrolysis cells. In-situ operation can involve some significant compromises, e.g. electron microscopy and spectroscopy techniques are traditionally thought to require vacuum conditions however ways around these limitations continue to be found, such that electron microscopy and photoelectron spectroscopy are now possible on samples at close to atmospheric conditions, and/or on samples covered by a thin liquid layer. Another emerging trend is towards combining information-rich spectroscopic techniques with sample imaging in both two and three dimensions. Raman Imaging, scanned probe and tomographic techniques provide images with unprecedented resolution containing highly informative localized information, for example on local elemental composition, water content, and substrate-metal homogeneity. Instrument developers continue to "push the envelope" on what is possible for *in-situ* sample characterization and imaging, and it seems likely that this trend will continue.

## Electrode Modeling

Computer modeling capabilities in all areas of science and engineering have undergone spectacular development over the past several decades, to the point where modeling is now an expected and often required part of any new research project. Recent federal initiatives aimed at "Materials by Design" and a "Materials Genome" speak to the power and prevalence of computational modeling in materials science and engineering. In the area of energy, system-level models that include multiple types of physics all at once are commercially available and may be easily used to model a wide range of energy storage and conversion devices. Ab-initio modeling at atomic/molecular scales has also advanced on many fronts and is available via both commercial and proprietary avenues. Application of atomic / molecular models to nanomaterials can be problematic due to the large size and large number of atoms that must be included in the simulation to achieve realistic results, however rapid advancements in supercomputer technology are making calculations routine that were once thought to be nearly impossible. This trend is almost certain to continue and will lead to advancements in understanding the relationship between structure and properties, and hopefully, to the design and ultimately the synthesis of new nanomaterials for energy.

## Energy Policy and Nanomaterials

Public policy is not normally thought of as a science and engineering area, but in the present national and global climate, with so many new paradigms being considered for energy harvesting, conversion, storage and consumption, it is only reasonable to think that informed policy decisions will be needed to allow for new technologies to safely enter the marketplace. The relationship between nanomaterials and energy policy is perhaps a circuitous one, but as nanomaterial-enabled technologies such as plug-in hybrid electric vehicles, fuel-cell-powered hydrogen vehicles, and grid-scale electrical energy storage systems become more well developed and widespread, it will become necessary to develop public policies that are forward-looking enough to enable new technologies to develop, while still allowing for the fair market competition that provides the selective pressure to determine which technologies will ultimately win the day. In fact, energy needs in society are so great and so diverse that there will likely be many winners for different markets. Predictions in the public policy area are particularly difficult to make because of the high uncertainty associated with global geopolitics, and also the price volatility that accompanies select fossil fuels such as petroleum and natural gas. Even so, it seems likely that over the next five to ten years, public policy will continue to evolve in areas relating to renewable energy and the electric grid, electrical energy distribution (charging) for electric cars, and hydrogen, bioethanol, biodiesel and natural gas as automotive transportation fuels.

In concluding, and keeping in mind the wise proverb about making predictions, we may speculate that the demand for energy will certainly increase as the number of people who have access to energy-consuming devices increases. The number of electric and green-energy automobiles and portable devices in the world is likely to increase, as is the incorporation of nanomaterials into offices and homes. Windows that generate electricity and energy-efficient appliances will become more widespread. Green-energy tax credits will mean that the children of the authors of this volume may well be compelled to write a companion piece outlying their challenges and their solutions with which they are faced. I imagine their approach will be innovative, far reaching and hypothesis-driven as are the solutions offered herein. The demand for food, energy and new materials for ancient and modern communities is as old as fire or the wheel and drives human desire to control their environment forward towards the next millennium.

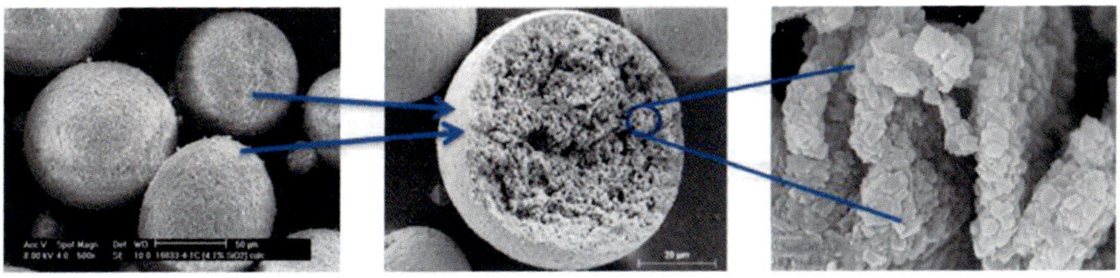

*Figure 1. Scanning Electron Microscopy (SEM) images of FCC catalyst particles from in situ synthesis where macropores are covered with nanoporous zeolite growth, leading to enhanced zeolite accessibility and activity. (Chapter 1)*

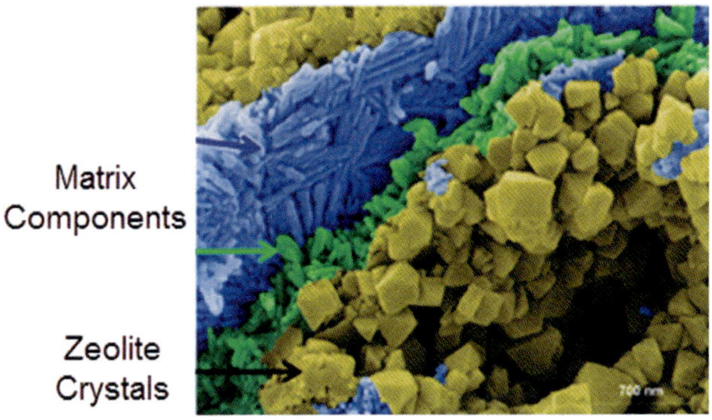


*Figure 2. Scanning Electron Microscopy (SEM) images depicting different matrix and zeolite layers in different colors. The exposed zeolite layer covering the macropore walls lead to selective cracking of large hydrocarbon molecules. (Chapter 1)*

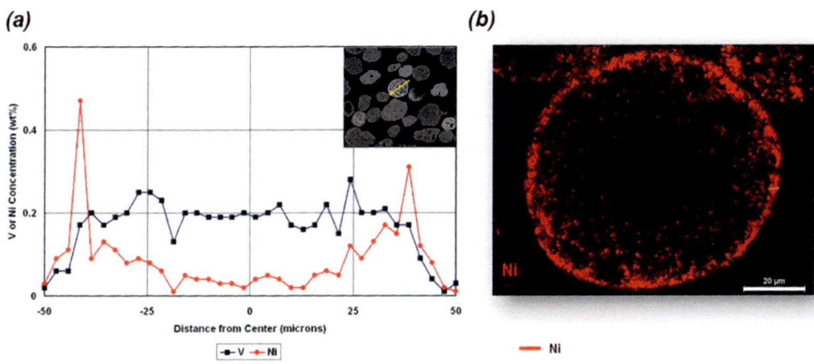


*Figure 3. (a) Deposition profiles for contaminant metals Ni and V on FCC catalysts measured by electron-probe microanalysis. (b) Nickel distribution on an FCC catalyst particle by energy dispersive X-ray spectroscopic analysis. (Chapter 1)*

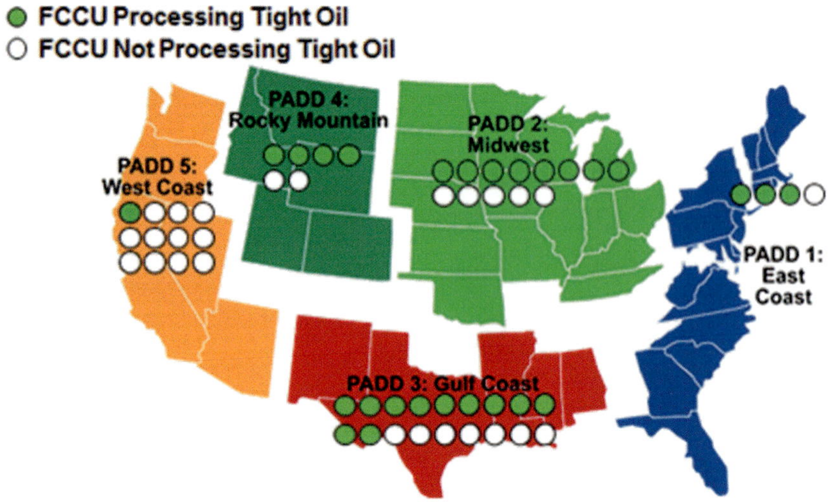


*Figure 4. A representation of the number of refineries that have moved to processing tight oils, either exclusively or partially, separated by PADD region. (Chapter 1)*

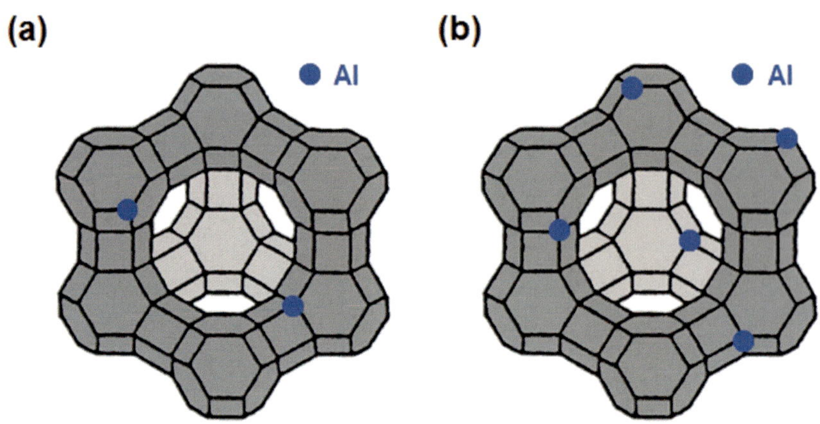


*Figure 5. Depiction of lower (a) and higher (b) unit cell size on zeolite Y framework within an FCC catalyst. Blue spots represent aluminum sites. (Chapter 1)*

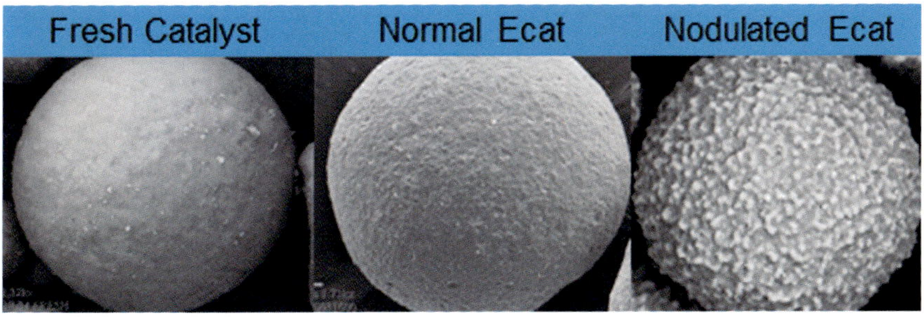


*Figure 6. Depiction of iron nodulation formation on the surface of FCC catalyst. (Chapter 1)*

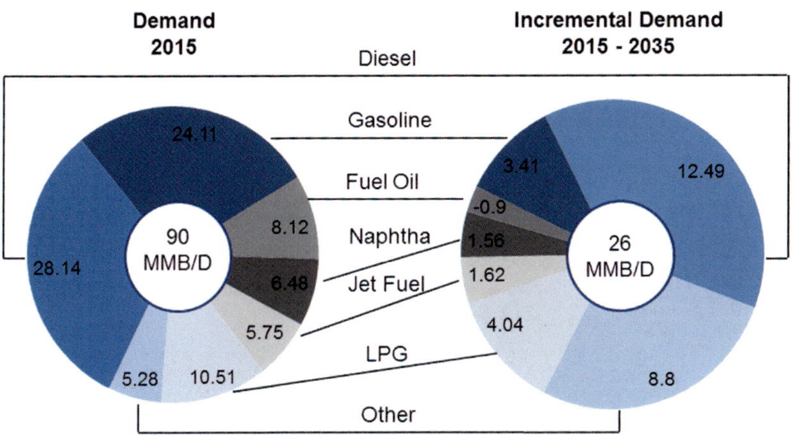


*Figure 7. Current demand and incremental demand growth for fuels. (Chapter 1)*

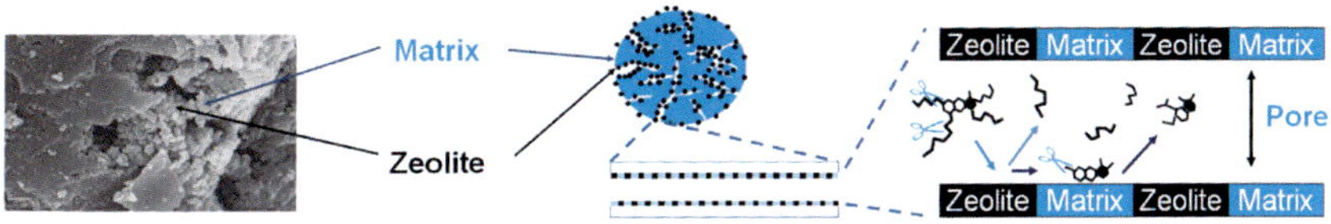


*Figure 8. SEM image and schematic depicting an FCC catalyst manufactured via in situ synthesis technology to maximize diesel yields from the refinery. (Chapter 1)*

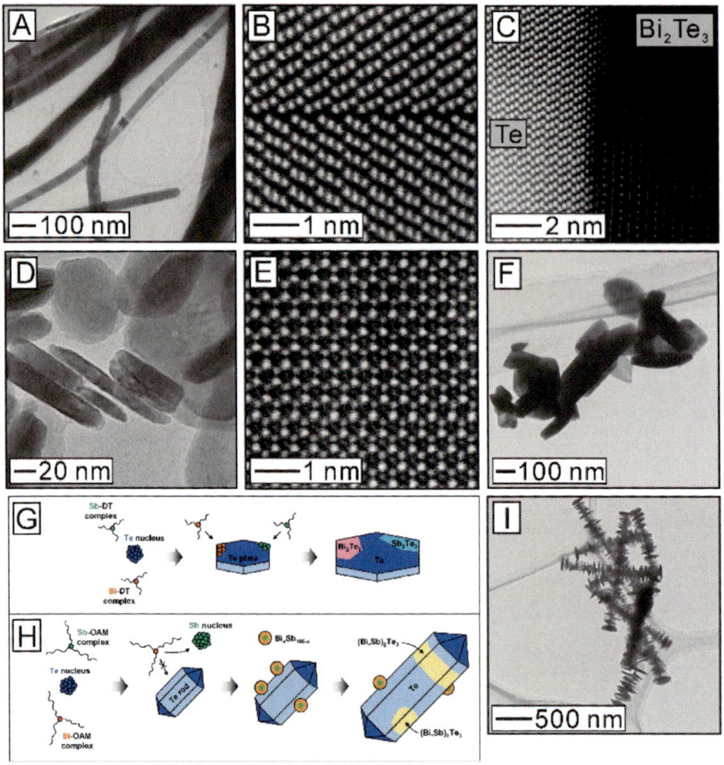


*Figure 2. TEM image of nanowires (A), STEM-HAADF image of tellurium segement (B) and transition from Te to $Bi_2Te_3$ (C), TEM image of nanodiscs (D), STEM-HAADF image of a nanodisc lying on face (E), TEM image of nanoflakes (F), scheme of synthetic formation pathway of nanodiscs (G), scheme of synthetic formation pathway of nanowires (H), TEM image of nanowires studded with nanodiscs (I). Reproduced with permission from references 13, 14, and 15. Copyright 2011, Wiley and Sons. Copyright 2011, American Chemical Society. Copyright 2011, Royal Society of Chemistry. (Chapter 3)*

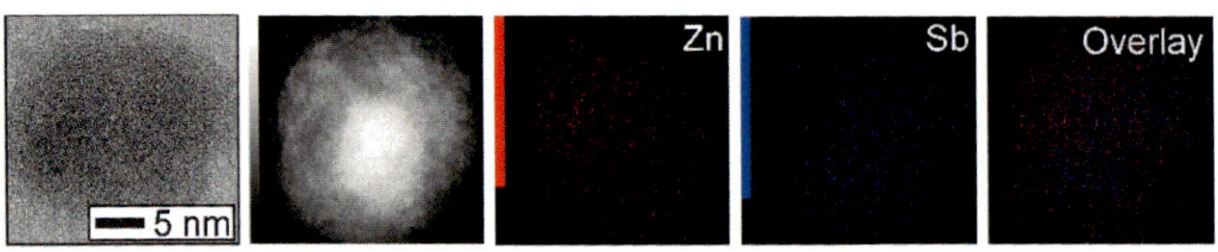


*Figure 3. HR-TEM image, STEM-HAADF image, EDS map of zinc, EDS map of antimony and an overlay of the zinc and antimony maps from left to right. The scale bar is the same for all images. (Chapter 3)*

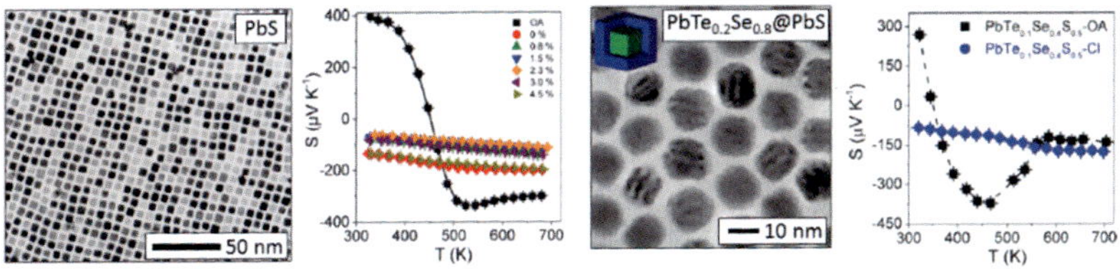


*Figure 6. TEM image of PbS nanoparticles, Seebeck measurements of PbS materials, TEM image of $PbTe_{0.2}Se_{0.8}$@PbS nanoparticles and Seebeck measurements of $PbTe_{0.1}Se_{0.4}S_{0.5}$ materials. Reproduced with permission from reference 8. Copyright 2015, American Chemical Society. (Chapter 3)*

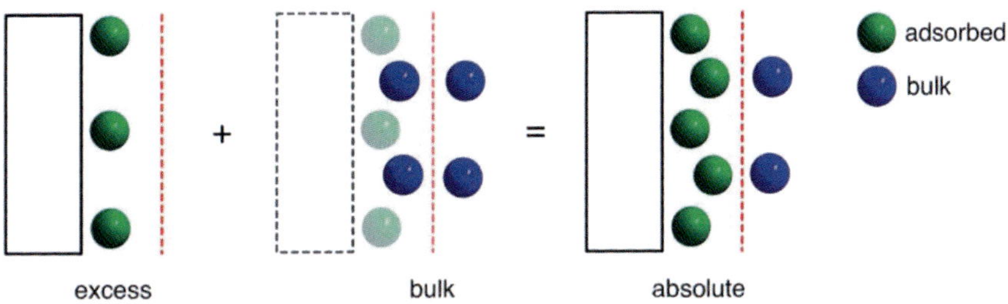


*Figure 1. The absolute adsorption is the sum of excess adsorption (green) and the bulk amount (blue). (Reproduced with permission from reference 19. Copyright 2014 The Royal Society of Chemistry). (Chapter 8)*

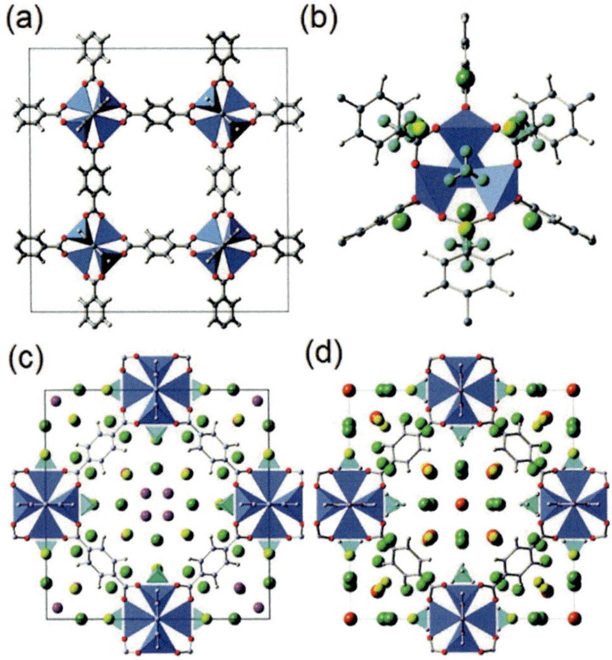


*Figure 2. (a) Crystal structure of MOF-5. C in gray and D in white. (b) The methane adsorption sites in MOF-5: cup sites (cyan) and secondary hex (yellow) and $ZnO_2$ sites (green) (c) [001] view of the I4/mmm structure of MOF-5 with more CD4 loading at 80 K, extra methane (pink) were observed to populate near the center of the pores. (d) [001] view of the P4mm structure of MOF-5 with higher CD4 loading below 60 K, confined methane sites (orange) were observed to align themselves along the c axis and further lower the symmetry to P4mm. Orientationally disordered methane molecules are shown as spheres for clarity. (Reproduced with permission from reference 48. Copyright 2009 American Chemical Society). (Chapter 8)*

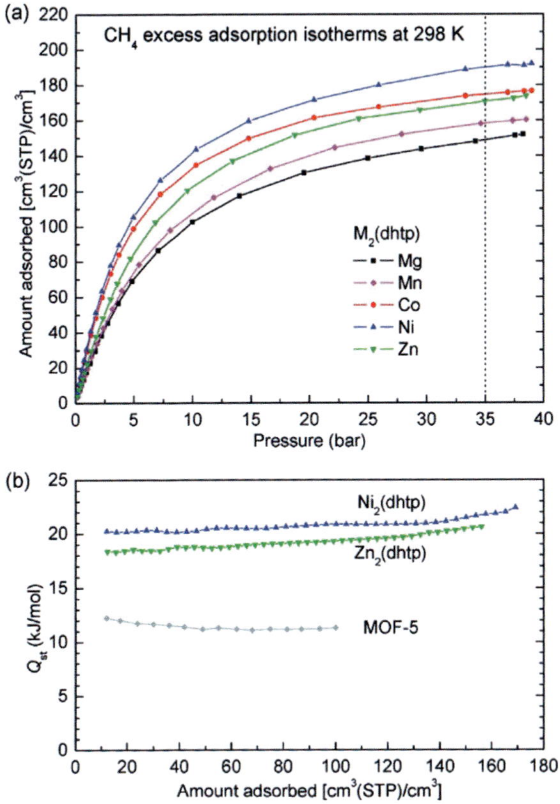


*Figure 3. (a) Excess $CH_4$ adsorption isotherms of $M_2$(dhtp) at 298 K. (b) The experimental $Q_{st}$ of $Ni_2$(dhtp) and $Zn_2$(dhtp). The $Q_{st}$'s of $Mg_2$(dhtp), $Mn_2$(dhtp), and $Co_2$(dhtp) fall between the two curves and, thus, are not shown for clarity. The $Q_{st}$'s of MOF-5 (from ref 58) are also plotted for comparison. (Reproduced with permission from reference 51. Copyright 2009 American Chemical Society). (Chapter 8)*

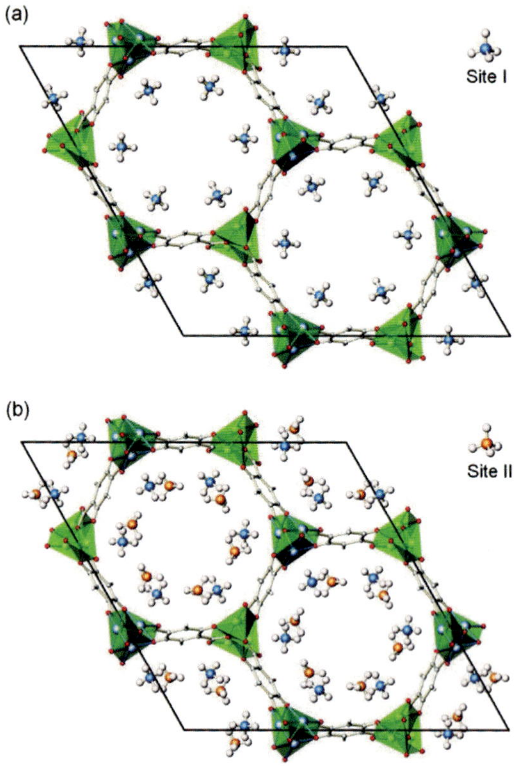


*Figure 4. Crystal unit cell of Mg-MOF-74 with methane adsorbed on site I (a) and site II (b). (Reproduced with permission from reference 51. Copyright 2009 American Chemical Society). (Chapter 8)*

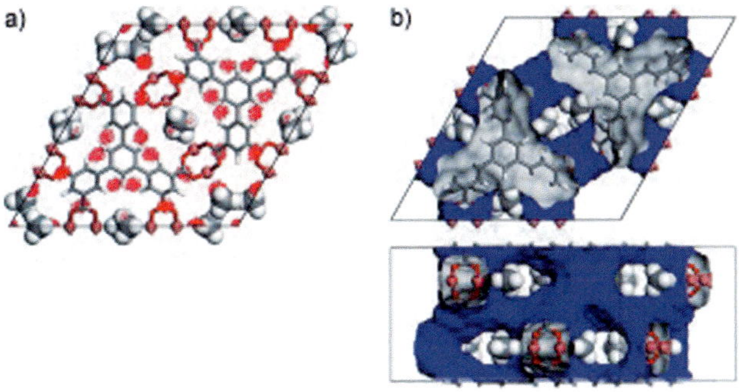


*Figure 6. (a) Probability distribution of the $CH_4$ center of mass in UTSA-20 ([0 0 1] view), obtained from Grand Canonical Monte Carlo simulation at 298 K and 10 bar. The red regions represent the places where methane molecules are heavily populated in the MOF structure. Note that the open Cu site is pre-occupied with $CH_4$ molecules to focus our effort on the search for other strong methane adsorption sites. b) The pore surface of the interconnected channel pores in UTSA-20 with adsorbed methane at the linker channel site. The channel width along the c axis matches well with the size of the adsorbed methane molecules, leading to enhanced van der Waals interaction. (Reproduced with permission from reference 63. Copyright 2011 John Wiley and Sons). (Chapter 8)*

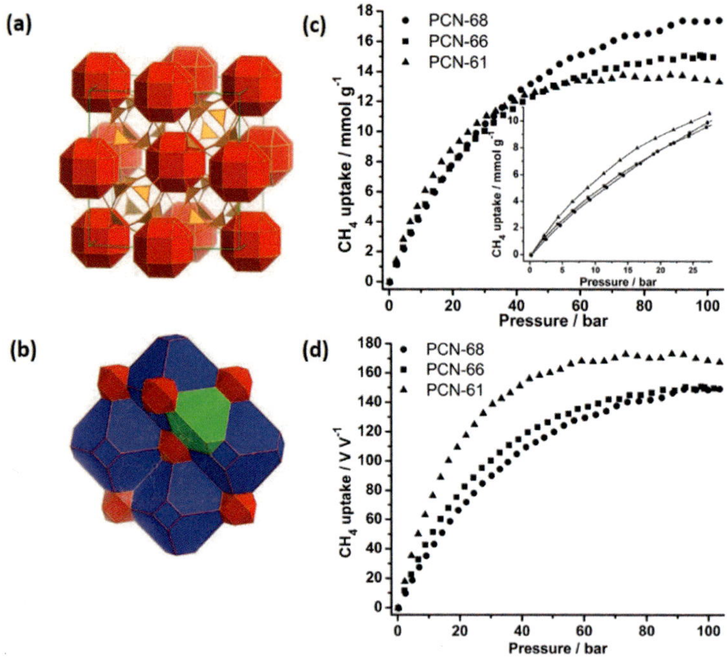


*Figure 7. (a) rht-type framework and (b) 3D polyhedra of PCN-68; (c) gravimetric and (d) volumetric methane capacities in PCN-6X at 298K. The inset (c) shows the medium-pressure region enlarged. (Reproduced with permission from reference 43. Copyright 2010 John Wiley and Sons). (Chapter 8)*

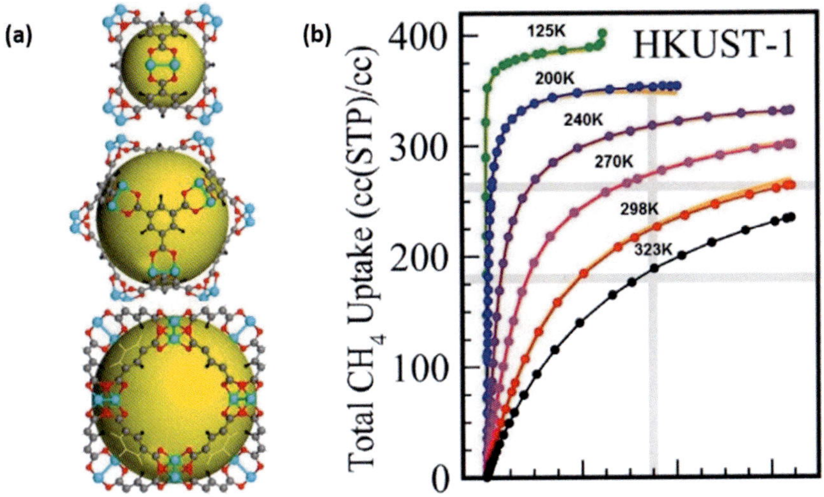


*Figure 8. (a) the nanocages and crystal structures HKUST-1. The gray, black, red, cyan, and blue spheres represent carbon, hydrogen, oxygen, copper, and nitrogen atoms, respectively. (b) Total volumetric uptake isotherms at various temperatures for HKUST-1. The gray horizontal lines indicate the DOE's old and new volumetric targets, respectively. (Reproduced with permission from reference 60. Copyright 2013 American Chemical Society). (Chapter 8)*

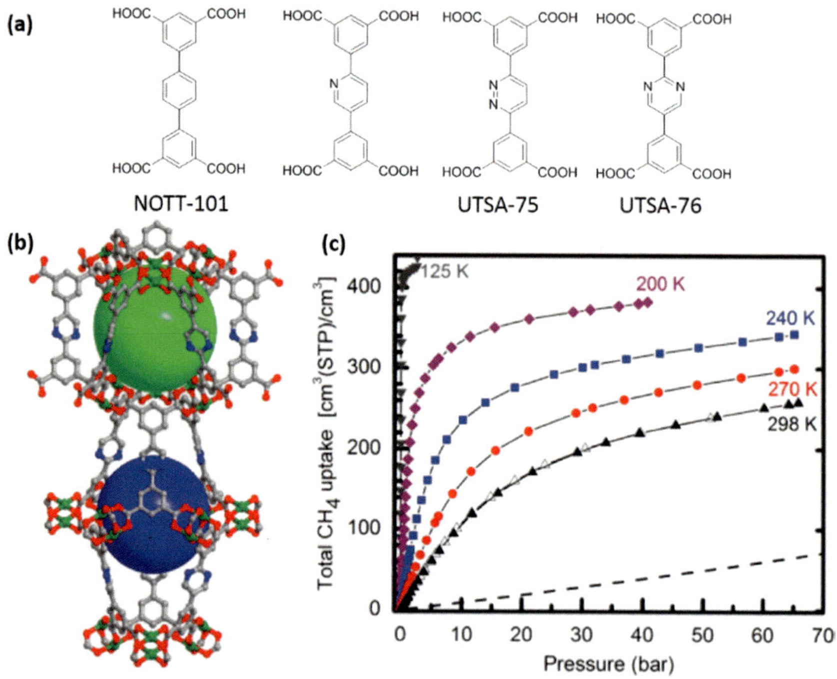


*Figure 9. (a) the schematic structure of the organic linker design for the construction of NOTT-101 and UTSA-7X; (b) Two cages of about 10.2 and 9.6×22.3Å, respectively in UTSA-76; (c) temperature-dependent high-pressure methane sorption isotherms of UTSA-76 (black dash curve represents the stored methane in a high pressure gas tank). (Reproduced with permission from reference 72. Copyright 2014 American Chemical Society). (Chapter 8)*

# Editors' Biographies

## Jingbo Louise Liu

Dr. Jingbo Louise Liu received her Ph.D. in Materials Science and Engineering from the University of Science and Technology Beijing in 2001 (Professor W. Li) and advanced training at the University of Calgary (Professor V. Birss). She was promoted to a tenured Associate Professor at Texas A&M University-Kingsville (TAMUK) due to her outstanding creativity and productivity for nanostructured materials preparation, characterization, and understanding of fundamental physical and chemical properties of nanoparticles, nanofilms and nanotubes, as well as applications of engineered nanomaterials in alternative energy and biological science. She established highest power density to advance performance of proton exchange membrane fuel cells and directed a new paradigm to apply metal-organic frameworks in the disinfection science. Dr. Liu has authored and co-authored textbook (2), book (3) and book chapters (5) and over 60 peer-reviewed journal articles. She chaired and organized professional conferences and reviewed dozens of journal articles. She has been hosting and co-hosting visiting scholars. During 9-year services in TAMUK, she trained more than 100 undergraduate students (90 % are Hispanic), 22 master (12 graduated) students to conduct leading edge research. She served as NSF panelist and Chaired the proposal review panel. She also served as journal editor/reviewer and reviewed NSF proposals.

Currently, Dr. Liu is the DEBI faculty fellow at the US Air Force Research Laboratory and user affiliate at the Lawrence Berkeley National Laboratory. She was awarded the "2012 to 2015 Annual foreign experts and talent from overseas project" supported by the State Administration of Foreign Experts Affairs, P.R. China; Japan Society for the Promotion of Science (JSPS) Invitation Fellow and worked at the Department of Materials Science, University of Tokyo (2010-2011). She has served as a "Faculty and Student Team" fellow, collectively funded by the National Science Foundation and US Department of Energy, Office of Science and worked at the Argonne National Laboratory (2009). She also received Faculty Fellowship Summer Institute in Israel (2008) and outstanding research and teaching awards at the university level. She directed and participated in the projects (>15) supported by the NSF (USA, CHINA), NSERC (CANADA), American Chemical Society Petroleum Research Funds (PRF), R. Welch Foundation (departmental grants since 2006), Department of Education, industrial and TAMUK as PI, Co-PI and senior personnel. She also received dozens of travel funds to attend QEM Workshops; NIH Faculty Grant Writing Workshop; Higher Education Consortium Workshop, Universities Space Research Association; and COACh (NSF women advancement) workshops.

## Sajid Bashir

Dr. Sajid Bashir was elected as a Royal Society Fellow, and Chartered Scientist and Chartered Chemist who has pioneered in laser dyes to improve the sensitivity of hydrophobic molecules by time-of-flight Mass Spectrometry. He explored characterization of macromolecules and their application in biomedical and analytical fields. He graduated from the University of Warwick in 2000/1 and conducted plant biological research in Cornell University from 2001-2004 (Professor J. Rose). His graduate theses concerned analysis of macromolecules by Fourier Transform Mass Spectrometry (Professor T. Wood) and the photo physics of electronically excited molecules under laser irradiation (Professor Derrick). In 2004, he joined Texas A&M University-Kingsville and was promoted to associate Professor due to his contribution to the bio-nanotechnology. He participated in and co-directed the NIH, NSF and USDA funded research as co-principle investigator and Department of Defense as principle investigator. He published more than five dozens of peer-reviewed papers and delivered dozens of invited presentations at professional societies and different universities. Dr. Bashir is certified by the Red Cross in basic first aid, use of automated external defibrillator, and cardiopulmonary resuscitation. He is also U.S Department of transport and International Air Transport Association certified in Dangerous Goods Regulations (hazardous chemicals).

Both Drs. Bashir and Liu have co-organized sessions at the American Chemical Society Division of Colloid and Surface Chemistry (COLL) that have been cited by the President of the society as thematic symposium, which have had robust attendance. Dr. Bashir and Liu have interned at Lawrence Berkeley National Laboratory Advanced Light Source and the U.S Airforce Research Laboratory (Dr. Bashir) and hold at least two discovery of inventions submitted through the A&M System office. Dr. Bashir also received the 2008/9 Olan Kruse Science Faculty Award at TAMUK for research related to his contributions regarding analysis of macromolecules to matrix-assisted laser desorption time-of-flight mass spectrometry.

# Indexes

# Author Index

# Subject Index

**O**



**P**

## T

Printed in the USA/Agawam, MA
June 7, 2016

635807.023